ERDŐS–KO–RADO THEOREMS: ALGEBRAIC APPROACHES

Aimed at graduate students and researchers, this fascinating text provides a comprehensive study of the Erdős–Ko–Rado (EKR) Theorem, with a focus on algebraic methods. The authors begin by discussing well-known proofs of the EKR bound for intersecting families of sets. The natural generalization of the EKR Theorem holds for many different objects that have a notion of intersection, and the bulk of this book focuses on algebraic proofs that can be applied to these different objects. The authors introduce tools commonly used in algebraic graph theory and show how these can be used to prove versions of the EKR Theorem. Topics include association schemes, strongly regular graphs, distance-regular graphs, the Johnson scheme, the Hamming scheme, and the Grassmann scheme. The book also gives an introduction to representation theory (aimed at combinatorialists) with a focus on the symmetric group. This theory is applied to orbital schemes, concentrating on the perfect matching scheme and other partitions, and to conjugacy class schemes, with an emphasis on the symmetric group.

Readers can expand their understanding at every step with the 170 end-of-chapter exercises. The final chapter discusses in detail 14 open problems, each of which would make an interesting research project.

Chris Godsil is a professor in the Combinatorics and Optimization department at the University of Waterloo, Ontario, Canada. He authored (with Gordon Royle) the popular textbook Algebraic Graph Theory. He started the Journal of Algebraic Combinatorics in 1992 and he serves on the editorial board of a number of other journals, including the Australasian Journal of Combinatorics and the Electronic Journal of Combinatorics.

Karen Meagher is an associate professor in the Department of Mathematics and Statistics at the University of Regina, Saskatchewan, Canada. Her research area is graph theory and discrete mathematics, in which she has published around 25 journal articles.

Erdős–Ko–Rado Theorems: Algebraic Approaches

CHRIS GODSIL
University of Waterloo, Ontario, Canada

KAREN MEAGHER
University of Regina, Saskatchewan, Canada

CAMBRIDGE
UNIVERSITY PRESS

CAMBRIDGE
UNIVERSITY PRESS

University Printing House, Cambridge CB2 8BS, United Kingdom

One Liberty Plaza, 20th Floor, New York, NY 10006, USA

477 Williamstown Road, Port Melbourne, VIC 3207, Australia

314-321, 3rd Floor, Plot 3, Splendor Forum, Jasola District Centre, New Delhi - 110025, India

103 Penang Road, #05-06/07, Visioncrest Commercial, Singapore 238467

Cambridge University Press is part of the University of Cambridge.

It furthers the University's mission by disseminating knowledge in the pursuit of education, learning and research at the highest international levels of excellence.

www.cambridge.org
Information on this title: www.cambridge.org/9781107128446

First published 2016

A catalogue record for this publication is available from the British Library

ISBN 978-1-107-12844-6 Hardback

To our families:

Gillian, Emily and Nicholas
Paul, Ada and Dash

Contents

Preface

There are 280 partitions of $\{1, \ldots, 9\}$ into three pairwise disjoint triples. Say two such partitions α and β are skew if each triple in α contains one point from each triple in β. We can now define a graph with our 280 partitions as vertices, where two partitions are adjacent if and only if they are skew. A coclique in this graph is a set of partitions such that no two are skew. We ask the innocent question: how large can a coclique be?

There is an easy lower bound. Let Ω be the set of partitions such that the points 1 and 2 lie in the same triple. (There are 70 of these.) Clearly no two partitions in Ω are skew and so we now have a lower bound of 70 on the size of a coclique. But now we have two questions. Can we do better? And, if not, are there any cocliques of size 70 that do not have this form?

Karen asked Chris the first question in 2001, and by a process of induction and complication, we were led to this book. The complications arise because there is a close connection to the Erdős–Ko–Rado Theorem, one of the fundamental results in combinatorics. This theorem provides information about systems of intersecting sets. A family $\mathcal{F}$ of subsets of a ground set – it might as well be $\{1, \ldots, n\}$ – is intersecting if any two sets in $\mathcal{F}$ have at least one point in common. More generally it is t-intersecting if any two elements of $\mathcal{F}$ have at least t points in common. The most commonly stated form of the EKR Theorem is the following.

0.0.1 Theorem. *If $\mathcal{F}$ is an intersecting family of k-subsets of $\{1, \ldots, n\}$, then*

$$|\mathcal{F}| \leq \binom{n-1}{k-1}.$$

If equality holds then $\mathcal{F}$ consists of the k-subsets that contain a given point of the underlying set. □

In our view this theorem has two parts: a bound and a characterization of families that meet the bound.

One reason this theorem is so important is that it has many interesting extensions. To address these, we first translate it to a question in graph theory. The Kneser graph $K(n, k)$ has all k-subsets of $\{1, \ldots, n\}$ as its vertices, and two k-subsets are adjacent if they are disjoint. (We assume $n \geq 2k$ to avoid trivialities.) Then an intersecting family of k-subsets is a coclique in the Kneser graph, and we see that the EKR Theorem characterizes the cocliques of maximum size in the Kneser graph. So we can seek to extend the EKR Theorem by replacing the Kneser graphs by other interesting families of graphs. The partition graphs just discussed provide an example.

There is a second class of extensions of the EKR Theorem. In their famous 1961 paper Erdős, Ko and Rado proved the following:

0.0.2 Theorem. *If $\mathcal{F}$ is a t-intersecting family of k-subsets of $\{1, \ldots, n\}$ and n is large enough, then*

$$|\mathcal{F}| \leq \binom{n-t}{k-t}.$$

If equality holds then $\mathcal{F}$ consists of the k-subsets that contain a given t-subset of the underlying set.   □

There are graph-theoretic analogs of this question too, but we have to work a little harder. In place of the Kneser graphs, we use the Johnson graphs $J(n, k)$. The vertices of $J(n, k)$ are the k-subsets of $\{1, \ldots, n\}$, but now we deem two k-subsets to be adjacent if they have exactly $k - 1$ points in common. Again we assume $n \geq 2k$. We can show by induction that $J(n, k)$ has diameter k and thus two k-subsets are adjacent in $K(n, k)$ if and only if they are at maximum possible distance in $J(n, k)$.

Define the width of a subset of the vertices of a graph to be the maximum distance between two vertices in the subset. Then our first version of the EKR Theorem characterizes the subsets of maximum size in $J(n, k)$ of width $k - 1$, and the second version the subsets of maximum size with width $k - t$. All known analogs of the EKR Theorem for t-intersecting sets can be stated naturally as characterizations of subsets of width $d - t$ in a graph of diameter d. However, such theorems have only been proved in cases where the distance graphs form an association scheme (which for now means that they fit together in a particularly nice way). To give one example, we can replace k-subsets of

$\{1, \ldots, n\}$ by subspaces of dimension k over a vector space of dimension n over $GF(q)$.

What of this book? One goal has been to show how the EKR Theorem can be tackled using tools from algebraic graph theory. But we are not zealots, and we begin by discussing most of the known proofs of the EKR bound for intersecting families; these are not algebraic. We go to develop many of the tools we need, and then we apply them to strongly regular graphs. We develop the basic properties of the Johnson scheme, and using these we offer two proofs of the EKR Theorem for intersecting families of k-sets.

We present a version of Wilson's proof of the EKR Theorem for t-intersecting families of k-subsets when $n \geq (t+1)(k-t+1)$. The main novelty is that in order to derive the characterization of the maximum families, we make explicit use of the concepts of width and dual width. (Here we are following important work by Tanaka [160].) It is comparatively easy to extend this approach to t-intersecting families of k-dimensional subspaces of a vector space over a finite field. We complete the first half of the book by treating EKR problems on words; here the Hamming schemes provide a natural framework.

In the second part of the book, we consider versions of EKR on sets of permutations and partitions. For this we need to make use of the fact that in these problems there is a natural action of the symmetric group, and this means we need information about the representation theory of the symmetric group. We treat this in some detail (although in many cases we refer the reader to the literature for proofs).

In the original version of the EKR Theorem there is a requirement that n be large relative to the size of the subsets and the size of the intersection; this condition cannot be dropped, and a significant body of work was required to determine the exact bound on n. For many of the analogs of the EKR Theorem a bound, analogous to this lower bound on n, is required. It may be possible to obtain strong results and the assumption that n (or its analog) is "sufficiently large." Such results are not the focus of this book; we are more interested in the combinatorial details involved in finding exact results.

We assume a working knowledge of graph theory, but otherwise we have tried to keep things self-contained. There are exercises of varying difficulty at the end of each chapter. In general, if there is a reference attached to the exercise, expect it to be more challenging.

This book is the culmination of many years of work, and there are many people whom we wish to thank for their assistance and encouragement in writing this book as well as many interesting and illuminating discussions. We specifically wish to thank following people: Bahman Ahmadi, Robert Bailey,

Adam Dyck, Ferdinand Ihringer, Bill Martin, Mike Newman, Alison Purdy and Pablo Spiga. Finally, we offer our gratitude to Brendan Rooney, who completed a detailed edit of the entire book. He corrected many errors, both mathematical and grammatical, and his contribution has significantly improved this book!

1

The Erdős–Ko–Rado Theorem

A family of subsets of a set is *intersecting* if any two elements of the family have at least one element in common. It is easy to find small intersecting families; the basic problem is to decide how large they can be, and to describe the structure of the families that meet whatever bound we can derive. The prototype of all these results is the following, which is often referred to as the *Erdős–Ko–Rado Theorem for intersecting families*.

1.0.1 Theorem. *Let k and n be integers with $n \geq 2k$. If $\mathcal{A}$ is a collection of k-subsets of the set $\{1, \ldots, n\}$, such that any two members of $\mathcal{A}$ have at least one point in common, then*

$$|\mathcal{A}| \leq \binom{n-1}{k-1}.$$

Moreover, if $n > 2k$, equality holds if and only if $\mathcal{A}$ consists of all the k-subsets that contain a given point from $\{1, \ldots, n\}$. □

This result provides both a bound and a characterization; the bound is easier to prove, but the characterization can be more useful. This theorem is essentially a corollary of a stronger result first proved by Erdős, Ko and Rado [58]. Define a *t-intersecting family* to be a collection of subsets such that any two have at least t points in common. Then the *Erdős–Ko–Rado Theorem for t-intersecting families* can be stated as follows:

1.0.2 Theorem. *Let n, k and t be positive integers with $0 \leq t \leq k$. There exists a function $f(k, t)$ such that if $n \geq f(k, t)$ and $\mathcal{A}$ is a t-intersecting family of k-sets of $\{1, \ldots, n\}$, then*

$$|\mathcal{A}| \leq \binom{n-t}{k-t}.$$

Moreover, equality holds if and only if A consists of all k-subsets that contain a specified t-subset of $\{1, \ldots, n\}$.

This theorem is clearly more general than our first theorem, but has the weakness that its conclusion holds only if n is greater than some unspecified lower bound.

We call a collection of subsets of $\{1, \ldots, n\}$ a *set system* with underlying set $\{1, \ldots, n\}$; if the subsets all have size k we refer to it as a *k-set system*. The easiest way to build a t-intersecting k-set system on an n-set is to simply take all k-subsets that contain a fixed t-set; clearly such a system has size

$$\binom{n-t}{k-t}.$$

We call a set system of this type a *canonical t-intersecting family*. The lower bound on n in Theorem 1.0.2 is necessary because when n is not large enough, an intersecting family of maximal size need not be canonical. Much work was devoted to determining the precise value of $f(n, k)$ needed. Examples show that we need $n \geq (t + 1)(k - t + 1)$ for the bound to hold, and in 1978 Frankl [62] proved that this constraint sufficed when t was large enough. In 1984, Wilson [172] proved that the bound in the EKR Theorem holds if $n \geq (t + 1)(k - t + 1)$, and the characterization holds provided $n > (t + 1)(k - t + 1)$. (We present his proof in Chapter 6.) In 1997 Ahlswede and Khachatrian [3] determined the largest t-intersecting k-set systems on an n-set, for all values of n. The result of this work is that, for each choice of n, k and t, we know that maximum size of the t-intersecting families and we know the structure of the families that reach this size.

In this chapter we present the original proof of the EKR Theorem, discuss the work of Ahlswede and Khachatrian, and then provide a number of different proofs of the EKR Theorem for intersecting families.

1.1 The original proof

In their paper, Erdős, Ko and Rado prove two statements. The first was the bound in Theorem 1.0.1, and the second was Theorem 1.0.2, with a very rough upper bound on the value of the function $f(k, t)$. In this section we present the original proofs of both statements. Both proofs use a simple yet useful operation on set systems called *shifting* or *set compression*. For a comprehensive overview of this method see the survey paper by Frankl [68].

Let $\mathcal{A}$ be a k-set system on an n-set. For any integers i, j from $\{1, \ldots, n\}$, define the (i, j)-*shift of a set A* in $\mathcal{A}$ by

$$S_{i,j}(A) = \begin{cases} (A \setminus \{j\}) \cup \{i\}, & \text{if } j \in A \text{ and } i \notin A \text{ and } (A \setminus \{j\}) \cup \{i\} \notin \mathcal{A}; \\ A, & \text{otherwise.} \end{cases}$$

Then the (i, j)-*shift of a set system* $\mathcal{A}$ is defined by

$$S_{i,j}(\mathcal{A}) = \{S_{i,j}(A) : A \in \mathcal{A}\}.$$

Obviously if $\mathcal{A}$ is a k-set system then $S_{i,j}(\mathcal{A})$ is also a k-set system of the same size. What is more interesting is that shifting a t-intersecting set system results in a system that is still t-intersecting.

1.1.1 Lemma. *If $\mathcal{A}$ is a t-intersecting system of k-sets, then for any distinct pair i, j the set system $S_{i,j}(\mathcal{A})$ is also t-intersecting.*

Proof. Let A and B be any two sets in $\mathcal{A}$. We will show that

$$|S_{i,j}(A) \cap S_{i,j}(B)| \geq t.$$

If both sets remain unchanged by the shifting operation or if both sets are changed by the shifting operation, then clearly $S_{i,j}(A)$ and $S_{i,j}(B)$ are still t-intersecting.

Thus we can assume that $S_{i,j}(A) = A$ and $S_{i,j}(B) \neq B$. Since the shifting operation changes the set B, we know that $S_{i,j}(B) = (B \setminus \{j\}) \cup \{i\}$. As the set A does not change, then at least one of the following holds:

(a) $j \notin A$,
(b) $i \in A$,
(c) $(A \setminus \{j\}) \cup \{i\} \in \mathcal{A}$.

Since A and B are t-intersecting, if either (a) or (b) holds, then it is easy to see that $S_{i,j}(A)$ (which equals A) and $S_{i,j}(B)$ are t-intersecting. If case (c) holds, then $(A \setminus \{j\}) \cup \{i\}$ is in $\mathcal{A}$, and it must be at least t-intersecting with B. Since j is not in $(A \setminus \{j\}) \cup \{i\}$ and i is not in B, neither i nor j are one of the t elements in the intersection of $(A \setminus \{j\}) \cup \{i\}$ and B. Thus, all of these t elements must also be in both $S_{i,j}(A) = A$ and $S_{i,j}(B)$. $\square$

If for all $i < j$ the set system $\mathcal{A}$ has the property that $S_{i,j}(\mathcal{A}) = \mathcal{A}$, then $\mathcal{A}$ is called a *left-shifted system* (or a *stable system*). Any set system $\mathcal{A}$ can be transformed into a left-shifted set system by shifting; in fact this can be done in as few as $\binom{n}{2}$ shifts [68, Proposition 2.2].

We now have all the tools to prove the bound in the Erdős–Ko–Rado Theorem for $t = 1$.

1.1.2 Theorem. *Let k and n be integers with $n \geq 2k$. If $\mathcal{A}$ is an intersecting
k-set system on an n-set, then*

$$|\mathcal{A}| \leq \binom{n-1}{k-1}.$$

Proof. First consider the case where $2k = n$. Since $\mathcal{A}$ is intersecting, it is not
possible that both a set and its complement are in $\mathcal{A}$, and thus

$$\mathcal{A} \leq \frac{1}{2}\binom{n}{k} = \binom{n-1}{k-1}.$$

Next consider the case where $2k < n$ and apply induction on n. If $n = 3$,
then $k = 1$, so clearly $|\mathcal{A}| = 1$ and the theorem holds. Assume the theorem
holds for $n_0 < n$. Let $\mathcal{A}$ be an intersecting k-set system on an n-set and further
assume that $\mathcal{A}$ is a left-shifted system. In particular, $S_{i,n}(\mathcal{A}) = \mathcal{A}$ for all $i < n$.

The system $\mathcal{A}$ can be split into two subsystems: the first is the collection
of all sets that do not contain n, and the second is all sets that do contain n.
The first subsystem is an intersecting k-set system on an $(n-1)$-set and by
induction has size no more than $\binom{n-2}{k-1}$. We claim that the size of the second
subsystem is no more than $\binom{n-2}{k-2}$, which implies that

$$|\mathcal{A}| \leq \binom{n-2}{k-2} + \binom{n-2}{k-1} = \binom{n-1}{k-1}$$

and completes the proof of the theorem.

To prove this claim, we show that if n is removed from each set in the second
subsystem, what remains is an intersecting $(k-1)$-set system on an $(n-1)$-
set, which by induction implies that the size of the second subsystem is no
more than $\binom{n-2}{k-2}$.

Thus, we only need to show for any two sets A_1 and A_2 from the second
subsystem that

$$(A_1 \setminus \{n\}) \cap (A_2 \setminus \{n\}) \neq \emptyset.$$

Since $n > 2k$, there is an $x \in \{1, \ldots, n\}$ that is not in $A_1 \cup A_2$. Further, the set
$\mathcal{A}$ is left-shifted, and hence the set

$$B_1 = (A_1 \setminus \{n\}) \cup \{x\}$$

is in $\mathcal{A}$. The set system $\mathcal{A}$ is intersecting, so this means that $|A_2 \cap B_1| \geq 1$ and
that the sets A_1 and A_2 have a point in common that is not n. $\square$

The bound $n \geq 2k$ is both necessary and the best possible, since if $n < 2k$
then any two k-sets from an n-set are intersecting.

Next we sketch the original proof given by Erdős, Ko and Rado that a t-intersecting k-set system that is not a canonical t-intersecting k-set system is strictly smaller than $\binom{n-t}{k-t}$, provided that n is sufficiently large relative to k and t.

1.1.3 Theorem. *Let $\mathcal{A}$ be a t-intersecting k-set system on an n-set. If $n \geq t + (k - t)\binom{k}{t}^3$ and $\mathcal{A}$ is not a canonical t-intersecting set system, then*

$$|\mathcal{A}| < \binom{n - t}{k - t}.$$

Proof. Set $r = |\cap_{A \in \mathcal{A}} A|$; since $\mathcal{A}$ is not a canonical t-intersecting set system we see that $r < t$.

Provided that $n \geq 2k - t$, a maximal t-intersecting set system is not a $(t + 1)$-intersecting system (this fact is left as an exercise), so there must be sets $A_1, A_2 \in \mathcal{A}$ such that $|A_1 \cap A_2| = t$. Since $\mathcal{A}$ is not a canonical t-intersecting system there also must be a set $A_3 \in \mathcal{A}$ such that

$$|A_1 \cap A_2 \cap A_3| < t.$$

Consider the collection T of triples (B_1, B_2, B_3) of t-subsets with $B_i \subset A_i$ for $i = 1, 2, 3$ such that

$$t < |B_1 \cup B_2 \cup B_3| \leq k.$$

For each triple (B_1, B_2, B_3) in T, define a system of sets by

$$\Phi_{(B_1, B_2, B_3)} = \{A : (B_1 \cup B_2 \cup B_3) \subseteq A \in \mathcal{A}\}.$$

Every set in $\mathcal{A}$ must have intersection of size at least t with each of A_1, A_2 and A_3; thus every set in $\mathcal{A}$ must contain at least one set $B_1 \cup B_2 \cup B_3$ where $(B_1, B_2, B_3) \in T$. Thus, the union of the sets $\Phi_{(B_1, B_2, B_3)}$ over all the triples from T is exactly $\mathcal{A}$. So by bounding the size of T and the size of any $\Phi_{(B_1, B_2, B_3)}$, we get a bound on the size of $\mathcal{A}$.

A very rough (but still effective) upper bound on the size of T is $\binom{k}{t}^3$. Further, if we assume that $s = |B_1 \cup B_2 \cup B_3|$, then an upper bound on the size of $\Phi_{(B_1, B_2, B_3)}$ is $\binom{n-s}{k-s}$. Since $s \geq t + 1$, this bound is no larger than $\binom{n-t-1}{k-t-1}$. Putting these together, we have the bound

$$|\mathcal{A}| = \left| \bigcup_{(B_1, B_2, B_3) \in T} \Phi_{(B_1, B_2, B_3)} \right| \leq \binom{n - t - 1}{k - t - 1} \binom{k}{t}^3.$$

Finally, if $n > t + (k - t)\binom{k}{t}^3$ then this upper bound is less than $\binom{n-t}{k-t}$ and the theorem holds. $\square$

In the previous proof the lower bound on the size of n is

$$n \geq t + (k - t)\binom{k}{t}^3.$$

Erdős, Ko and Rado state in their paper that this is clearly not the optimal lower bound on n, but that a bound on n is needed. They give the following example of a 2-intersecting 4-set system on an 8-set system that is strictly larger than the canonical intersecting set system:

$$\{\{1,2,3,4\}, \quad \{1,2,3,5\}, \quad \{1,2,3,6\}, \quad \{1,2,3,7\}, \quad \{1,2,3,8\},$$
$$\{1,2,4,5\}, \quad \{1,2,4,6\}, \quad \{1,2,4,7\}, \quad \{1,2,4,8\}, \quad \{1,3,4,5\},$$
$$\{1,3,4,6\}, \quad \{1,3,4,7\}, \quad \{1,3,4,8\}, \quad \{2,3,4,5\}, \quad \{2,3,4,6\},$$
$$\{2,3,4,7\}, \quad \{2,3,4,8\} \ \}$$

Figure 1.1 A 2-intersecting 4-set system on an 8-set

This system has size 17, whereas the canonical 2-intersecting set system has only $\binom{8-2}{4-2} = 15$ sets. There is a simple pattern to this system: it is the collection of every 4-set that has at least 3 elements from the set $\{1, 2, 3, 4\}$. Such systems are discussed in more detail in the next section.

1.2 Non-canonical intersecting set systems

In Figure 1.1 we give an example of a 2-intersecting set system that is larger than the canonical 2-intersecting set system. This example can be generalized to provide a large family of t-intersecting set systems that are strictly larger than the canonical t-intersecting set system. Consider the system of all $2r$-subsets from a $4r$-set that have at least $r + 1$ elements from the set $\{1, \ldots, 2r\}$. It is not hard to see that this is a 2-intersecting set system of size

$$\sum_{i=r+1}^{2r} \binom{2r}{i}\binom{2r}{2r-i} = \frac{1}{2}\left(\binom{4r}{2r} - \binom{2r}{r}^2\right).$$

This is larger than the size of a canonical 2-intersecting system for all $r > 1$.

In fact if $n < (k - t + 1)(t + 1)$ and $t > 1$, there is a t-intersecting k-set system on an n-set that is strictly larger than the canonical system. It is easy to describe these systems. Assume that n, k and t are integers with $t \leq k \leq n$. For $i \in \{0, \ldots, k - t\}$ define

$$\mathcal{F}_i = \{A \ : \ |A| = k, \ |A \cap \{1, \ldots, 2i + t\}| \geq i + t\}.$$

If $i \in \{0, \ldots, k - t\}$, the system $\mathcal{F}_i$ is a t-intersecting k-set system on an n-set and $\mathcal{F}_0$ is a canonical t-intersecting system. The size of $\mathcal{F}_i$ is

$$|\mathcal{F}_i| = \sum_{j=t+i}^{t+2i} \binom{t + 2i}{j} \binom{n - (t + 2i)}{k - j}. \tag{1.2.1}$$

Ahlswede and Khachatrian [3] determined bounds on n (relative to k and t) for when $\mathcal{F}_{i+1}$ is larger than $\mathcal{F}_i$. Further, they prove that for all n one of the systems $\mathcal{F}_i$ is the largest intersecting system. This impressive result is stated as Theorem 1.3.1 in the next section; we give one case of this theorem in the following lemma.

1.2.1 Lemma. *If $n = (k - t + 1)(t + 1)$, then $|\mathcal{F}_0| = |\mathcal{F}_1|$ and if $n < (k - t + 1)(t + 1)$, then $|\mathcal{F}_0| < |\mathcal{F}_1|$.*

Proof. The system $\mathcal{F}_1$ is the collection of all k-subsets from an n-set that contain at least $t + 1$ elements from the set $\{1, \ldots, t + 2\}$. From (1.2.1),

$$|\mathcal{F}_1| = \binom{t + 2}{t + 1}\binom{n - (t + 2)}{k - (t + 1)} + \binom{t + 2}{t + 2}\binom{n - (t + 2)}{k - (t + 2)}.$$

Since

$$|\mathcal{F}_0| = \binom{n - t}{k - t} = \binom{n - t - 2}{k - t} + 2\binom{n - t - 2}{k - t - 1} + \binom{n - t - 2}{k - t - 2}$$

it follows that

$$\binom{n - t}{k - t} - |\mathcal{F}_1| = \binom{n - t - 2}{k - t} - t\binom{n - t - 2}{k - t - 1}$$

$$= \binom{n - t - 2}{k - t - 1}\left(\frac{n - k - 1}{k - t} - t\right)$$

$$= \binom{n - t - 2}{k - t - 1}\left(\frac{n - (t + 1)(k - t + 1)}{k - t}\right). \qquad \square$$

With this simple example we can see that the lower bound of $f(k, t) = (k - t + 1)(t + 1)$ on n in Theorem 1.0.2 cannot be improved. It is considerably more difficult to prove for $n > (k - t + 1)(t + 1)$ that the only systems of maximum size are the canonical intersecting set systems.

1.3 The complete Erdős–Ko–Rado Theorem

Theorem 1.0.2 describes the size and the structure of the largest intersecting set systems, provided that the size of the underlying set is large relative to the size of the subsets and the size of the intersection. Ahlswede and Khachatrian [3] proved in 1997 what they called the "Complete Erdős–Ko–Rado Theorem." This theorem gives the size and structure of the maximum t-intersecting k-set systems for all values of n.

The characterization is up to isomorphism: two set systems on an n-set are *isomorphic* if one system can be can be obtained from the other by a permutation of the underlying n-set. Any two of the canonical t-intersecting k-set systems are isomorphic (and these are the only systems isomorphic to the canonical t-intersecting systems).

1.3.1 Theorem. *Let t, k, n be positive integers with $1 \leq t \leq k \leq n$ and r be an integer with $r \leq k - t$. If*

$$(k - t + 1)\left(2 + \frac{t-1}{r+1}\right) < n < (k - t + 1)\left(2 + \frac{t-1}{r}\right), \qquad (1.3.1)$$

then $\mathcal{F}_r$ is the unique (up to isomorphism) t-intersecting k-set system with maximal cardinality. (By convention, $\frac{t-1}{r} = \infty$ for $r = 0$.) If

$$n = (k - t + 1)\left(2 + \frac{t-1}{r+1}\right), \qquad (1.3.2)$$

then $|\mathcal{F}_r| = |\mathcal{F}_{r+1}|$, and these systems are the unique (up to isomorphism) t-intersecting k-set systems with maximal cardinality.

This theorem covers all relevant values of n, since if $r = k - t + 1$ then $n < 2k - t + 1$ and any two k-sets would be t-intersecting. Further, this theorem includes Theorem 1.0.2 with the exact lower bound on n. In particular, if $n > (k - t + 1)(t + 1)$ then $r = 0$ in inequality (1.3.1), and the canonical t-intersecting system $\mathcal{F}_0$ is the largest system.

The proof of the Erdős–Ko–Rado Theorem for t-intersecting families is much more complicated than the proof of the Erdős–Ko–Rado Theorem for intersecting families. The only known proof of the general result, given by Ahlswede and Khachatrian [3], also uses the shifting operation defined in Section 1.1. We do not include the proof here, but we refer the reader to the book *Lectures on Advances in Combinatorics* by Ahlswede and Blinovsky [2], where the proof of this theorem is presented in Lecture 1. We give two examples to help illustrate what this theorem means.

The example from Figure 1.1 is a 2-intersecting 4-set system on an 8-set (so $t = 2$, $k = 4$ and $n = 8$). The value of r that satisfies

$$(k - t + 1)\left(2 + \frac{t-1}{r+1}\right) < n < (k - t + 1)\left(2 + \frac{t-1}{r}\right)$$

for these values of n, k and t is $r = 1$. This means that the largest 2-intersecting 4-set system on an 8-set is isomorphic to the system of 4-sets

$$\mathcal{F}_1 = \{A : |A \cap \{1, 2, 3, 4\}| \geq 3\},$$

which has size

$$\binom{4}{3}\binom{4}{1} + \binom{4}{4} = 17.$$

If $t = 1$, the inequality (1.3.1) in Theorem 1.3.1 reduces to

$$2k < n < 2\left(k + \frac{0}{r}\right).$$

If $n > 2k$ this is only satisfied when $r = 0$ (with the convention that $\frac{0}{0} = \infty$) and the unique maximum system is $\mathcal{F}_0$, the canonical intersecting system.

On the other hand, if $n = 2k$ then (1.3.2) in the complete Erdős–Ko–Rado theorem reduces to

$$n = k\left(2 + \frac{0}{r+1}\right)$$

which is satisfied for any value of $r \in \{0, \ldots, k-1\}$. In this case there are many non-isomorphic systems that meet the bound of $\binom{n-1}{k-1}$, since the system $\mathcal{F}_r$ is a maximum system for all $r \in \{0, \ldots, k-1\}$. To see that all these systems have the same size, consider

$$
\begin{aligned}
|\mathcal{F}_r| &= \sum_{i=r+1}^{2r+1} \binom{2r+1}{i}\binom{2k-(2r+1)}{k-i} \\
&= \frac{1}{2}\sum_{i=0}^{2r+1} \binom{2r+1}{i}\binom{2k-2r-1}{k-i} \qquad (1.3.3) \\
&= \frac{1}{2}\binom{2k}{k} \\
&= \binom{2k-1}{k-1}.
\end{aligned}
$$

We see that (1.3.3) holds since

$$
\begin{aligned}
\sum_{i=r+1}^{2r+1} \binom{2r+1}{i}\binom{2k-2r-1}{k-i} &= \sum_{i=r+1}^{2r+1} \binom{2r+1}{2r+1-i}\binom{2k-2r-1}{k+i-2r-1} \\
&= \sum_{i=0}^{r} \binom{2r+1}{i}\binom{2k-2r-1}{k-i}.
\end{aligned}
$$

In particular, the collection of all k-subsets that contain 1 form an intersecting set system of this size; this is isomorphic to the set system $\mathcal{F}_0$. Also, the collection of all k-subsets that contain at least two elements from the set $\{1, 2, 3\}$ is of this size. Further, the system of all sets that do not contain 1 also form an intersecting set system that is isomorphic to the system $\mathcal{F}_{k-1}$.

1.4 The shifting technique

Shortly after the original proof of the Erdős–Ko–Rado theorem appeared, there was a rush to find shorter and more elegant proofs of the result, particularly for the case when $t = 1$. In this, and the following section, two of these proofs are outlined (and a third is outlined in the Exercises). In Section 2.14, Section 6.6 and Section 7.1, three more more proofs of this result are given.

Our first alternate proof of the Erdős–Ko–Rado theorem appears in Frankl's paper on the shifting technique in extremal set theory [68]. This proof only establishes the bound in the theorem, but this method can be generalized to a nice proof of the Hilton–Milner Theorem (see Section 1.6.)

1.4.1 Lemma. *Suppose $\mathcal{A}$ is a left-shifted intersecting k-set system. Then for all $A, B \in \mathcal{A}$,*

$$A \cap B \cap \{1, \ldots, 2k - 1\} \neq \emptyset.$$

Proof. If this statement does not hold, consider sets A and B with

$$A \cap B \cap \{1, \ldots, 2k - 1\} = \emptyset,$$

where $|A \cap \{1, \ldots, 2k - 1\}|$ is maximized. Then there exists a $j \in A \cap B$ with $j > 2k - 1$. Further, since $|A \setminus \{j\}| = k - 1$ and $|B \setminus \{j\}| = k - 1$, there is an $i \leq 2k - 1$ with $i \notin A \cup B$. Since the system $\mathcal{A}$ is left-shifted, the set $(A \setminus \{j\}) \cup \{i\}$ is in $\mathcal{A}$ and has

$$((A \setminus \{j\}) \cup \{i\}) \cap B \cap \{1, \ldots, 2k - 1\} = \emptyset,$$

but this contradicts the maximality of $|A \cap \{1, \ldots, 2k - 1\}|$ among all such sets. $\qquad\square$

This proposition can be used to prove the bound in the Erdős–Ko–Rado theorem when $t = 1$ and $n \geq 2k$. We apply induction on the size of k. Clearly, if $k = 1$ then the statement is trivial, and if $n = 2k$ it follows easily since $\mathcal{A}$ cannot contain both a set and its complement.

For $n > 2k$, assume that $\mathcal{A}$ is a left-shifted intersecting k-set system. Define

$$\mathcal{A}_i = \{A \cap \{1, \ldots, 2k\} : A \in \mathcal{A}, |A \cap \{1, \ldots, 2k\}| = i\}.$$

By Lemma 1.4.1 for each i in $\{1, \ldots, k\}$ the system $\mathcal{A}_i$ is an intersecting i-set system. By induction, for each $i = 1, 2, \ldots, k - 1$,

$$|\mathcal{A}_i| \leq \binom{2k - 1}{i - 1}$$

and since $\mathcal{A}_k$ is an intersecting k-set system on a $2k$ set,

$$|\mathcal{A}_k| \leq \frac{1}{2}\binom{2k}{k} = \binom{2k - 1}{k - 1}.$$

For each set $B \in \mathcal{A}_i$ there are at most $\binom{n-2k}{k-i}$ sets A in $\mathcal{A}$ such that $B \subseteq A$, and thus we have

$$|\mathcal{A}| \leq \sum_{i=1}^{k} \binom{2k - 1}{i - 1}\binom{n - 2k}{k - i} = \binom{n - 1}{k - 1}.$$

1.5 The Kruskal–Katona Theorem

Our next proof of the Erdős–Ko–Rado Theorem follows Katona [105] and Daykin [49]. It is based on the Kruskal–Katona Theorem, which gives a lower bound on the size of a *shadow of a set system*. The shadow $\partial(\mathcal{A})$ of the k-set system $\mathcal{A}$ is defined by

$$\partial(\mathcal{A}) = \{B : |B| = k - 1 \text{ and } B \subseteq A \text{ for some } A \in \mathcal{A}\},$$

and the gth shadow $\partial^g(\mathcal{A})$ of a k-set system is the collection

$$\partial^g(\mathcal{A}) = \{B : |B| = k - g \text{ and } B \subseteq A \text{ for some } A \in \mathcal{A}\}.$$

A set system in which no set is a subset of another in the system is called a *Sperner set system*. In the Exercises you are invited to prove that if $\mathcal{S}$ is a Sperner set system on a set of size n, then

$$|\mathcal{S}| \leq \binom{n}{\lfloor \frac{n}{2} \rfloor}.$$

This result is known as *Sperner's Theorem*. The next theorem is called the *LYM inequality* (named after Lubell [121], Yamamoto [174] and Mešalkin [134]); it is a refinement of Sperner's Theorem.

1.5.1 Theorem. *If $\mathcal{A}$ is a Sperner set system on an n-set, then*

$$\sum_{A \in \mathcal{A}} \binom{n}{|A|}^{-1} \leq 1. \qquad \square$$

Using this inequality, we can get a lower bound on the size of the shadow of a set system.

1.5.2 Corollary. *If $\mathcal{A}$ is a k-set system on an n-set, then*

$$|\mathcal{A}| \binom{n}{k-1} \leq |\partial(\mathcal{A})| \binom{n}{k}.$$

Proof. Let $\mathcal{B}$ be the collection of all $(k-1)$-subsets of $\{1, \ldots, n\}$ that are not in $\partial(\mathcal{A})$. Then the set system $\mathcal{A} \cup \mathcal{B}$ is a Sperner set system, and by the previous theorem and the LYM inequality,

$$\sum_{A \in \mathcal{A} \cup \mathcal{B}} \binom{n}{|A|}^{-1} = \frac{|\mathcal{A}|}{\binom{n}{k}} + \frac{\binom{n}{k-1} - |\partial(\mathcal{A})|}{\binom{n}{k-1}} = 1 + \frac{|\mathcal{A}|}{\binom{n}{k}} - \frac{\partial(\mathcal{A})}{\binom{n}{k-1}} \leq 1$$

and the result follows.   □

Katona [105] extended this result to gth shadows of intersecting set systems, as follows.

1.5.3 Theorem. *Let $0 \leq g \leq k-1$ and $1 \leq t \leq k$ and $g \leq t$. If $\mathcal{A}$ is a t-intersecting k-set system on an n-set, then*

$$|\mathcal{A}| \binom{2k-t}{k-g} \leq |\partial^g(\mathcal{A})| \binom{2k-t}{k},$$

and equality holds if and only if $t = g$ or $g = 0$.   □

Katona's Theorem can be used to prove the bound in the Erdős–Ko–Rado Theorem. Assume $\mathcal{A}$ is an intersecting k-set system on the n-set $\{1, \ldots, n\}$. If we set

$$\mathcal{B} = \mathcal{A}^c = \{\{1, \ldots, n\} \setminus A : A \in \mathcal{A}\},$$

then each $B \in \mathcal{B}$ has cardinality $n - k$. Further, any distinct pair $B_1, B_2 \in \mathcal{B}$ with $B_i = \{1, \ldots, n\} \setminus A_i$ for $i = 1, 2$ has

$$|B_1 \cap B_2| = |\{1, \ldots, n\} \setminus (A_1 \cup A_2)| \geq n - (2k - 1).$$

Thus the system $\mathcal{B}$ is an $(n - 2k + 1)$-intersecting $(n - k)$-set system.

Consider the $(n - 2k)$-shadow of $\mathcal{B}$. This system is denoted $\partial^{n-2k}(\mathcal{B})$ and is a k-set system. Further, $\partial^{n-2k}(\mathcal{B}) \cap \mathcal{A} = \emptyset$, which implies that

$$|\partial^{n-2k}(\mathcal{B}) \cup \mathcal{A}| = |\partial^{n-2k}(\mathcal{B})| + |\mathcal{A}| \leq \binom{n}{k}.$$

Then we have the following:

$$|\mathcal{B}| \frac{\binom{n-1}{k}}{\binom{n-1}{k-1}} = |\mathcal{B}| \frac{\binom{2(n-k)-(n-2k+1)}{k}}{\binom{2(n-k)-(n-2k+1)}{n-k}} \leq |\partial^{n-2k}(\mathcal{B})|.$$

The last inequality follows from applying Katona's Theorem (Theorem 1.5.3) to $\mathcal{B}$.

Finally, since $|\mathcal{A}| = |\mathcal{B}|$,

$$\binom{n}{k} \geq |\mathcal{A}| + |\partial^{n-2k}(\mathcal{B})| \geq |\mathcal{A}| + |\mathcal{A}|\frac{\binom{n-1}{k}}{\binom{n-1}{k-1}} = |\mathcal{A}|\frac{\binom{n}{k}}{\binom{n-1}{k-1}},$$

and $|\mathcal{A}| \leq \binom{n-1}{k-1}$. $\qquad\qquad\square$

Before we can proceed with the proof that only a canonical intersecting system can meet this bound, we need to introduce an ordering on k-sets. There are several different possible orderings of the collection of all k-subsets of an n-set; the one we will use is the *co-lexicographic order*. In the co-lexicographic order, we have $A < B$ if and only if A and B are distinct and the largest element in $(A \cup B) \setminus (A \cap B)$ is in B. In this ordering, if $k = 4$, the first sets are

$$\{1, 2, 3, 4\}, \{1, 2, 3, 5\}, \{1, 2, 4, 5\}, \{1, 3, 4, 5\}, \{2, 3, 4, 5\}, \{1, 2, 3, 6\}, \ldots$$

Note that the first elements are the same for all n. For any k-set system, let $\partial(|\mathcal{A}|)$ represent the number of $(k - 1)$-sets in the shadow of the first (in the co-lexicographic order) $|\mathcal{A}|$ sets of size k of $\{1, \ldots, n\}$.

We also need the following result which is known as the Kruskal–Katona Theorem [103, 107].

1.5.4 Theorem. *Assume $\mathcal{A}$ is a k-set system on an n-set. Then*

$$\partial(|\mathcal{A}|) \leq |\partial(\mathcal{A})|,$$

and if $|\mathcal{A}| = \binom{n_k}{k}$ for some $k \leq n_k \leq n$, then equality holds if and only if $\mathcal{A}$ is isomorphic to the collection of all k-subsets of $\{1, \ldots, n_k\}$. $\qquad\square$

The Kruskal–Katona Theorem implies that if

$$|\mathcal{A}| = \binom{a_k}{k} + \binom{a_{k-1}}{k-1} + \cdots + \binom{a_s}{s}$$

with $a_k > a_{k-1} > \cdots > a_s$, then

$$|\partial(\mathcal{A})| = \binom{a_k}{k-1} + \binom{a_{k-1}}{k-2} + \cdots + \binom{a_s}{s-1}$$

(the proof of this implication is left as an exercise).

If we define for a real number x

$$\binom{x}{k} = \frac{x(x-1)(x-2)\cdots(x-k+1)}{k!},$$

then for any k-set system $\mathcal{A}$ there is a real number $x \leq n$ such that

$$|\mathcal{A}| = \binom{x}{k}.$$

Lovász [117, Problem 13.31] proved that if $|\mathcal{A}| = \binom{x}{k}$, then $|\partial(\mathcal{A})| \geq \binom{x}{k-1}$; applying this bound repeatedly, we have that

$$|\partial^g(\mathcal{A})| \geq \binom{x}{k-g} \qquad (1.5.1)$$

for $1 \leq g \leq k$.

We can now characterize all the 1-intersecting k-set systems of maximum size. Assume for $n > 2k$ that $\mathcal{A}$ is an intersecting k-set system on an n-set with size $\binom{n-1}{k-1}$.

Again consider the $(n - k)$-set system

$$\mathcal{B} = \mathcal{A}^c = \{X \setminus A : A \in \mathcal{A}\}.$$

Since

$$|\mathcal{B}| = |\mathcal{A}| = \binom{n-1}{k-1} = \binom{n-1}{n-k},$$

the first $|\mathcal{B}|$ elements of X of size $n - k$ in the co-lexicographic order are exactly all the sets of size $n - k$ that do not contain n.

By the Kruskal–Katona Theorem (specifically by (1.5.1)),

$$|\partial^{n-2k}(\mathcal{B})| \geq \binom{n-1}{n-k-(n-2k)} = \binom{n-1}{k}.$$

Since

$$|\mathcal{A}| + |\partial^{n-2k}(\mathcal{B})| \leq \binom{n}{k}$$

and $|\mathcal{A}| = \binom{n-1}{k-1}$, it must be the case that $|\partial^{n-2k}(\mathcal{B})| = \binom{n-1}{k}$. This implies that

$$|\partial^{n-2k-i}(\mathcal{B})| = \binom{n-1}{k+i}$$

for all $i \in \{1, \ldots, n - 2k - 1\}$. In particular we have that $|\mathcal{B}| = \binom{n-1}{n-k}$ and

$$|\partial(\mathcal{B})| = \binom{n-1}{n-k-1} = \partial(|\mathcal{B}|).$$

By the Kruskal–Katona Theorem, these two facts imply that $\mathcal{B}$ is isomorphic to the collection of all $(n - k)$-sets that miss the element n, and this happens only when $\mathcal{A}$ is isomorphic to the collection of all k-sets that contain that element.   □

The Kruskal–Katona Theorem and the Erdős–Ko–Rado Theorem are perhaps the two most central theorems in combinatorial set theory. It is therefore quite interesting that we can derive the EKR Theorem (for intersecting families) from the Kruskal–Katona Theorem. No derivation of the Kruska–Katona Theorem from the EKR Theorem is known.

1.6 The Hilton–Milner Theorem

Erdős, Ko and Rado conjectured that the largest 1-intersecting system that was not a subset of the canonical intersecting set system is the set of all k-subsets that contain at least two elements from a fixed set of three elements (this is the set $\mathcal{F}_1$ from Section 1.2). It turns out that this conjecture is not true; the actual maximum sets were given by Hilton and Milner [95].

Hilton and Milner proved that the largest intersecting system that is not a subset of a canonical intersecting system can be constructed as follows. Let $\mathcal{F}_0$ be the set system of all k-sets that contain the element 1, and let $A = \{2, 3, \ldots, k + 1\}$. Define $\mathcal{F}'$ to be the system of all the sets in $\mathcal{F}_0$ that intersect A, together with the set A. This system is intersecting and has size

$$\binom{n-1}{k-1} - \binom{n-k-1}{k-1} + 1.$$

The next result is known as the Hilton–Milner Theorem [95]. Hilton and Milner's original proof is rather involved and actually proves a more general statement, so we give the simpler proof by Frankl and Füredi that can be found in [66]. This is very similar to Frankl's proof of the Erdős–Ko–Rado Theorem given in Section 1.4.

1.6.1 Theorem. *Let k and n be positive integers with $2 \leq k \leq \frac{n}{2}$. Let $\mathcal{A}$ be an intersecting k-set system on an n-set such that $\cap_{A \in \mathcal{A}} A = \emptyset$. Then*

$$|\mathcal{A}| \leq \binom{n-1}{k-1} - \binom{n-k-1}{k-1} + 1.$$

Proof. We prove this theorem by induction on k. If $k = 2$, then $\mathcal{A}$ must be isomorphic to the system $\{\{1, 2\}, \{1, 3\}, \{2, 3\}\}$, so the result holds. Assume the bound holds for all $k_0 < k$.

Let $\mathcal{A}$ be a maximal collection of intersecting k-sets that is not a canonical intersecting system or a subset of one. Apply the shifting operation $S_{i,j}$ for $i, j \in \{1, \ldots, n\}$ to the system $\mathcal{A}$ either until the system is a canonical system, or until it is a left-shifted system.

In the first case, we can assume that $\mathcal{A}$ is not a canonical intersecting system, but that $S_{x,y}(\mathcal{A})$ is. In this case all the sets in $S_{x,y}(\mathcal{A})$ contain x, this means that every set in $\mathcal{A}$ contains either x or y. Since $\mathcal{A}$ is maximal, we can further assume that every k-subset of $\{1, \ldots, n\}$ that contains both x and y is in $\mathcal{A}$.

Apply all (i, j)-shifts with $i, j \in \{1, \ldots, n\} \setminus \{x, y\}$ to $\mathcal{A}$ and call this new system $\mathcal{B}$. Every set in $\mathcal{B}$ contains at least one of x or y (since the (i, j)-shifts used to obtain $\mathcal{B}$ will never remove either of these elements). Further, this set is left-shifted in the sense that $S_{i,j}(\mathcal{B}) = \mathcal{B}$ whenever

$$i, j \in \{1, \ldots, n\} \setminus \{x, y\}.$$

Set Y to be the first $2k - 2$ elements of $\{1, \ldots, n\} \setminus \{x, y\}$ together with $\{x, y\}$. Then for any $A, B \in \mathcal{B}$,

$$A \cap B \cap Y \neq \emptyset. \tag{1.6.1}$$

(The proof of this is very similar to the proof of Lemma 1.4.1, so we omit it.)

Define

$$\mathcal{B}_i = \{B \cap Y \; : \; B \in \mathcal{B}, \; |B \cap Y| = i\}.$$

It is clear that $|\mathcal{B}_1| = 0$ since $\mathcal{B}$ is not a canonical intersecting system. It is also not difficult to see that

$$\mathcal{B}_k \leq \frac{1}{2} \binom{2k}{k} = \binom{2k - 1}{k - 1}.$$

We claim that for all $i \in \{1, \ldots, k - 1\}$

$$|\mathcal{B}_i| \leq \binom{2k - 1}{i - 1} - \binom{k - 1}{i - 1}.$$

Assume $i \in \{2, \ldots, k - 1\}$. Then by (1.6.1), $\mathcal{B}_i$ is an intersecting i-set system with elements from a set of size $2k$ (namely, Y). Thus, if the claim does not hold, then

$$|\mathcal{B}_i| > \binom{2k - 1}{i - 1} - \binom{k - 1}{i - 1} \geq \binom{2k - 1}{i - 1} - \binom{2k - i - 1}{i - 1} + 1.$$

By the induction hypothesis, $\mathcal{B}_i$ must be a canonical intersecting system (or the subset of a canonical intersecting system). This implies that every set in $\mathcal{B}_i$ contains a common element a. But, since $\mathcal{B}$ is not a canonical intersecting system, there is a k-set in $\mathcal{B}$ that does not contain a. We can bound the size of $\mathcal{B}_i$ by counting the number of i-sets from Y that contain the element a and subtracting the number of these sets that do not intersect with the k-set in $\mathcal{B}$ that does not contain a. Thus we have that

$$|\mathcal{B}_i| \leq \binom{2k - 1}{i - 1} - \binom{k - 1}{i - 1} \tag{1.6.2}$$

and this proves our claim.

For each i-set B in $\mathcal{B}_i$, there are at most $\binom{n-2k}{k-i}$ k-sets B' in $\mathcal{B}$ with $B' \cap Y = B$; thus the total number of sets in $\mathcal{B}$ is no more than

$$\binom{2k-1}{k-1} + \sum_{i=1}^{k-1} \left(\binom{2k-1}{i-1} - \binom{k-1}{i-1} \right) \binom{n-2k}{k-i}$$
$$\leq \binom{n-1}{k-1} - \binom{n-k-1}{k-1} + 1.$$

We still need to consider the situation when applying the shifting operation to $\mathcal{A}$ produces a left-shifted set that is not canonical. In this case the proof is the same from (1.6.1), except that we define the set Y to be the set $\{1, \ldots, 2k\}$. $\square$

For $k > 3$, this proof can also be used to show that equality holds if and only if $\mathcal{A}$ is isomorphic to the system of all sets in $\mathcal{F}_0$ that intersect $\{2, \ldots, k+1\}$ together with the set $\{2, \ldots, k+1\}$. The key to this is to note that equality must hold for every i in (1.6.2) and in particular for $i = 2$. Finally, if $k = 3$ there are two non-isomorphic systems that meet this bound, the one described above and $\mathcal{F}_1$.

1.7 Cross-intersecting sets

There are many ways to create variants of the EKR Theorem; in this section we consider a different type of intersection and in the following section we add a restriction on the sets in the set systems that we consider.

A pair of set systems $(\mathcal{A}, \mathcal{B})$ is *cross intersecting* if every set from $\mathcal{A}$ has nontrivial intersection with every set from $\mathcal{B}$. Similarly, $(\mathcal{A}, \mathcal{B})$ is cross t-intersecting if every set from $\mathcal{A}$ is t-intersecting with every set from $\mathcal{B}$. We consider only the case where the sets in $\mathcal{A}$ are restricted to k-sets and the sets in $\mathcal{B}$ are ℓ-sets. The values of k and ℓ are not required to be equal, but if they are, the systems $\mathcal{A}$ and $\mathcal{B}$ need not be disjoint.

Setting $\mathcal{A}$ to be the collection of all k-sets that contain a fixed element and $\mathcal{B}$ to be all ℓ-sets that contain the same fixed element, we construct a cross-intersecting set with

$$|\mathcal{A}|\,|\mathcal{B}| \leq \binom{n-1}{k-1} \binom{n-1}{\ell-1}.$$

The next result states that, if $n \geq 2k$ and $n \geq 2\ell$, this is the largest possible value for $|\mathcal{A}|\,|\mathcal{B}|$. This result was first proved by Pyber [142] (with a weaker lower bound on n). Pyber's proof uses a combination of the Kruskal–Katona Theorem and a proof similar to Katona's cycle proof outlined in Exercise 1.7 of this chapter. Pyber's bound on n was later improved by Matsumoto and

Tokushige [126]. We give only a brief outline of their proof, since it involves careful and detailed manipulations with binomial coefficients.

1.7.1 Theorem. *Let n, k and ℓ be integers with 2k ≤ n and 2ℓ ≤ n. Let $\mathcal{A}$ be a k-set system and $\mathcal{B}$ be an ℓ-set system, and assume that both systems have $\{1, \ldots, n\}$ as their underlying set. If $(\mathcal{A}, \mathcal{B})$ is cross intersecting, then*

$$|\mathcal{A}|\,|\mathcal{B}| \leq \binom{n-1}{k-1}\binom{n-1}{\ell-1}.$$

Moreover, if 2k < n and 2ℓ < n, equality holds if and only if there is an $i \in \{1, \ldots, n\}$ such that $\mathcal{A}$ consists of all k-sets that contain i and $\mathcal{B}$ consists of all ℓ-sets containing i.

Proof. If both $|\mathcal{A}| < \binom{n-1}{k-1}$ and $|\mathcal{B}| < \binom{n-1}{\ell-1}$, then the theorem obviously holds; thus we will assume that $|\mathcal{A}| \geq \binom{n-1}{k-1}$. Since $|\mathcal{A}| \geq \binom{n-1}{k-1}$, we can express

$$|\mathcal{A}| = \binom{n-1}{k-1} + \binom{n-2}{k-1} + \cdots + \binom{n-j}{k-1} + \binom{x}{n-k-j}, \qquad (1.7.1)$$

where j is an integer in $\{1, \ldots, n-k\}$ and x is some real number with $n - k - j \leq x \leq n - j - 1$.

The pair $(\mathcal{A}, \mathcal{B})$ is cross intersecting if and only if the sets $\partial^{n-k-\ell}(\mathcal{A}^c)$ and $\mathcal{B}$ are disjoint; thus

$$|\mathcal{A}|\,|\mathcal{B}| \leq |\mathcal{A}|\left(\binom{n}{\ell} - |\partial^{n-k-\ell}(\mathcal{A}^c)|\right).$$

This is maximized when $|\partial^{n-k-\ell}(\mathcal{A}^c)|$ is minimized, by the Kruskal–Katona Theorem (Theorem 1.5.4). This happens when $\mathcal{A}^c$ consists of the first $|\mathcal{A}|$ sets of size $(n - k)$ in the co-lexicographic order. From the expression of $|\mathcal{A}|$ given in (1.7.1) and the comments following the Kruskal–Katona Theorem, we have that

$$|\partial^{n-k-\ell}(\mathcal{A}^c)| \geq \binom{n-1}{\ell} + \binom{n-2}{\ell-1} + \cdots + \binom{n-j}{\ell-j+1} + \binom{x}{\ell-j}.$$

Since

$$\binom{n}{\ell} - \left(\binom{n-1}{\ell} + \cdots + \binom{n-j}{\ell-j+1} + \binom{x}{\ell-j}\right) = \binom{n-j}{\ell-j} - \binom{x}{\ell-j},$$

the bound in this theorem can be proven by showing that

$$\left(\binom{n-1}{k-1} + \cdots + \binom{n-j}{k-1} + \binom{x}{n-k-j}\right)\left(\binom{n-j}{\ell-j} - \binom{x}{\ell-j}\right)$$

is no more than

$$\binom{n-1}{k-1}\binom{n-1}{\ell-1}.$$

Since $\binom{n-i}{k-1} \le \binom{n-1}{k-1}$ for $1 \le i \le j$, and $\binom{x}{n-k-j} \le \binom{n-j-1}{n-k-j} = \binom{n-j-1}{k-1}$, a very rough upper bound on the left-hand side is

$$\left((j+1)\binom{n-1}{k-1}\right)\binom{n-j}{\ell-j}.$$

If $j \ge 3$, the inequality is straightforward using these bounds since

$$(j+1)\binom{n-1}{k-1}\binom{n-j}{\ell-j} = (j+1)\binom{n-1}{k-1}\frac{\ell-j+1}{n-j+1}\cdots\frac{\ell-1}{n-1}\binom{n-1}{\ell-1}$$

$$\le \frac{j+1}{2^{j-1}}\binom{n-1}{k-1}\binom{n-1}{\ell-1}$$

$$\le \binom{n-1}{k-1}\binom{n-1}{\ell-1}.$$

This result also holds for $j = 1$ and $j = 2$, but, since in both of these cases the calculations are much more involved, we omit the details and refer the reader to [126]. $\square$

Theorem 1.7.1 actually implies the EKR Theorem for intersecting sets. If $\mathcal{A}$ is any intersecting k-set system, then $(\mathcal{A}, \mathcal{A})$ is a cross-intersecting family. By Theorem 1.7.1,

$$|\mathcal{A}|^2 \le \binom{n-1}{k-1}^2$$

and equality holds only if $\mathcal{A}$ is a canonical intersecting system.

1.8 Separated sets

In the final section of this chapter, we give a different variation of the EKR Theorem; in this case we determine the largest intersecting set system in which all the sets satisfy an additional condition.

Let X be the circulant graph on $\mathbb{Z}_n$ with connection set $\{\pm 1, \pm 2, \ldots, \pm r\}$. A k-subset $A = \{a_1, a_2, \ldots, a_k\}$ of $\mathbb{Z}_n$ is an *r-separated set* if for any i and j the numbers a_i and a_j are not adjacent in X. To be consistent with the rest of this chapter, we use n rather than 0 in our sets. If the numbers 1 to n are arranged in a cycle, then any two elements in an r-separated set are separated by at least r numbers in the cycle. If $r = 1$, then an r-separated set is simply called a *separated* set.

It was conjectured by Holroyd and Johnson that if $n \geq (r + 1)k$, then the largest system of intersecting separated k-sets from an n-set is the collection of all separated sets that contain a fixed element (see [173] or [96, Problem BCC 16.25]). Such a collection will, not surprisingly, be called a *canonical intersecting separated set system*. Talbot [159] showed that this conjecture is true for $r = 1$ and, provided that $n \neq 2k + 2$, the only maximum sets are the canonical ones. Talbot actually proved this result for all r, but we only present a brief outline of the result for separated sets (the proof for $r > 1$ is similar).

1.8.1 Theorem. *Let* $n \geq 2k$ *and* $n \neq 2k - 2$. *If* $\mathcal{A}$ *is a system of intersecting separated k-sets from an n-set, then the size of* $\mathcal{A}$ *is no larger than the size of a canonical intersecting separated k-set system and equality holds if and only if* $\mathcal{A}$ *is a canonical system.* □

The proof of this result uses a shifting operation similar to the one used in the original proof of the EKR Theorem, but in the case of separated sets this shifting needs to be done more carefully. Define the shifting operation f on a separated set $A = \{a_1, a_2, \ldots, a_k\}$ by

$$f(A) = \begin{cases} \{1, a_2 - 1, \ldots, a_k - 1\}, & \text{if } a_1 = 1; \\ \{a_1 - 1, a_2 - 1, \ldots, a_k - 1\}, & \text{otherwise.} \end{cases}$$

For a system of separated sets $\mathcal{A}$, define

$$f(\mathcal{A}) = \{f(A) : A \in \mathcal{A}\}.$$

If A is a separated k-set, then $f(A)$ will be a k-subset of $\{1, \ldots, n - 1\}$, but it may not be a separated set; specifically, if the set A contains either the pair $\{1, 3\}$ or the pair $\{2, n\}$, then $f(A)$ will not be a separated set. Further, if the symmetric difference between two separated sets A and B is exactly $\{1, 2\}$, then $f(A) = f(B)$. So, unlike the compression operation for set systems, we do not have that $f(\mathcal{A})$ has the same size as $\mathcal{A}$ and, moreover, it may not be an intersecting separated k-set system.

To deal with these two problems, two systems of separated k-sets are constructed. The first, called $\mathcal{A}_1$, consists of the sets that are "problem" sets; these are the sets A for which $f(A)$ is not separated, and the sets A, with $1 \in A$, for which there exists another set $B \in \mathcal{A}$ with $f(A) = f(B)$. The second system, called $\mathcal{A}_2$, consists of the remaining sets. It can be shown that the system $\mathcal{A}'_1$ formed by removing the element 1 from every set in $f(\mathcal{A}_1)$ is an intersecting separated $(k - 1)$-set system from the set $\{1, \ldots, n - 2\}$. It can also be shown (in fact it is easier to see) that $\mathcal{A}'_2 = f(\mathcal{A}_2)$ is an intersecting separated k-set system from $\{1, \ldots, n - 1\}$. Then

$$|\mathcal{A}| = |\mathcal{A}'_1| + |\mathcal{A}'_2|.$$

Using induction on n gives a bound on each of $\mathcal{A}'_1$ and $\mathcal{A}'_2$, which gives the desired bound on $|\mathcal{A}|$. The characterization of when equality holds requires careful consideration the structure that $\mathcal{A}_1$ must have so that the induction bound holds with equality on $\mathcal{A}'_1$.

This result can be considered as the EKR Theorem for separated sets; it is our first example of an EKR type result for objects other than k-sets. This direction is one that we continue throughout the book. In future chapters, we consider EKR Theorems for vector spaces over a finite field, integer sequences, permutations and partitions. In general, if there is a collection of objects for which there is some notion of intersection, then we can ask what is the largest set of these objects such that any two are intersecting? If the answer is the largest collection in which any two intersect in the same place, or in some "canonical" fashion, then we say a version of the EKR Theorem holds.

1.9 Exercises

1.1 Assume that $n \geq 2k - t$. Show that a maximal t-intersecting k-set system on an n-set is not also a $(t + 1)$-intersecting system.

1.2 Show that the maximum size of a Sperner set system on an n-set is $\binom{n}{\lfloor \frac{n}{2} \rfloor}$, without using the LYM inequality! (See either [23] or [153].)

1.3 The original proof of the EKR Theorem did not require all sets to have size k; rather it required that any set in the intersecting system have size at most k, and that no set in the system be contained in another set. Prove that these conditions yield the same bound on the size of the system.

1.4 Let $\mathcal{A}$ be a canonical intersecting k-set system on an n-set. Prove that for all i, j in $\{1, \ldots, n\}$, the system $S_{i,j}(\mathcal{A})$ is also a canonical intersecting system.

1.5 Let $\mathcal{A}$ be a set system that is isomorphic to $\mathcal{F}_i$ for some i. Then $\mathcal{A}$ can be shifted until it is left-shifted; show that this left-shifted system is $\mathcal{F}_i$. Does the shifting operation preserve isomorphism? Specifically, can you find two non-isomorphic set systems that can be shifted to the same system?

1.6 Prove that the Kruskal–Katona Theorem implies that if

$$|\mathcal{A}| = \binom{a_k}{k} + \binom{a_{k-1}}{k-1} + \cdots + \binom{a_s}{s}$$

with $a_k > a_{k-1} > \cdots > a_s$, then

$$|\partial(\mathcal{A})| = \binom{a_k}{k-1} + \binom{a_{k-1}}{k-2} \cdots \binom{a_s}{s-1}.$$

1.7   There is another popular proof for the EKR Theorem given by Katona in [104], known as *Katona's cycle proof*. The point of this exercise is to develop this proof. (The cycle proof is really a version of the proof given in Section 2.14.) Let $\mathcal{A}$ be an intersecting k-set system on an n-set (with $n \geq 2k$). Consider the pairs of $(\pi, \mathcal{A}')$ where π is a cyclic arrangement of $\{1, \ldots, n\}$ and $\mathcal{A}'$ is a subset of k-sets from $\mathcal{A}$ that occur as a consecutive subsequence of the arrangement. Prove the bound in the EKR Theorem by counting these pairs in two different ways.

1.8   Prove for $k > 3$ the only sets that meet the bound in Theorem 1.6.1 are isomorphic to the system of all sets in $\mathcal{F}_0$ which intersect $\{2, \ldots, k+1\}$ together with the set $\{2, \ldots, k+1\}$.

1.9   Describe all the systems of intersecting separated k-sets that are of maximum size from $\{1, \ldots, n\}$ for $n = 2k + 2$.

Notes

There are many proofs of the EKR Theorem that we have not included, and new proofs using different ideas, or different combinations of ideas, continue to appear. For example, we have not included proofs by Friedgut [72], by Füredi et al. [75], and by Frankl and Füredi [70].

We point out that deriving the bound on the size of an intersecting set system is comparatively easy – it is the characterization of the systems that meet the bound that is challenging. The EKR Theorem for t-intersecting set systems seems to lie deeper than the theorem for intersecting systems. The only successful approach to the latter has been the work of Ahlswede and Khachatrian [3], based on shifting, and the work of Wilson [172] using association schemes. Wilson's approach only applies when $n \geq (t+1)(k-t+1)$, but does extend to other cases. The EKR Theorem for intersecting systems has generalizations to a surprisingly wide range of discrete structures, as will become clear in the course of this book. By comparison, the theorem for t-intersecting systems appears to generalize only to association schemes that are both metric and cometric.

In [164] a result similar to Theorem 1.7.1 for cross-t-intersecting families is proved (with some restrictions on the size of t and the size of n relative to t and k). This result uses the *random walk* method introduced by Frankl in [62] and [68]. Various authors have proven versions of Theorem 1.7.1 for intersecting objects other than sets; for some examples, see [25, 26, 27, 71, 101, 110].

There are many other variations of the EKR Theorem. For example there is a vast amount of literature on the maximum size of a k-set system in which any r sets have nontrivial intersection, or have intersection of size at least t – such set systems are called *r-wise t-intersecting k-set systems*. Frankl proved that a version of the EKR Theorem holds for r-wise intersecting k-sets in [61], and he also determined the largest size of a r-wise t-intersecting set system in which there is no restriction on the size of the sets [63]. In 2011 Tokushige [165] published a version of the EKR Theorem for r-wise t-intersecting k-set systems.

The proof of Theorem 1.8.1 that is outlined in the final section of this chapter is interesting, since it shows some of the complexities involved in shifting operations. There are many generalizations and variations of the shifting operation that we do not discuss in this book. It is an operation that is very useful (it is key to the proof of Theorem 1.3.1). However, it is very easy to get subtle details wrong when using it – for example, there are published proofs using shifting on permutations [19] and vector spaces [48, 53] that are incorrect.

2

Bounds on cocliques

The problem of determining the largest system of intersecting sets can be rephrased as a problem on graphs. For any k and n, define a graph whose vertex set is all k-subsets of $\{1, \ldots, n\}$ and two vertices are adjacent if and only if the k-sets are disjoint. These graphs have been widely studied; they are denoted by $K(n, k)$ and are called *Kneser graphs*. Any two nonadjacent vertices in a Kneser graph are a pair of intersecting sets, and a collection of vertices in which no two vertices are adjacent is an intersecting set system.

A *clique* in a graph is a set of vertices in which any two vertices are adjacent, and a *coclique* (often called an *independent set*) is a set of vertices in which no two vertices are adjacent. The EKR Theorem for 1-intersecting sets is equivalent to determining the size and the structure of the largest cocliques in $K(n, k)$. We follow the standard notation and use $\omega(X)$ to denote the size of the largest clique in a graph X, and $\alpha(X)$ the size of the largest coclique.

Our aim in this chapter is to derive bounds on the size of cocliques and cliques that we will use to provide new proofs of the EKR Theorem. We have chosen proofs that illustrate generally useful techniques; in some cases these may not be the simplest. We focus on bounds that do not make use of the theory of association schemes – we will see how to apply the theory of association schemes in the following chapter. We prove our results in general since we aim to establish analogs of the EKR not just for sets, but also for sequences, partitions, matchings and permutations. (We call these *EKR-type theorems*.)

Background for much of this chapter appears in Godsil and Royle [87]; in particular the missing proofs will be found there.

2.1 A clique-coclique bound

Suppose the graph X on v vertices has a partition into vertex-disjoint cliques of size c. Since any coclique intersects a clique in at most one vertex, we deduce that

$$\alpha(X) \le \frac{v}{c}.$$

The results in this section allow us to derive this bound in cases where there is no partition of $V(X)$ into vertex-disjoint cliques.

We may view a set S of subsets of the vertices of a graph X as an incidence structure, with the vertices of X as its points and the elements of S as its blocks. We say the structure is *block regular* if all subsets in S have the same size, and *point regular* each point lies in the same number of blocks. The *incidence matrix* of the structure is the matrix with the characteristic vectors of the blocks as its columns. In the cases of interest to us, the set S consists of cliques or cocliques.

In the proof that follows, and in the rest of the book, we use **1** to denote the all-ones vector; we use J to denote the all-ones matrix.

2.1.1 Theorem. *Let X be a graph on v vertices and let C be a set of maximum cliques in X. If the vertices of X and the cliques in C form a point- and block-regular incidence structure, then*

$$\alpha(X)\omega(X) \le |V(X)|.$$

If equality holds, then any coclique of maximum size meets each clique from C in exactly one vertex.

Proof. Assume $v = |V(X)|$. Let $\omega(X)$ be the size of the cliques in C, and let r be the number of cliques in C that contain a given vertex of X. Finally, let N be the incidence matrix for the incidence structure given by C.

Let y be the characteristic vector of a coclique S in X, since S intersects each clique in C in at most one vertex

$$y^T N \le \mathbf{1}^T$$

(where the inequality is taken entry-wise). Therefore, since each vertex of X is in exactly r cliques from C,

$$|C| = \mathbf{1}^T\mathbf{1} \ge y^T N\mathbf{1} = ry^T\mathbf{1} = r|S|. \tag{2.1.1}$$

As each clique in C has size $\omega(X)$, we also have the following equality:

$$|C|\omega(X) = r|V(X)|. \tag{2.1.2}$$

Putting (2.1.1) and (2.1.2) together, we see that $\frac{r|V(X)|}{\omega(X)} \ge r|S|$ and the theorem follows. $\square$

If C is a maximum clique in a vertex-transitive graph, then the images of C under $\mathrm{Aut}(X)$ form a point- and block-regular incidence structure. Therefore we have the following corollary, which is known as the clique-coclique bound.

2.1.2 Corollary. *If X is vertex-transitive, then $\alpha(X)\omega(X) \leq |V(X)|$.* $\square$

From this corollary, there is a very easy proof of the bound in the EKR Theorem in the special case where k divides n (details are given in Section 6.1). The characterization in the EKR Theorem can also be proven using these cliques (this is left as an exercise).

Another (and perhaps better) example of a graph for which equality holds in the clique-coclique bound is the *d-cube*. The vertices of this graph are binary sequences of length d where two sequences are adjacent if they differ in just one position. The d-cube is vertex transitive, and because any clique in it has at most two vertices, the set of all edges forms the block incidence structure required in Theorem 2.1.1. By the clique-coclique bound, the size of the maximum coclique is no more than 2^{n-1}. The set of all sequences that contain an even number of entries equal to 0 is a coclique of this size. Note that any edge (so any maximum clique) will have one sequence with an even number of entries equal to 0 and one with an odd number. The d-cube is a graph in the Hamming scheme $H(d, 2)$, which we consider at length in Chapter 10.

2.1.3 Corollary. *Let X be a vertex-transitive graph such that*

$$\alpha(X)\omega(X) = |V(X)|.$$

Let S and C be a coclique and a clique of maximum size, respectively. Let M and N denote the incidence matrices of the incidence structures formed by the distinct translates (under $\mathrm{Aut}(X)$) of S and C, respectively. Then

$$N^T M = J.$$ $\square$

We say that a vector in $\mathbb{R}^n$ is *balanced* if it is orthogonal to the all-ones vector $\mathbf{1}$. If v_S is the characteristic vector of a subset S of the set V, then we say that

$$v_S - \frac{|S|}{|V|}\mathbf{1}$$

is the *balanced characteristic vector* of S. A short calculation shows that the previous corollary is equivalent to saying the balanced characteristic vector of a coclique of size $\alpha(X)$ is orthogonal to the balanced characteristic vector of a clique of size $\omega(X)$.

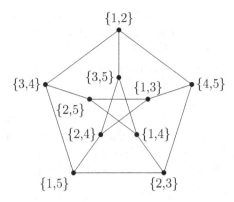

Figure 2.1 The Petersen graph $K(5, 2)$.

2.2 Equitable partitions

Any graph can be represented by a matrix in which the rows and columns correspond to the vertices of the graph and the entries in the matrix are 1 if the vertices are adjacent and 0 otherwise – this matrix is called the *adjacency matrix* of the graph. The eigenvalues of this matrix are called the *eigenvalues of the graph*. In this section we show a method to find eigenvalues of a graph using *equitable partitions*; in the following section we see how eigenvalues can be used to bound the size of a coclique in some graphs.

Let π be a partition of the vertices of the graph X and let $C_1, \ldots, C_r$ be the cells of this partition. This partition is *equitable* if the number of vertices in cell C_j that are adjacent to a vertex in C_i is determined only by i and j, and not by the specific vertex from C_i used. If π is an equitable partition of X, we define the *quotient graph* X/π to be the directed multi-graph with the cells of π as its vertices, and if a vertex in C_i has exactly ν neighbors in C_j, then X/π has ν arcs from C_i to C_j. In general X/π will have loops and parallel arcs.

We consider a simple example. Let X be $K(5, 2)$. This is the graph in Figure 2.1. Its vertices are the unordered pairs of elements of $\{1, 2, 3, 4, 5\}$, and two pairs are adjacent if and only if they are disjoint (this graph is better known as the *Petersen graph*).

The four pairs that contain the element 1 form a coclique S, and each pair that does not contain 1 is disjoint from exactly two pairs in S. Thus the partition $\{S, \bar{S}\}$ is equitable and the adjacency matrix of the corresponding quotient matrix is

$$\begin{pmatrix} 0 & 3 \\ 2 & 1 \end{pmatrix}.$$

Since the rows of this matrix sum to 3, we see that any nonzero constant vector is an eigenvector for the quotient matrix with eigenvalue 3. Since the trace of this matrix is 1, we conclude that the second eigenvalue is -2. It will follow from Lemma 2.2.2 that 3 and -2 are both eigenvalues of the Petersen graph, and for each there is a corresponding eigenvector that is constant on the cells of the partition.

If π is a partition, we use $|\pi|$ to denote the number of cells of π. We can represent π by a $|V(X)| \times |\pi|$ matrix whose ith column is the characteristic vector of the ith cell of π. This matrix is called the *characteristic matrix of the partition*, and it is a 01-matrix whose columns sum to $\mathbf{1}$ – this implies that its columns are linearly independent.

2.2.1 Theorem. *Let π be a partition of the vertices of the graph X with characteristic matrix P, and let A be the adjacency matrix of X. Then the following are equivalent:*

   (a) π *is equitable.*
   (b) *The column space of P is A-invariant.*
   (c) *There is a matrix B such that $AP = PB$.*
   (d) *If B is the adjacency matrix $A(X/\pi)$ of X/π, then $AP = PB$.* □

Assuming that B is the adjacency matrix $A(X/\pi)$, if z is an eigenvector for B with eigenvalue θ, then

$$APz = PBz = \theta Pz.$$

Thus Pz is an eigenvector for A with eigenvalue θ; moreover, it is an eigenvector that is constant on the cells of π. The vector Pz is call the *lift* of z. Note that B is the matrix that represents the action of A on the A-invariant subspace col(P), and hence its characteristic polynomial must divide the characteristic polynomial of A.

2.2.2 Lemma. *If π is an equitable partition of the vertices of the graph X, the characteristic polynomial of X/π divides the characteristic polynomial of X.* □

2.2.3 Corollary. *If π is an equitable partition of the vertices of the graph X, then the eigenvalues of X/π are eigenvalues of X.* □

It is not difficult to prove that the set of orbits of any group of automorphisms of X forms an equitable partition. For example, consider the group of permutations of $\{1, \ldots, n\}$ that map 1 to 1. This group is a subgroup of the automorphism of any Kneser graph $K(n, k)$. It has exactly two orbits; the first is the set of all k-subsets that contain 1, and the second is all sets that do not contain 1. The first orbit is a coclique of size $\binom{n-1}{k-1}$ and the quotient graph for this equitable partition is

$$\begin{pmatrix} 0 & \binom{n-k}{k} \\ \binom{n-k-1}{k-1} & \binom{n-k-1}{k} \end{pmatrix}.$$

The eigenvalues of this matrix are

$$\binom{n-k}{k}, \quad -\binom{n-k-1}{k-1}.$$

By Corollary 2.2.3, these are two eigenvalues of $K(n,k)$.

2.3 Interlacing

Let $\theta_i(M)$ denote the ith largest eigenvalue of a matrix M. If A is an $n \times n$ matrix and B is an $m \times m$ matrix where $m \le n$, we say that the eigenvalues of B *interlace* the eigenvalues of A if the eigenvalues of A and B are all real and the following inequalities hold:

$$\theta_i(B) \le \theta_i(A), \quad i = 1, \ldots, m \tag{2.3.1}$$

and

$$\theta_i(-B) \le \theta_i(-A), \quad i = 1, \ldots, m. \tag{2.3.2}$$

The interlacing is *tight* if there is some s such that equality holds for $i = 1, \ldots, s$ in (2.3.1) and for $i = 1, \ldots, m - s$ in (2.3.2).

The case of interest to us is when $m = 2$. Then interlacing means that

$$\theta_1(B) \le \theta_1(A), \quad \theta_n(A) \le \theta_2(B)$$

and the interlacing is tight if both inequalities are equalities.

Let π be a partition of the vertices of X with cells C_i where $i = 1, \ldots, r$. If π is not equitable, then the number of vertices in the cell C_i adjacent to a fixed vertex of C_j may not be the same for all vertices in C_j, so the quotient graph may not be well defined. In this case, we define $A(X/\pi)$ to be the $r \times r$ matrix with ij-entry equal to the average number of neighbors that a vertex in the cell C_i has in C_j.

2.3.1 Theorem. *Let π be a partition of the vertices of the graph X. Then the eigenvalues of $A(X/\pi)$ interlace the eigenvalues of A; if the interlacing is tight, then π is equitable.* $\square$

By way of example, suppose S is a coclique of size s in a k-regular graph X on v vertices. If $\pi = \{S, \bar{S}\}$, then

$$A(X/\pi) = \begin{pmatrix} 0 & k \\ \frac{sk}{v-s} & k - \frac{sk}{v-s} \end{pmatrix}.$$

The eigenvalues of $A(X/\pi)$ are k and $-sk/(v-s)$, and interlacing implies that

$$\theta_n(A) \leq -\frac{sk}{v-s}.$$

From this we obtain that

$$s \leq \frac{v}{1 - \frac{k}{\theta_n}}.$$

We call this the *ratio bound* for cocliques. This bound will be used frequently throughout; it can be used to prove the bound in many versions of the EKR Theorem.

2.4 The ratio bound for cocliques

In this section we offer a self-contained proof of the ratio bound for cocliques. The proof is useful because it allows us to extract extra information when equality holds. This extra information is the key to proving the characterization in the EKR Theorem, as we will see in Section 6.6.

2.4.1 Theorem. *Let X be a k-regular graph with least eigenvalue τ. Then*

$$\alpha(X) \leq \frac{|V(X)|}{1 - \frac{k}{\tau}}$$

and, if equality holds for some coclique S with characteristic vector v_S, then

$$v_S - \frac{|S|}{|V(X)|}\mathbf{1}$$

is an eigenvector with eigenvalue τ.

Proof. Set $v = |V(X)|$ and $s = |S|$ and denote $A(X)$ by A. If we define

$$M = A - \tau I - \frac{k - \tau}{v}J$$

then M is positive semidefinite. Hence for any vector x we have

$$0 \leq x^T M x = x^T A x - \tau x^T x - \frac{k - \tau}{v}x^T J x. \tag{2.4.1}$$

If x is the characteristic vector of a coclique of size s, then $x^T A x = 0$ and (2.4.1) simplifies to

$$0 \leq 0 - \tau s - \frac{k - \tau}{v} s^2.$$

Hence

$$\tau s \leq \frac{\tau - k}{v} s^2$$

from which the inequality follows.

Now suppose that equality holds, then $x^T M x = 0$ and, since M is positive semidefinite, it follows that $Mx = 0$. Therefore

$$(A - \tau I)x = \frac{k - \tau}{v} J x, \tag{2.4.2}$$

and setting $x = v_S - \frac{s}{v}\mathbf{1}$, we see that our second claim is an easy consequence of this. $\qquad\square$

In the proof of this theorem, the fact that A is a 01-matrix was not used. All that is required is that A be symmetric with constant row sum, and the (i, j)-entry of A be zero if the corresponding vertices are not adjacent in X. We say a matrix B is *compatible* with a graph X on n vertices if it is symmetric with order $n \times n$, and, for all vertices u and v, we have $B_{u,v} = 0$ if u is not adjacent to v. Note that this implies that the diagonal entries of B are all zero. Compatible matrices are also called *weighted adjacency matrices*. Hence, using the same proof used to show that Theorem 2.4.1 holds, we get the following result for compatible matrices with constant row sum.

2.4.2 Theorem. *Let A be a matrix, with constant row sum d, that is compatible with a graph X. If the least eigenvalue of A is τ, then*

$$\alpha(X) \leq \frac{|V(X)|}{1 - \frac{d}{\tau}}$$

and, if equality holds for some coclique S with characteristic vector v_S, then

$$v_S - \frac{|S|}{|V(X)|}\mathbf{1}$$

is an eigenvector with eigenvalue τ. $\qquad\square$

Any coclique in a k-regular graph that meets the ratio bound is called *ratio tight*. A ratio tight coclique has some additional properties.

2.4.3 Corollary. *Let X be a k-regular graph on v vertices with least eigenvalue τ. If S is a coclique in X such that*

$$|S| = \frac{v}{1 - \frac{k}{\tau}}$$

then the partition $\{S, \overline{S}\}$ of $V(X)$ is equitable and each vertex not in S has exactly $-\tau$ neighbors in S.

Proof. Let v_S be the characteristic vector of a coclique S. From (2.4.2) it follows that the subspace spanned by v_S and $\mathbf{1}$ is A-invariant, and therefore the span of v_S and $\mathbf{1} - v_S$ is A-invariant. Now Theorem 2.2.1(b) implies that the partition $\{S, \overline{S}\}$ is equitable. If the adjacency matrix of the quotient is

$$\begin{pmatrix} 0 & k \\ a & k-a \end{pmatrix}$$

then its eigenvalues are k and $-a$. By counting edges between S and $\overline{S}$ we find that $a = \frac{|S|k}{v - |S|} = -\tau$; thus each vertex in $\overline{S}$ has exactly $-\tau$ neighbors in S. $\square$

2.4.4 Corollary. *Let X be a k-regular graph on v vertices and let S be a coclique in X with size s. If the least eigenvalue of X is*

$$\frac{ks}{v - s}$$

then the partition $\{S, \overline{S}\}$ is equitable and S is a maximal coclique. $\square$

2.4.5 Lemma. *Let X be a k-regular graph on v vertices and let S be a coclique in X with size s. Then $\{S, \overline{S}\}$ is an equitable partition if and only if $s = \frac{v}{1 - \frac{k}{\lambda}}$ for some eigenvalue λ of X.*

Proof. First, assume that $\{S, \overline{S}\}$ is an equitable partition. The adjacency matrix of the quotient is

$$\begin{pmatrix} 0 & k \\ a & k-a \end{pmatrix}$$

where $a = \frac{k(s)}{v-s}$. The eigenvalues of this quotient graph are k and $\lambda = -\frac{k(s)}{v-s}$; since the partition is equitable, these are also eigenvalues of X.

Next, assume that $s = \frac{v}{1 - \frac{k}{\lambda}}$ for some eigenvalue λ of X. Then the eigenvalues of the quotient graph are k and λ and the interlacing is tight, so $\{S, \overline{S}\}$ is an equitable partition. $\square$

2.5 Application: Erdős–Ko–Rado

In this section, the ratio bound is used to prove the bound in the EKR Theorem when $t = 1$. This proof uses the eigenvalues of the Kneser graphs. Recall that the vertices of the Kneser graph $K(n, k)$ are all the k-sets from an n-set, and two are adjacent if and only if they are disjoint. The Kneser graph $K(n, 1)$ is better known as the complete graph K_n, and $K(n, 2)$ is the complement of the line graph of K_n.

As stated in the introduction to this chapter, a coclique in the Kneser graph $K(n, k)$ is a collection of k-subsets of $\{1, \ldots, n\}$ such that any two k-subsets have at least one point in common; hence it is a 1-intersecting family. The eigenvalues of the Kneser graph $K(n, k)$ have been determined (this result is proved in Section 6.5). They are the integers

$$(-1)^i \binom{n - k - i}{k - i} \tag{2.5.1}$$

where $i = 0, \ldots, k$ and the ith eigenvalue has multiplicity

$$m_i = \binom{n}{i} - \binom{n}{i - 1} \tag{2.5.2}$$

(with the understanding that $\binom{n}{-1} = 0$).

If $n < 2k$, then $K(n, k)$ is the empty graph and the eigenvalues are all 0 (and the size of the maximum coclique is known without using the ratio bound). If $n = 2k$, the graph is a perfect matching and the eigenvalues are 1 and -1 (and the size of a maximum coclique is exactly half the number of vertices). So we consider only Kneser graphs with $n \geq 2k$.

2.5.1 Lemma. *If $n \geq 2k$, then $\alpha(K(n, k)) = \binom{n-1}{k-1}$.*

Proof. It is easy to see that $K(n, k)$ has $\binom{n}{k}$ vertices and that its degree is $\binom{n-k}{k}$. The least eigenvalue occurs when $i = 1$ in (2.5.1), which equals

$$-\binom{n - k - 1}{k - 1}$$

and so the ratio bound yields that

$$\alpha(K(n, k)) \leq \frac{\binom{n}{k}}{1 + \frac{n-k}{k}} = \binom{n - 1}{k - 1}.$$

The set of all vertices that contain 1 (or any fixed element) is clearly a coclique of this size. $\qquad\square$

If x is the characteristic vector of a coclique of size $\binom{n-1}{k-1}$, then the corresponding balanced characteristic vector

$$x - \frac{k}{n}\mathbf{1}$$

lies in the eigenspace of X with eigenvalue $-\binom{n-1}{k-1}$. The characterization of the coclique of maximum size can be done by considering the eigenvectors in the $-\binom{n-1}{k-1}$-eigenspace.

Define $W_n(1, k)$ to be the $n \times \binom{n}{k}$ matrix with the characteristic vectors of the k-subsets of $\{1, \ldots, n\}$ as its columns. Since

$$W_n(1, k)(W_n(1, k))^T = \left(\binom{n-1}{k-1} - \binom{n-2}{k-2}\right) I + \binom{n-2}{k-2} J,$$

this product is invertible and hence the rows of $W_n(1, k)$ are linearly independent.

Therefore the subspace of $\mathrm{col}(W_n(1, k)^T)$ orthogonal to $\mathbf{1}$ has dimension exactly $n - 1$. Since each column of $W_n(1, k)^T$ is the characteristic vector of a coclique of size $\binom{n-1}{k-1}$, we conclude that this subspace is exactly the eigenspace of the Kneser graph corresponding to the eigenvalue $-\binom{n-k-1}{k-1}$. So the balanced characteristic vector of any maximum coclique is in this subspace. In Section 6.6 we will see how to use this to prove that a coclique of size $\binom{n-1}{k-1}$ must consist of the k-subsets that contain a given point, that is, the only maximum cocliques in $K(n, k)$ are the canonical intersecting sets. (This illustrates the point that deriving the bound for the EKR Theorem can be significantly easier than characterizing the cocliques that meet the bound.)

At this point we know exactly what the maximum cocliques in the Kneser graph are. Using these cocliques (assuming that $n \geq 2k$), it is straightforward to color the Kneser graph $K(n, k)$ with $n - 2k + 2$ colors. For each i in $\{1, \ldots, n - (2k - 1)\}$, color the sets that contain the element i with the color i (if a set contains more than one of these elements, then any one of the colors can be used). Then the only sets that remain uncolored are the k-subsets of $\{n - (2k - 1) + 1, \ldots, n\}$; since any two such k-sets must be intersecting, they can all be assigned the same color.

What is not so straightforward is to show that there is no proper coloring with fewer colors. This was conjectured to be true by Kneser in 1955 and proven to be so by Lovász [116] (a proof can be found in [87, Section 7.11]). The proof given by Lovász is particularly interesting as it uses Borsuk's Theorem, which is a result from topology.

2.5.2 Theorem. *If* $n \geq 2k$ *the* $\chi(K(n, k)) = n - 2k + 2$. $\square$

2.6 A ratio bound for cliques

We can always get a bound on the size of a clique in a regular graph by applying the ratio bound for cocliques to its complement. However, there is another bound, which can be sharper in some cases. The version we offer here, and its proof, comes from unpublished work of Mike Newman.

2.6.1 Theorem. *Let X be a k-regular graph with least eigenvalue τ, and suppose $\mathcal{C}$ is a set of cliques of X each with size $\omega(X)$ that gives a point- and block-regular incidence structure on $V(X)$. If there is a constant λ such that each edge of X lies in exactly λ cliques from $\mathcal{C}$, then*

$$\omega(X) \leq 1 - \frac{k}{\tau}.$$

If equality holds and N is the vertex-clique incidence matrix, then $NN^T = \lambda(A - \tau I)$, where A is the adjacency matrix of X.

Proof. Let A be the adjacency of X and let N be the incidence matrix for the cliques in $\mathcal{C}$. Assume that each vertex of X lies in exactly r cliques from $\mathcal{C}$ and each edge lies in exactly λ cliques. Then

$$NN^T = rI + \lambda A.$$

The matrix NN^T is positive semidefinite, so all of its eigenvalues are non-negative and consequently

$$0 \leq r + \lambda\tau. \tag{2.6.1}$$

If we let $\omega = \omega(X)$, then

$$|\mathcal{C}|\omega = |V(X)|r, \qquad |\mathcal{C}|\binom{\omega}{2} = \lambda|E(X)|$$

whence

$$\omega - 1 = \frac{2\lambda\,|E(X)|}{|\mathcal{C}|\omega} = \frac{2\lambda\,|E(X)|}{r\,|V(X)|} = k\frac{\lambda}{r} \leq \frac{k}{-\tau}.$$

This implies the stated bound; further, if equality holds, then equality holds in (2.6.1) and $NN^T = \lambda(A - \tau I)$. $\qquad\square$

We see that if the bound is tight, then the τ-eigenspace of A is equal to $\ker(NN^T)$, which is equal to $\ker(N^T)$ (to see this, consider that $\|Nx\| = x^T NN^T x$ for any x). Thus the τ-eigenspace is equal to the set of vectors that are orthogonal to the characteristic vectors of the cliques in $\mathcal{C}$. If S is a coclique of size $|V(X)|/\omega$ with characteristic vector v_S, then $v_S^T N = \mathbf{1}$ and consequently

$$\left(v_S - \frac{1}{\omega}\mathbf{1}\right)^T N = 0.$$

Thus the balanced characteristic vectors of cocliques with size $|V(X)|/\omega(X)$ are eigenvectors for A with eigenvalue τ.

There is a class of graphs where the set of all maximum cliques is a point- and block-regular incidence structure on $V(X)$.

2.6.2 Lemma. *If X is a k-regular arc-transitive graph with least eigenvalue τ then*

$$\omega(X) \leq 1 - \frac{k}{\tau}. \qquad \square$$

It is not hard to see that the Kneser graphs are arc-transitive and by this lemma we have the (obvious) inequality

$$\omega(K(n, k)) \leq 1 - \frac{\binom{n-k}{k}}{-\binom{n-k-1}{k-1}} = \frac{n}{k},$$

which holds with equality if and only if $k|n$. More interesting examples where the ratio bound for cliques is tight are given in Chapter 5.

2.7 Ratio bound for cross cocliques

For a graph X, we say that a pair of subsets (S, T) of $V(X)$ is a *cross coclique* (or *cross independent set*) if no edge of X joins a vertex of S to a vertex of T. This definition is similar to the cross-intersecting sets that were defined in Section 1.7. Indeed, if $(\mathcal{A}, \mathcal{B})$ is a cross-intersecting set and all the sets in both $\mathcal{A}$ and $\mathcal{B}$ are k-sets, then $(\mathcal{A}, \mathcal{B})$ forms a cross coclique in a Kneser graph.

Theorem 1.7.1 gives an upper bound on the product of the sizes of the two sets of vertices in a cross coclique in a Kneser graph. In this section we show another way to prove this bound that uses the eigenvalues of the graph. We actually give a more general result that bounds the size of cross cocliques in regular graphs in terms of the two largest (in absolute value) eigenvalues of the graph. The result, and the proof we present, is due to Ellis [54].

For vectors u and v in $\mathbb{R}^n$ we denote the ith entry of u by $u(i)$ and use the standard inner product: namely

$$\langle u, v \rangle = \frac{1}{n} \sum_{i=1}^{n} u(i)v(i).$$

2.7.1 Theorem. *Let X be a k-regular graph on n vertices with adjacency matrix A. Let $k \geq \lambda_2 \geq \lambda_3 \geq \cdots \geq \lambda_n$ be the eigenvalues of A. Set $\mu = \max\{|\lambda_2|, |\lambda_n|\}$. Assume (S, T) is a cross coclique in X; then*

$$|S|\,|T| \leq \left(\frac{n}{1 + \frac{k}{\mu}} \right)^2.$$

Proof. Set

$$v'_S = \frac{|S|}{\sqrt{|V(X)|}} v_S, \qquad v'_T = \frac{|S|}{\sqrt{|V(X)|}} v_T$$

(where v_S is the characteristic vector of S). Let $\{\frac{1}{\sqrt{n}}\mathbf{1}, u_2, u_3, \ldots u_n\}$ be an orthonormal set of eigenvectors where u_i has eigenvalue λ_i. Set $\xi_i = \langle v'_S, u_i \rangle$ and $\eta_i = \langle v'_T, u_i \rangle$; then

$$v'_S = \sum_{i=1}^n \xi_i u_i, \qquad v'_T = \sum_{i=1}^n \eta_i u_i$$

and, since $\|v'_S\| = 1$ and $\|v'_T\| = 1$, we also have

$$\sum_{i=1}^n |\xi_i|^2 = 1, \qquad \sum_{i=1}^n |\eta_i|^2 = 1.$$

Note that

$$\xi_1 = \sqrt{\frac{|S|}{n}}, \qquad \eta_1 = \sqrt{\frac{|T|}{n}}. \tag{2.7.1}$$

Since (S, T) is a cross coclique,

$$\sum_{i=1}^n \lambda_i \xi_i \eta_i = (v'_s)^T A v'_T = 0,$$

and thus

$$k\xi_1\eta_1 = \left| \sum_{i=2}^n \lambda_i \xi_i \eta_i \right|$$

$$\leq \sum_{i=2}^n |\lambda_i| \, |\xi_i| \, |\eta_i|$$

$$\leq \mu \sum_{i=2}^n |\xi_i| \, |\eta_i|$$

$$\leq \mu \sqrt{\sum_{i=2}^n |\xi_i|^2 \sum_{i=2}^n |\eta_i|^2}$$

$$= \mu\sqrt{(1 - \xi_1^2)(1 - \eta_1^2)}.$$

This can be rewritten as

$$\frac{\xi_1^2 \eta_1^2}{(1 - \xi_1^2)(1 - \eta_1^2)} \leq \left(\frac{\mu}{k}\right)^2.$$

By the arithmetic-geometric mean inequality, the left-hand side of this is bounded below by

$$\frac{\xi_1^2 \eta_1^2}{(1 - \xi_1 \eta_1)^2}$$

so we have that

$$\xi_1 \eta_1 \leq \frac{\mu}{k + \mu}.$$

Using the values of ξ_1 and η_1 from (2.7.1), the result follows.   □

As with the ratio bound for cocliques, we can actually say more when equality holds in the foregoing equation. In this case, ξ_i and η_i must be equal to zero for all $i \geq 1$ with $\lambda_i \neq \mu$; this implies that $v_S - \frac{|S|}{n}\mathbf{1}$ and $v_T - \frac{|T|}{n}\mathbf{1}$ are both μ-eigenvalues.

Finally, this theorem can be applied to the Kneser graph to get an algebraic proof of Theorem 1.7.1 when both set systems in the cross-intersecting set are k-set systems.

2.7.2 Theorem. *Assume $n \geq 2k$ and let (S, T) be a cross coclique in the Kneser graph $K(n, k)$. Then*

$$|S|\,|T| \leq \binom{n-1}{k-1}^2.$$

Proof. From (2.5.1) the eigenvalue of $K(n, k)$ that is the second largest in absolute value is $\mu = -\binom{n-k-1}{k-1}$. By Theorem 2.7.1,

$$|S|\,|T| \leq \left(\frac{\binom{n-k-1}{k-1}\binom{n}{k}}{\binom{n-k}{k} + \binom{n-k-1}{k-1}}\right)^2 = \binom{n-1}{k-1}^2.   □$$

2.8 Cocliques in neighborhoods

We derive another bound for the size of a coclique. Although the proof is quite elementary, the result is still useful.

2.8.1 Theorem. *Let X be a k-regular graph on v vertices, and let α_1 denote the maximum size of a coclique in the neighborhood of a vertex in X. Then*

$$\alpha(X) \leq \frac{v}{1 + \frac{k}{\alpha_1}}.$$

If equality holds for a coclique S, then the partition $\{S, \bar{S}\}$ is equitable and $-\alpha_1$ is an eigenvalue of X.

Proof. Suppose S is a coclique in X of size s. Since X is k-regular, the number of edges that join a vertex in S to a vertex in $\bar{S}$ is ks. On the other hand, if $u \notin S$ then the neighbors of u in S form a coclique in the neighborhood of u. So u has at most α_1 neighbors in S, and we conclude that

$$sk \leq \alpha_1(v - s).$$

This implies the given bound. We also see that if equality holds, each vertex in $\bar{S}$ has exactly α_1 neighbors in S, and therefore the partition $\{S, \bar{S}\}$ is equitable with quotient matrix

$$\begin{pmatrix} 0 & k \\ \alpha_1 & k - \alpha_1 \end{pmatrix}.$$

Its eigenvalues are k and $-\alpha_1$. $\qquad\qquad\qquad\qquad\qquad\qquad\qquad\qquad\qquad\square$

The neighborhood of a vertex in the Kneser graph $K(n, k)$ is isomorphic to the Kneser graph $K(n - k, k)$. By the EKR Theorem, if $n - k \geq 2k$ the size of the maximum coclique in the neighborhood of a vertex is $\binom{n-k-1}{k-1}$ (which is the absolute value of the least eigenvalue of $K(n, k)$).

2.9 The inertia bound

We introduce another bound on the size of a coclique due to Cvetković. Recall that a matrix B is a compatible matrix for a graph X on n vertices if it is symmetric with order $n \times n$ and the (u, v)-entry of B is 0 whenever u is not adjacent to v.

For a symmetric matrix B, all the eigenvalues are real-valued, and we use $n^+(B)$ and $n^-(B)$ to denote respectively the number of positive and negative eigenvalues of B. The triple

$$\left(n^+(B),\ n - n^+(B) - n^-(B),\ n^-(B)\right)$$

is sometimes called the *inertia* of B.

2.9.1 Theorem. *If X is a graph and B is compatible with X, then*

$$\alpha(X) \leq \min\{n - n^-(B), n - n^+(B)\}.$$

Proof. Let S be a coclique in X of size s, and let B be compatible with X. Although it may seem silly, let B_0 denote the submatrix of B with rows and columns indexed by the vertices in S. Then, by interlacing,

$$0 = \theta_s(B_0) \leq \theta_s(B)$$

and

$$\theta_{n+1-s}(B) \leq \theta_1(B_0) = 0.$$

The first inequality implies that the number of non-negative eigenvalues of B is at least s, whereas from the second it follows that the number of non-positive eigenvalues is at least s. As the first number is $n - n^-(B)$ and the second is $n - n^+(B)$, the theorem follows.   □

The obvious matrix compatible with X is its adjacency matrix, but often better bounds can be found using different matrices. For a first example, the inertia bound is tight for complete graphs; another, more interesting, example is given in the next section.

2.10 Inertia bound: Kneser graphs

Our first real use of the inertia bound is due to Calderbank and Frankl [37]. We begin with the following binomial identity:

$$\sum_{i=0}^{k} (-1)^i \binom{v}{k-i} = \binom{v-1}{k}. \tag{2.10.1}$$

Once we see that

$$\binom{v}{k} - \binom{v-1}{k} = \binom{v-1}{k-1},$$

it is very easy to prove this identity by induction. By (2.5.2), depending on the parity of k, (2.10.1) is either the number of positive or the number of negative eigenvalues in the Kneser graph $K(n, k)$.

The inertia bound applied to the adjacency matrix of Kneser graph now yields

$$\alpha(K(n, k)) \le \min\left\{ \binom{v-1}{k}, \binom{v-1}{k-1} \right\} = \binom{v-1}{k-1}.$$

This provides a second proof of the EKR Theorem. It is irritating that in this case we are unable to derive any information about the case of equality!

2.11 Inertia bound: folded cubes

We now consider a second class of examples, the *folded cubes*. The *folded d-cube* is the graph that results when we quotient the d-cube over the equitable partition formed by the antipodal pairs of vertices. Equivalently it is the graph we get from the $(d-1)$-cube by joining each antipodal pair of vertices by an edge. If d is even, the folded d-cube is bipartite and there is not much of interest we can say about its maximal cocliques

The eigenvalues of the folded $(2r + 1)$-cubes are known [31] to be

$$2r + 1 - 4i,$$

where $i = 0, 1, \ldots, r$, with respective multiplicities

$$\binom{2r + 1}{2i}.$$

If $i \leq \frac{r}{2}$ the eigenvalues are positive, so if r is odd the number of positive eigenvalues is

$$\sum_{i=0}^{\frac{r-1}{2}} \binom{2r + 1}{2i}.$$

Similarly, if r is even, the number of negative eigenvalues is

$$\sum_{i=\frac{r}{2}+1}^{r} \binom{2r + 1}{2i}.$$

In both cases this gives an upper bound on the size of a coclique. This bound is tight in both cases; if r is odd, a coclique of this size can be constructed by taking all the vertices with the small part having an even number of elements, and if r is even, by taking all vertices with the large part having an even number of vertices.

Note that the ratio bound is not tight for folded $(2d + 1)$-cubes.

2.12 Fractional chromatic number

In this section we develop some results on fractional chromatic number, in part using graph homomorphisms. Since the results are standard, we omit some proofs. For more detail, and extra background, we refer the reader to [87, Chapter 7].

Let X be a graph and let N be the incidence matrix of the incidence structure with the vertices of X as points and the cocliques of X as blocks. Any proper coloring of a graph of X can be specified by a 01-vector z such that

$$Nz \geq \mathbf{1}. \tag{2.12.1}$$

The *chromatic number* of X is the minimum value of $\mathbf{1}^T z$ given that z is a 01-vector and (2.12.1) holds; this number is denoted by $\chi(X)$.

We define the *fractional chromatic number* $\chi_f(X)$ of X to be the solution to the following linear program (or LP)

$$\min \mathbf{1}^T z, \quad Nz \geq \mathbf{1}, \quad z \geq 0. \tag{2.12.2}$$

Clearly $\chi_f(X) \leq \chi(X)$. We will say that a non-negative vector z such that $Nz \geq 1$ is a *fractional coloring* of X. The dual of this linear program is

$$\max y^T \mathbf{1}, \quad y^T N \leq \mathbf{1}, \quad y \geq 0.$$

The constraints here are satisfied by the characteristic vectors of cliques, and optimal solutions to this linear programming problem are called *fractional cliques*. The value of an optimal solution is denoted by $\omega_f(X)$.

Since a proper coloring of a graph X is a collection of $\chi(X)$ pairwise disjoint cocliques, we have the trivial inequality $\chi(X)\alpha(X) \leq |V(X)|$. Surprisingly enough, this inequality still holds when we replace χ by χ_f.

2.12.1 Lemma. *For any graph X,*

$$\chi_f(X) \geq |V(X)|/\alpha(X)$$

and equality holds if X is vertex transitive.

Proof. The vector y in the above dual LP can be considered to be a function on $V(X)$, and for any $g \in \text{Aut}(X)$, we can define

$$y^g(v) = y(v^g).$$

We see that if y is an optimal solution to our dual LP, then y^g is also an optimal solution for any $g \in \text{Aut}(X)$, as is

$$\hat{y} = \frac{1}{|\text{Aut}(X)|} \sum_g y^g.$$

Here $\hat{y}$ is constant on the orbits of $\text{Aut}(X)$, so if $\text{Aut}(X)$ is vertex transitive, $\hat{y}$ must be constant. Now

$$\mathbf{1}^T N \leq \alpha(X)\mathbf{1}$$

and so the optimal constant solution to the dual LP is $y = \alpha(X)^{-1}\mathbf{1}$ and its value is

$$\frac{|V(X)|}{\alpha(X)}.$$

Consequently,

$$\frac{|V(X)|}{\alpha(X)} \leq \omega_f(X) = \chi_f(X)$$

with equality if X is vertex-transitive. $\qquad\qquad\qquad\qquad\qquad\square$

By Lemma 2.12.1 and the EKR Theorem, for any Kneser graph $K(n, k)$ the fractional chromatic number is

$$\frac{\binom{n}{k}}{\binom{n-1}{k-1}} = \frac{n}{k}.$$

2.13 Homomorphisms

An important feature of fractional chromatic number is that it behaves well with respect to homomorphisms. Recall that a *homomorphism* from graph Y into X is a map from $V(Y)$ to $V(X)$ that maps edges in Y to edges in X. We may write $X \to Y$ to denote that there is a homomorphism from X to Y. If $\psi : X \to Y$ is a homomorphism and $y \in V(Y)$, then

$$\psi^{-1}(y) = \{x \in V(X) : \psi(x) = y\}$$

is the *fiber* of ψ at y. Any fiber is necessarily a coclique, from which we see that a graph X is m-colorable if and only if there is a homomorphism from X to K_m. Thus we can define chromatic number in terms of homomorphisms; it is interesting that we can also define fractional chromatic number in this way. (This result is due to Bondy and Hell [24], a proof appears also in [87, Chapter 7].)

2.13.1 Theorem. *For any graph X*

$$\chi_f(X) = \inf \left\{ \frac{n}{k} : X \to K(n, k) \right\}. \qquad \Box$$

We note one useful consequence of this result (which is not hard to prove directly).

2.13.2 Lemma. *If there is a homomorphism from Y into X, then*

$$\chi_f(Y) \le \chi_f(X). \qquad \Box$$

This implies the inequality in the following lemma, which is equivalent to the "no-homomorphism lemma" of Albertson and Collins [9].

2.13.3 Lemma. *If Y is vertex transitive and there is a homomorphism from X to Y, then*

$$\frac{|V(X)|}{\alpha(X)} \le \frac{|V(Y)|}{\alpha(Y)}.$$

If X is vertex transitive and equality holds, the preimage in X of a coclique of maximum size in Y is a coclique of maximum size in X. $\qquad \Box$

If X is actually an induced subgraph of Y, this result implies that if S is a coclique of maximal size in Y, then $S \cap V(X)$ is a coclique of maximum size in X.

As another simple application, if $\omega(X) = \chi_f(X) = r$ and X is vertex transitive, the lemma implies that a coclique S of maximum size must have non-trivial intersection with any r-clique.

2.14 Cyclic intervals

We present a short proof that $\alpha(K(n, k)) = \binom{n-1}{k-1}$, using ideas related to fractional colorings of graphs. One object we need is the *cyclic interval graph* $C(n, k)$, which we can define as the Cayley graph for $\mathbb{Z}_n$ relative to the connection set

$$C = \{k, k+1, \ldots, n-k\}.$$

Alternatively we can view it as the graph whose vertices are the "cyclic intervals"

$$\{i, i+1, \ldots, i+k-1\}$$

(for $i \in \mathbb{Z}_n$) where two intervals are adjacent if they are disjoint. From this definition we see that $C(n, k)$ is an induced subgraph of the Kneser graph $K(n, k)$. The other properties we need are summarized in our next result.

2.14.1 Lemma. *If $n \geq 2k$ then $\alpha(C(n, k)) = k$, and if $n > 2k$ then a coclique of size k consists of k intervals that contain a common element.* $\qquad\square$

We now prove the EKR Theorem for intersecting sets. Our proof follows closely the proof in [87, Theorem 7.8.1].

Since $C(n, k)$ is vertex transitive, its fractional chromatic number is n/k; since $C(n, k)$ is an induced subgraph of $K(n, k)$, by Lemma 2.13.3 we have

$$\frac{n}{k} = \chi_f(C(n, k)) \leq \chi_f(K(n, k)) \leq \frac{\binom{n}{k}}{\alpha(K(n, k))}$$

whence we deduce that $\alpha(K(n, k)) \leq \binom{n-1}{k-1}$. As we have cocliques in $K(n, k)$ of size $\binom{n-1}{k-1}$, we have the bound in the EKR Theorem.

It remains to prove the characterization of the cocliques of maximum size. Assume S is a coclique of size $\binom{n-1}{k-1}$ in $K(n, k)$. By Lemma 2.13.3, the intersection of S with any cyclic interval graph must consist of k cyclic shifts of some set of k consecutive elements in some cyclic ordering of $\{1, \ldots, n\}$.

We consider first the natural ordering and note that, by relabeling if needed, we may assume that S contains the sets

$$\{1, 2, \ldots, k\}, \ldots, \{k, k + 1, \ldots, 2k - 1\}.$$

We complete the proof by showing that S must contain all k-subsets that contain k. Note that since S contains exactly k vertices from any $C(n, k)$, it does not contain any k-subset of the form $\{x, 1, \ldots, k - 1\}$, where $2k \leq x \leq n$.

Let g be any element of $\mathrm{Sym}(n)$ that fixes $\{1, \ldots, k - 1\}$ as a set and consider a cyclic ordering that starts

$$x, 1^g, \ldots, (k - 1)^g, k, \ldots, \tag{2.14.1}$$

where $2k \leq x \leq n$. Then S contains

$$\beta = \{1^g, \ldots, (k - 1)^g, k\},$$

but since it does not contain $\{x, 1^g, \ldots, (r - 1)^g\}$, it contains the k cyclic right shifts of β. For any k-subset α that contains k, there is some cyclic ordering of the form in (2.14.1) that contains α, unless α contains each element of $\{2k, \ldots, n\}$ (in which case there is no way to choose x.)

Suppose that $y \in \{k + 1, \ldots, 2k - 1\}$ and consider the ordering obtained from the natural cyclic ordering by swapping y and $2k$. Reapplying the argument of the previous paragraph, we see that S contains all k-subsets that contain k, possibly excepting those that contain $\{y, 2k, \ldots, n\}$. Varying y, we conclude that a k-subset that contains k and is not in S must contain each element of $\{k + 1, \ldots, n\}$. Since $n > 2k$, there are no such k-subsets, and the result follows. $\qquad\square$

2.15 Exercises

2.1 Using an equitable partition, show that $\pm n$ are eigenvalues for the complete bipartite graph $K_{n,n}$. Similarly, for a semiregular bipartite graph in which the vertices in one part have degree k, and the vertices in the other part have degree ℓ, show that $\pm\sqrt{k\ell}$ are eigenvalues.

2.2 Describe the equitable partition formed by the set of orbits of

$$\mathrm{Sym}(\{1, 2, \ldots, k\}) \times \mathrm{Sym}(\{k + 1, k + 2, \ldots, n\})$$

acting on the vertices of $K(n, k)$.

2.3 Find the quotient graph of $K(n, 2)$ where π is the partition formed by the set of orbits of

$$\mathrm{Sym}(\{1, 2\}) \times \mathrm{Sym}(\{3, 4, \ldots, n\})$$

acting on the vertices of $K(n, 2)$.

2.4 Prove that the only maximum cocliques in the n-hypercube are the set of all sequences with either an even number of zeros or an odd number of zeros.

2.5 Recall that in Section 1.8 we defined the separated sets. Let $SG(n, k)$ be the subgraph of $K(n, k)$ induced by the separated sets (this graph is called the *stable Kneser graph*).

   (a) Show that $SG(n, k)$ is not, in general, either vertex transitive or regular.

   (b) Using the Borsuk-Ulam Theorem, it can be shown that $\chi(SG(n, k)) = n - 2k + 2$ [149]. Show that $SG(n, k)$ is also *vertex-critical*. (Show that if a single vertex is removed, then the chromatic number of the resulting graph is strictly less thatn $n - 2k + 2$.)

   (c) Assume that $n \geq 2k + 1$ and $k \geq 2$. Show that the automorphism group of the stable Kneser graph is the diherdral group of order $2n$.

2.6 Let A be the adjacency matrix of a graph with least eigenvalue τ. If the matrix $A - \tau I - \gamma J$ is positive semidefinite for some positive real γ, show that $\alpha(X) \leq -\tau/\gamma$.

2.7 Show that Theorem 2.8.1 holds with equality for any regular bipartite graph.

2.8 Give an example of a Kneser graph for which the inequality in Theorem 2.8.1 is strict.

2.9 Prove that if A and B are real symmetric matrices, there is an invertible matrix X such that $XAX^T = B$ if and only A and B have the same inertia.

2.10 For graphs X and Y, the *Cartesian product* of X and Y (denoted by $X \, \Box \, Y$) is the graph on vertex set $V(X) \times V(Y)$ with (x_1, y_1) adjacent to (x_2, y_2) if either $x_1 = x_2$ and y_1 is adjacent to y_2, or x_1 is adjacent to x_2 and $y_1 = y_2$. First show that a coclique in $X \, \Box \, K_r$ is equivalent to an induced r-partite graph in X. Next, prove for any n that

$$\binom{n-1}{k-1} + \binom{n-2}{k-1} \leq \alpha(K(n, k) \, \Box \, K_2).$$

2.11 Using the inertia bound, prove that

$$\alpha(K(2k+1, k) \, \Box \, K_2) \leq \binom{2k}{k} + \binom{2k}{k-2}.$$

2.12 Let k be even. Construct a coclique in $K(2k + 1, k) \square K_2$ of size

$$\sum_{i=k/2}^{k-1} \binom{k-1}{i}\binom{k+2}{k-i} + \sum_{i=k/2+1}^{k} \binom{k+1}{i}\binom{k}{k-i}.$$

For k an odd integer construct a coclique in $K(2k + 1, k) \square K_2$ of size

$$2 \sum_{i=(k+1)/2}^{k} \binom{k}{i}\binom{k+1}{k-i}.$$

(Note that this is larger than the coclique constructed in Exercise 2.10, but smaller than the bound in Exercise 2.11.)

2.13 The *Andrásfai graph* And(m) is the circulant graph on $3m - 1$ vertices, with connection set consisting of the m elements of $\mathbb{Z}_{3m-1}$ that are congruent to 1 modulo 3. Use the inertia bound on the adjacency matrix to show that $\alpha(\text{And}(m)) = m$.

2.14 Use the inertia bound to show that the folded 11-cube is not 3-colorable. (Weight the edges in the Cartesian product of the folded 11-cube with K_3 so that the edges within a copy of the folded cube have weight 1 and the edges between the copies of the graph have weight 1.1.)

2.15 Define a graph whose vertex set is the set of all 2-subsets of $\{1, 2, 3, 4, 5\}$ and two vertices are adjacent if and only if the subsets have exactly one point in common. (This is the line graph on the complete graph on 5 vertices, also known as the Johnson graph $J(5, 2)$). Show that the size of a maximum coclique in this graph is no more than 2 using the inertia bound. (Hint: consider three cliques (i) the set of all vertices containing 1, (ii) the set of all vertices containing 2 but not 1, and (iii) the remaining vertices. Use a weighted adjacency matrix in which the edges in these three cliques all have weight t and the remaining edges in the graph have weight 1. Calculate the eigenvalues of this matrix using some mathematics software.)

2.16 Use Lemma 2.13.3 to prove that the clique-coclique bound (Theorem 2.1.1) holds for all vertex-transitive graphs.

2.17 Use Lemma 2.13.3 to show that if C_k is the k-cycle, then $C_k \rightarrow C_\ell$ if and only if either k is even or ℓ is odd with $\ell \le k$.

2.18 Let X be a vertex-transitive graph such that $\alpha(X)\omega(X) = |V(X)|$. If M and N are the respective incidence matrices for cocliques and cliques, prove that $MN = NM$.

Notes

Exercises 2.10, 2.11 and 2.12 are all related to finding the largest induced r-partite subgraph in the Kneser graph $K(n, k)$. Erdős in [57] proved that the largest such subgraph has size $\sum_{i=1}^{r} \binom{n-i}{k-1}$ when n is sufficiently large, and he conjectured that the result held for $n \geq 2k + t - 1$. This lower bound on n is not sufficient, as is shown by Frankl and Füredi in [65]. The value of the exact lower bound on n for which this result holds is an open problem.

The proof that the stable Kneser graph, $SG(n, k)$, has chromatic number $n - 2k + 2$ can be found in [149] (this proof is very similar to the proof that the chromatic number of $K(n, k)$ is $n - 2k + 2$). This paper also gave the first proof that $SG(n, k)$ is vertex-critical (this is Exercise 2.5(b)). Exercise 2.5(c) is due to Braun [29].

3

Association schemes

In the previous chapter, we defined the Kneser graphs and saw that a coclique in this graph is a 1-intersecting set system. Similarly, we can define a graph to have the set of all k-subsets of $\{1, \ldots, n\}$ as its vertex set, where two k-subsets are adjacent if and only if they have at most $t - 1$ elements in common. A coclique in this graph is a t-intersecting set system. In fact, there is an entire family of graphs that can be defined on the set of all k-sets of an n-set based on the size of the intersection of the sets. For $2k \leq n$, define $J(n, k, i)$ to be the graph in which the vertices are adjacent if the sets have intersection size exactly $k - i$ (this set of graphs forms the *Johnson scheme*, which is discussed in detail in Chapter 6). With this definition $J(n, k, 0)$ is the empty graph (alternately, this could be viewed as the graph with $\binom{n}{k}$ vertices and a loop at each vertex) and $J(n, k, k)$ is the Kneser graph $K(n, k)$.

For each of these graphs, the adjacency matrix is a $\binom{n}{k} \times \binom{n}{k}$ 01-matrix. Since each pair of sets is adjacent in exactly one of these graphs, the sum of these matrices is the all-ones matrix. More importantly, it is possible to calculate the product of any two of these matrices. If we use $A(n, k, i)$ to denote the adjacency matrix for the graph $J(n, k, i)$, then the product of $A(n, k, i)$ and $A(n, k, j)$ is

$$\sum_{x=0}^{k} \left(\sum_{y=0}^{x} \binom{x}{y} \binom{k-x}{i-y} \binom{k-x}{j-y} \binom{n-2k+x}{k-i-j+y} \right) A(n, k, x). \quad (3.0.1)$$

The important thing to note in this equation is that the entries in the product only depend on the size of the intersection of the corresponding sets. Thus the algebra generated by these adjacency matrices is actually spanned by these matrices; since the matrices are linearly independent they form a basis for the algebra they generate. A set of matrices with these properties is called an *association scheme*. Nearly all the graphs that we consider belong to an association scheme; this chapter provides the basic concepts and theory. The subsequent chapters go into detail on the properties of some important association schemes.

3.1 Definitions

An association scheme can be viewed as a highly structured set of graphs, but it is easier to define association schemes in terms of matrices, so this is what we do.

An *association scheme* is a set of $v \times v$ 01-matrices $\mathcal{A} = \{A_0, \ldots, A_d\}$, with the following properties:

(a) $A_0 = I$;
(b) $\sum_{i=0}^{d} A_i = J$;
(c) $A_i^T \in \mathcal{A}$, for all $i = 1, \ldots, d$;
(d) for all $i, j \in \{0, \ldots, d\}$, the product $A_i A_j$ lies in the span of $\mathcal{A}$;
(e) $A_i A_j = A_j A_i$ for all i and j.

We view the matrices $A_1, \ldots, A_d$ as the adjacency matrices of graphs $\{X_1, \ldots, X_d\}$, and we call these graphs the *graphs of the scheme*. These are a set of graphs, all on the same vertex set, and two distinct vertices are *i-related* if they are adjacent in the ith graph. Further, every vertex is 0-related to itself; this is the *identity relation* as it corresponds to the identity matrix. The number of nonidentity relations is the number of classes in the association scheme. Since the sum of the matrices in an association scheme is the all-ones matrix, every pair of vertices is i-related for exactly one i. In this sense, an association scheme is a partition of the ordered pairs of the vertices.

If all matrices in the association scheme are symmetric, then all of the graphs of the scheme are undirected; such an association scheme is called a *symmetric association scheme*. Note that if a scheme is symmetric, then property (e) is implied by the other properties. In many situations only symmetric schemes need be considered.

Since the matrices in an association scheme commute, every matrix in the association scheme commutes with the all-ones matrix. This implies that the row and column sums are all equal for each matrix, or equivalently, that each of the graphs of the scheme is a regular graph. An association scheme is called *primitive* if each of these graphs is connected.

Define the *Schur product* of two matrices $A = [a_{i,j}]$ and $B = [b_{i,j}]$ to be the matrix $A \circ B = [a_{i,j} b_{i,j}]$. Then for any two matrices A_i and A_j in an association scheme,

$$A_i \circ A_j = \begin{cases} 0, & i \neq j; \\ A_i, & i = j. \end{cases}$$

This equation implies that the matrices A_i are linearly independent and that each matrix A_i is a *Schur idempotent*. Further, this equation also shows that any two distinct matrices A_i and A_j are Schur orthogonal, meaning that the Schur product of any two distinct matrices is the zero matrix.

The graph of an association scheme with one class is simply the complete graph (this scheme has exactly two matrices, I and $J - I$). We will see in Section 5.1 that strongly regular graphs correspond to symmetric schemes with two classes.

The vector space consisting the complex linear combinations of the matrices in an association scheme $\mathcal{A}$ is denoted by $\mathbb{C}[\mathcal{A}]$; it is called the *Bose-Mesner algebra* of the scheme. Since the Bose-Mesner algebra is closed under matrix multiplication, it is a matrix algebra. Further, the set of matrices $\mathcal{A} \cup \{0\}$ is closed under Schur multiplication, so the Bose-Mesner algebra is as well. Thus the Bose-Mesner algebra is a Schur-closed algebra. We define a *graph in the Bose-Mesner algebra* to be any graph whose adjacency matrix is a matrix in $\mathbb{C}[\mathcal{A}]$.

3.1.1 Lemma. *The Bose-Mesner algebra of an association scheme is a commutative matrix algebra with identity, which is closed under transposition and contains the all-ones matrix.* □

The matrices in an association scheme form a basis for the Bose-Mesner algebra of the association scheme. This is a very strong property; it can be thought of as meaning that the algebra generated by the matrices is as small (in dimension) as possible. Such an algebra has some nice properties that we will use to determine properties of the graphs of the scheme.

3.2 Commutants

Many of the examples of association schemes that we are interested in arise as commutants of permutation groups. The *commutant* of a set of matrices $\mathcal{M}$ in $\mathrm{Mat}_{v \times v}(\mathbb{C})$ is the set of matrices A such that $AM = MA$ for all M in $\mathcal{M}$. The proof of the following is easy and has been left as a exercise—the only part that requires effort is verifying that the commutant of a set of permutation matrices is Schur-closed.

3.2.1 Lemma. *The commutant of a set of $v \times v$ permutation matrices is a Schur-closed matrix algebra that contains I and J and is closed under taking transposes.* □

Every Schur-closed matrix algebra has a basis of 01-matrices, and such a basis can be chosen so that the matrices are Schur orthogonal. This leads us to the following result. (The proof is left as an exercise.)

3.2.2 Theorem. *The commutant of a set of permutation matrices is commutative if and only if it is the Bose-Mesner algebra of an association scheme.* □

It is not the case that the commutant of a set of permutation matrices always gives rise to an association scheme, since the commutant need not be commutative. (An example of such a set is given in the Exercises.) We give one illustration where the commutant is commutative. In the introduction, we described the *Johnson scheme*, which has the k-subsets of a fixed n-set as its vertices, and two k-subsets are i-related if and only if the size of their intersection is exactly $k - i$; this scheme has k classes. We will see that the Bose-Mesner algebra of this association scheme is the commutant for a set of permutation matrices.

The symmetric group $\mathrm{Sym}(n)$ acts on the k-subsets of $\{1, \ldots, n\}$. Using this action, each permutation σ in $\mathrm{Sym}(n)$ can be represented as a permutation matrix, M_σ of size $\binom{n}{k} \times \binom{n}{k}$. Further, this action preserves the relations on k-subsets, so the adjacency matrices for the relations in the Johnson scheme commute with M_σ for all $\sigma \in \mathrm{Sym}(n)$. In fact, it is not difficult to show that the adjacency matrices for the relations are a basis of the commutant for the set of matrices M_σ. Finally, (3.0.1) shows that the adjacency matrices commute, and we conclude that the Johnson scheme is an association scheme with k classes.

3.3 The conjugacy class scheme

Let G be any group and assume that $C_0, \ldots, C_d$ are the conjugacy classes of the group G, where $C_0 = \{1\}$. The *conjugacy class scheme* on G has the elements of G as its vertices, and two group elements g and h are i-related if and only if $hg^{-1} \in C_i$. Specifically, the matrices A_i of the scheme are the $|G| \times |G|$ matrices with rows and columns indexed by the elements of G; and, the entry in the (g, h) position is 1 if $hg^{-1} \in C_i$ and 0 otherwise.

For each i, the matrix A_i is a sum of permutation matrices. Consider for each element x of G the permutation matrix $P(x)$ that represents the action of x on G by right multiplication. Then the matrices in the conjugacy class scheme are the matrices

$$A_i = \sum_{x \in C_i} P(x).$$

For each conjugacy class C_i (other than C_0), the graph X_i in the conjugacy class scheme is the Cayley graph for G relative to the connection set C_i. If C_i is not closed under taking inverses, then this is a directed graph, and the matrix A_i is not symmetric. In this case, take i' to be the index for which $A_{i'} = A_i^T$. If $x \in C_i$, then $C_{i'}$ is the conjugacy class that contains x^{-1}.

3.3.1 Lemma. *For any group G, the conjugacy class scheme is an association scheme.*

Proof. All the conditions for an association scheme are easily seen to hold for the conjugacy class scheme, except the final two. First we show that the product of any two matrices in the conjugacy class scheme lies in the span of the other matrices.

For each $x, y \in G$ the (x, y)-entry of $A_i A_j$ is equal to the number of elements $z \in G$ with $zx^{-1} \in C_i$ and $yz^{-1} \in C_j$. For every such z, and for any $g \in G$, it follows that $(gz)(gx)^{-1} \in C_i$ and $(gy)(gz)^{-1} \in C_j$. This means that the (gx, gy)-entry of $A_i A_j$ is equal to the (x, y)-entry. Thus the (x, y)-entry of $A_i A_j$ depends only on the conjugacy class that yx^{-1} belongs to, and not which specific x and y are considered.

The final point to check in verifying that we have an association scheme is that $A_i A_j = A_j A_i$ for all i and j. This follows from the fact that $C_i C_j = C_j C_i$, which we leave as an exercise. $\square$

Each element in the product $G \times G$ acts on the vertices of the conjugacy class scheme of G according to the rule

$$(g, h) : x \mapsto gxh^{-1}.$$

This action is an automorphism of each graph in the association scheme, so $G \times G$ is a subgroup of the automorphism of the association scheme. Further, the subgroups $1 \times G$ and $G \times 1$ both act as regular groups of automorphisms of the association scheme. Define the *diagonal subgroup* to be

$$D = \{(g, g) : g \in G\}.$$

Clearly, this group is the stabilizer in $G \times G$ of the identity. Further, the diagonal element (g, g) sends x to gxg^{-1}, and so the orbits of D are exactly the conjugacy classes of G. It follows that the Bose-Mesner algebra of the conjugacy class scheme is the commutant of the permutation group arising from the action of $G \times G$ on the cosets of D. More details about this association scheme are given in Section 11.11.

3.4 A basis of idempotents

We have seen that the matrices from an association scheme form a basis for the Bose-Mesner algebra of the association scheme. Moreover, the matrices in the association scheme are Schur idempotents. In this section, we show that there is a second basis for this algebra that is composed of idempotents.

Let $\mathcal{A} = \{A_0, \ldots, A_d\}$ be an association scheme with d classes on v vertices. If the scheme is symmetric, then the matrices A_i are Hermitian and commute, and they can be simultaneously diagonalized. In this case, we can express $\mathbb{C}^v$ as an orthogonal direct sum of subspaces; on each of these subspaces the matrices A_i all act as scalars. It can be shown that there are exactly $d+1$ subspaces in this decomposition, and we call them the *eigenspaces* of the scheme. Associated to the ith eigenspace is the idempotent Hermitian matrix E_i, which represents orthogonal projection onto the eigenspace. The main result of this section is that a basis of idempotents exists for all schemes, not just symmetric schemes.

The next three results are also given in *Association Schemes* by Godsil [81].

3.4.1 Theorem. *Let $\mathcal{M}$ be a commutative matrix algebra with identity over $\mathbb{C}$. Assume $\mathcal{M}$ has the additional property that for all $N \in \mathcal{M}$ it is the case that $N^2 = 0$ only if $N = 0$. Then each matrix in $\mathcal{A}$ can be expressed as a linear combination of pairwise orthogonal idempotents.*

Proof. Suppose $A \in \mathcal{M}$. Let $\psi(t)$ be the minimal polynomial of A and assume that

$$\psi(t) = \prod_{i=1}^{k} (t - \theta_i)^{m_i}.$$

Define

$$\psi_i(t) = \frac{\psi(t)}{(t - \theta_i)^{m_i}},$$

so that the polynomials $\psi_1, \ldots, \psi_k$ are coprime. Therefore, there are polynomials $f_1(t), \ldots, f_k(t)$ such that

$$1 = \sum_i f_i(t) \psi_i(t).$$

Replacing t with the matrix A gives that

$$I = \sum_i f_i(A) \psi_i(A). \qquad (3.4.1)$$

Define the matrices $E_i = f_i(A)\psi_i(A)$. We show that these E_i are pairwise orthogonal idempotents. If $i \neq j$, then ψ divides $\psi_i \psi_j$, and we have that

$$\psi_i(A)\psi_j(A) = 0.$$

This implies that $E_i E_j = 0$ when $i \neq j$. Multiplying both sides of (3.4.1) by $f_i(A)\psi_i(A)$, we see that this also implies that E_i is an idempotent.

Finally, we show that A is a linear combination of the matrices E_i. Since $\psi(t)$ divides $(t - \theta_i)^{m_i} f_i(t)\psi_i(t)$, we have

$$(A - \theta_i I)^{m_i} E_i = 0,$$

and, consequently,

$$((A - \theta_i I)E_i)^{m_i} = 0.$$

Given our hypothesis, it follows that $(A - \theta_i I)E_i = 0$, or $AE_i = \theta_i E_i$ (so the columns of E_i are eigenvectors for A with eigenvalue θ_i). Finally, since (3.4.1) can be rewritten as

$$I = E_1 + \cdots + E_k,$$

we can express A as the following linear combination of idempotents:

$$A = AE_1 + \cdots + AE_k = \theta_1 E_1 + \cdots + \theta_k E_k. \qquad \square$$

Our next major result is that there is a basis for any Bose-Mesner algebra that consists of idempotent matrices. Before stating this result, we need two lemmas. The first is a technical result that shows that the condition that $N^2 = 0$ only if $N = 0$ holds in the Bose-Mesner algebra of an association scheme.

3.4.2 Lemma. *If N and N^* commute, then $N^2 = 0$ implies $N = 0$.*

Proof. Suppose N and N^* commute and $N^2 = 0$. Then

$$0 = (N^*)^2 N^2 = (N^* N)^2$$

and hence

$$0 = \mathrm{tr}((N^* N)^2) = \mathrm{tr}((N^* N)^*(N^* N)).$$

Using the fact that $\mathrm{tr}(H^* H) = 0$ if and only if $H = 0$, we deduce that $N^* N = 0$. Using this fact again with $\mathrm{tr}(N^* N) = 0$, we conclude that $N = 0$. $\qquad \square$

Before stating the next lemma we need to define a partial ordering on the idempotents of a commutative algebra. Suppose E and F are idempotents in the algebra. We write $E \leq F$ if $FE = E$ (this implies that the column space of E is a subspace of the column space of F). This relation is reflexive, antisymmetric and transitive; therefore it is a partial order. A *minimal idempotent* is a minimal element of the set of nonzero idempotents with respect to this order.

3.4.3 Lemma. *For any set of idempotents in an algebra, there is a set of minimal idempotents of the set and any two distinct minimal idempotents are orthogonal.*

Proof. Suppose that E and F are distinct idempotents and $E \leq F$. Then

$$F(I - E) = F - E \neq 0$$

but $E(I - E) = 0$. Hence the column space of E must be a proper subspace of the column space of F. Therefore, if $E_1, \ldots, E_m$ are distinct idempotents and

$$E_1 \leq \cdots \leq E_m,$$

then m is bounded above by the number of columns in the matrices, and each chain in the partial order has a minimal element.

Further, if E and F are minimal idempotents, then $EF \leq E$ and $FE \leq F$. It follows that if E and F are distinct minimal idempotents, then they are orthogonal. $\square$

3.4.4 Theorem. *Let $\mathcal{A}$ be an association scheme and let $\mathbb{C}[\mathcal{A}]$ be the Bose-Mesner algebra of the association scheme. Then $\mathbb{C}[\mathcal{A}]$ has a basis of pairwise orthogonal idempotents $\{E_0, \ldots, E_d\}$ that satisfies the following conditions:*

    (a) $E_0 = \frac{1}{v}J$;

    (b) $\sum_{i=0}^{d} E_i = I$;

    (c) $E_i^T \in \{E_0, \ldots, E_d\}$ *for all* $i = 1, \ldots, d$;

    (d) for all $i, j \in \{0, \ldots, d\}$, *the Schur product* $E_i \circ E_j$ *lies in the span of* $\mathcal{A}$;

    (e) $E_i E_j = \delta_{i,j} E_i$, *for all* i *and* j.

Proof. From Lemma 3.1.1, Theorem 3.4.1 and Lemma 3.4.2, there is a set of idempotents that span $\mathbb{C}[\mathcal{A}]$; call this set $\mathcal{E}$. The essential problem that remains is to show that there exists a subset of these idempotents that is pairwise orthogonal and is a spanning set for $\mathbb{C}[\mathcal{A}]$.

Consider the subset $\mathcal{E}'$ of minimal idempotents from $\mathcal{E}$. By Lemma 3.4.3 the set $\mathcal{E}'$ is non-empty and is a set of orthogonal idempotents. To show that $\mathcal{E}'$ is a basis, we only need to prove that $\mathcal{E}'$ is a spanning set for $\mathbb{C}[\mathcal{A}]$. To do this, we show that each idempotent in $\mathcal{E}$ is a sum of the minimal idempotents from $\mathcal{E}'$.

Suppose $F \in \mathcal{E}$ and let F_0 be the sum of the distinct minimal idempotents E such that $E \leq F$. Clearly F_0 is an idempotent, and we claim that $F_0 = F$. If this was not the case, then $F - F_0$ would be an idempotent and there would be a minimal idempotent below it. But this would contradict our choice of F_0. Thus we conclude that $F = F_0$ and that $\mathcal{A}$ is spanned by minimal idempotents.

The all-ones matrix J is in $\mathbb{C}[\mathcal{A}]$, so by the construction given in Theorem 3.4.1 the idempotent E_0 is in the set $\mathcal{E}$ (specifically, in the notation of the theorem, $f_1(J)\psi_1(J) = E_0$). The dimension of the column space of J is 1, which implies that E_0 is minimal in $\mathcal{E}$, so it is in the basis $\mathcal{E}'$.

The identity matrix is in $\mathbb{C}[\mathcal{A}]$, so it can be expressed a linear combination of $E_0, \ldots, E_d$:

$$I = \sum_i a_i E_i.$$

Since the scalars a_i are eigenvalues for I, they must all equal 1 and (b) holds.

From Theorem 3.4.1, the set $\mathcal{E}$ is composed of matrices of the form $E_i = f_i(A)\psi_i(A)$. By taking transposes we have that $E_i^T = f_i(A^T)\psi_i(A^T)$; since $A^T \in \mathcal{A}$, the matrix E_i^T is one of the idempotents in $\mathcal{E}$. If $E \in \mathcal{E}'$ then $E^T \in \mathcal{E}$. To show that $\mathcal{E}'$ is closed under transpositions, we only need to show that E^T is also minimal. If E^T was not minimal, then there would exist a minimal idempotent $F \neq E$ such that $E^T F = F$. But then $EF^T = F^T E = F^T$, which implies that $F^T = E$, since E is minimal.

Finally, condition (d) holds since $\mathbb{C}[\mathcal{A}]$ is closed under Schur multiplication. $\qquad\square$

3.4.5 Lemma. *The idempotents E_i in the basis of a Bose-Mesner algebra are Hermitian.*

Proof. Since $\mathbb{C}[\mathcal{A}]$ is closed under transposition and complex conjugation, $E_i^* \in \mathbb{C}[\mathcal{A}]$. Thus there are scalars $a_0, \ldots, a_d$ such that

$$E_i^* = \sum_j a_j E_j,$$

and for any $j \in \{1, \ldots d\}$

$$E_i^* E_j = a_j E_j.$$

As in the proof in the previous theorem, we can show that E_i^* is a minimal idempotent; thus $a_j = 0$ for all j except one. Since $\mathrm{tr}(E_i^* E_i) > 0$, it follows that $a_i \neq 0$, which implies that E_i^* is a scalar multiple of E_i. But $\mathrm{tr}(E_i) = \mathrm{tr}(E_i^*)$, and therefore $E_i^* = E_i$. $\qquad\square$

Since $E_i^T \in \{E_0, \ldots, E_d\}$, this result implies that $\overline{E_i} \in \{E_0, \ldots, E_d\}$.

3.5 Some fundamental identities

There are many identities relating the Schur idempotents A_i, and the matrix idempotents E_j. In this section we prove three that we will use in this chapter.

In the previous section we saw that the set $\{E_0, \ldots, E_d\}$ forms a basis of $\mathbb{C}[\mathcal{A}]$. Thus there are scalars $p_i(j)$ such that

$$A_i = \sum_{j=0}^{d} p_i(j)E_j. \tag{3.5.1}$$

Using (e) from Theorem 3.4.4, this implies that

$$A_i E_j = p_i(j)E_j,$$

whence the scalars $p_i(0), \ldots, p_i(d)$ are the eigenvalues of A_i; this also shows that the columns of E_j are eigenvectors for A. The scalars $p_i(j)$ are called the *eigenvalues* of the association scheme. We note that $p_i(0)$ is the valency of the ith graph in the association scheme and is denoted by v_i (and $v_0 = p_0(0) = 1$).

Since the set $\{A_0, \ldots, A_d\}$ also forms a basis of $\mathbb{C}[\mathcal{A}]$, there are scalars $q_i(j)$ such that

$$E_j = \frac{1}{v} \sum_{i=0}^{d} q_j(i)A_i. \tag{3.5.2}$$

The scalars $q_j(i)$ are known as the *dual eigenvalues* of the association scheme. We see that $\mathrm{tr}(E_j) = q_j(0)$, which is equal to the dimension of the jth eigenspace, and is denoted by m_j.

The matrix P defined by

$$P_{i,j} = p_j(i)$$

is the *matrix of eigenvalues* of the association scheme. As we will see, it completely determines the algebraic structure of the scheme. The matrix Q given by

$$Q_{i,j} = q_j(i)$$

is the *matrix of dual eigenvalues*. Substituting the expression for A_j given in (3.5.1) into (3.5.2), we find that

$$\sum_{k} p_k(i)q_j(k) = \begin{cases} 0, & i \neq j; \\ v, & i = j. \end{cases}$$

In other words, we find that

$$PQ = vI. \tag{3.5.3}$$

To derive the next two identities, we use an inner product on the space of complex $v \times v$ matrices. If M and N are $v \times v$ complex matrices, we define a bilinear form $\langle M, N \rangle$ by

$$\langle M, N \rangle = \mathrm{tr}(M^*N).$$

It is well known that this is an inner product – note that it is linear in the second coordinate. We define a linear functional $\mathrm{sum}(M)$ to be the sum of the entries of the matrix M. Using this, we have an alternate expression for our inner product:

$$\langle M, N \rangle = \mathrm{sum}(\overline{M} \circ N).$$

From the first definition it is immediate that the matrix idempotents E_i form an orthogonal basis for the Bose-Mesner algebra $\mathbb{C}[\mathcal{A}]$. From the second we see that the Schur idempotents, A_i, form a second orthogonal basis.

Using the second definition of the inner product we have that

$$\mathrm{sum}(E_i) = v \, \mathrm{sum}(E_i \circ E_0) = v \langle E_i, E_0 \rangle.$$

This implies that $\mathrm{sum}(E_i) = 0$ unless $i = 0$, in which case it is v. A similar argument shows that $\mathrm{tr}(A_i) = 0$ unless $i = 0$ (but this fact is clear from the definition of an association scheme!).

For an association scheme with Schur idempotents $A_0, \ldots, A_d$ and matrix idempotents $E_0, \ldots, E_d$, it is possible to calculate $\langle A_i, E_j \rangle$ in two different ways. On one hand,

$$\langle A_i, E_j \rangle = \mathrm{tr}(A_i^T E_j) = \mathrm{tr}(\overline{p_i(j)}E_j) = \overline{p_i(j)}m_j;$$

while on the other hand,

$$\langle A_i, E_j \rangle = \mathrm{sum}(E_j \circ A_i) = q_j(i)v_i.$$

Thus we have

$$\frac{\overline{p_i(j)}}{v_i} = \frac{q_j(i)}{m_j}.$$

If Δ_v and Δ_m denote respectively the diagonal matrix of valencies and the diagonal matrix of multiplicities, then we can express the preceding equation as

$$\Delta_m P = Q^* \Delta_v. \tag{3.5.4}$$

We can also use these two forms of the inner product to prove the following fundamental identity.

3.5.1 Theorem. *Suppose $\mathcal{A}$ is an association scheme on v vertices with Schur idempotents $A_0, \ldots, A_d$ and matrix idempotents $E_0, \ldots, E_d$. Then for any x*

in $\mathbb{C}^v$,

$$\sum_{i=0}^{d} \frac{x^T A_i x}{v v_i} A_i = \sum_{j=0}^{d} \frac{x^T E_j x}{m_j} E_j.$$

Proof. Suppose M is a $v \times v$ Hermitian matrix. Since we have two orthogonal bases for $\mathbb{C}[\mathcal{A}]$, we can compute the orthogonal projection $\widehat{M}$ of M onto $\mathbb{C}[\mathcal{A}]$ in two ways. With respect to the A_i's we find that

$$\widehat{M} = \sum_{i=0}^{d} \frac{\langle M, A_i \rangle}{\langle A_i, A_i \rangle} A_i, \tag{3.5.5}$$

and with respect to the E_j's we have

$$\widehat{M} = \sum_{i=0}^{d} \frac{\langle M, E_j \rangle}{\langle E_j, E_j \rangle} E_j. \tag{3.5.6}$$

To derive the stated identity, we set $M = xx^T$ and calculate the projection of M onto $\mathbb{C}[\mathcal{A}]$. Observe that

$$\langle M, A_i \rangle = \text{tr}(xx^T A_i) = x^T A_i x, \qquad \langle M, E_j \rangle = \text{tr}(xx^T E_j) = x^T E_j x.$$

Further, the facts that

$$\langle A_i, A_i \rangle = \text{sum}(A_i \circ A_i) = v v_i,$$

and that

$$\langle E_j, E_j \rangle = \text{tr}(E_j^2) = m_j$$

yield the theorem. $\qquad \qquad \square$

In the identity in Theorem 3.5.1, each side is a matrix. The left side of the equation gives the entries of the matrix (these are the coefficients of the matrices A_i) while the right side of the equation gives the eigenvalues of the matrix (these are the coefficients of the matrices E_j).

3.6 An example

In this section we consider an important example. Fix n and k with $n \geq 2k$. For $r \in \{0, 1, \ldots, k\}$ let A_r denote the matrix with the rows and columns indexed by the k-subsets of $\{0, 1, \ldots, n\}$ with

$$(A_r)_{\alpha,\beta} = \begin{cases} 1, & \text{if } |\alpha \cap \beta| = k - r; \\ 0, & \text{otherwise.} \end{cases}$$

These are the matrices in the Johnson scheme. Let Ω denote the set of all k-subsets of $\{1, \ldots, n\}$ that contain some fixed t-subset. If x is the characteristic vector of Ω and $\alpha \in \Omega$, then

$$(A_r x)_\alpha = \binom{k-t}{k-r-t}\binom{n-k}{r}.$$

Therefore the projection of $x x^T$ onto the Bose-Mesner algebra of $J(n, k)$ is

$$
\sum_r \frac{x^T A_r x}{\binom{n}{k}\binom{k}{r}\binom{n-k}{r}} A_r = \sum_r \frac{|\Omega|\binom{k-t}{k-r-t}\binom{n-k}{r}}{\binom{n}{k}\binom{k}{r}\binom{n-k}{r}} A_r
$$
$$
= \frac{\binom{n-t}{k-t}}{\binom{n}{k}} \sum_r \frac{\binom{k-t}{k-r-t}}{\binom{k}{r}} A_r
$$
$$
= \frac{\binom{n-t}{k-t}}{\binom{n}{k}} \sum_r \frac{\binom{k-r}{t}}{\binom{k}{t}} A_r
$$
$$
= \frac{1}{\binom{n}{t}} \sum_r \binom{k-r}{t} A_r.
$$

There is a second expression for this sum, which we will need later. Let $W_{t,k}$ be the 01-matrix with rows indexed by t-subsets and columns by k-subsets of a fixed n set, where

$$
(W_{t,k})_{\alpha,\beta} = \begin{cases} 1, & \text{if } \alpha \subseteq \beta; \\ 0, & \text{otherwise.} \end{cases}
$$

Then for two k-subsets α and β,

$$
(W_{t,k}^T W_{t,k})_{\alpha,\beta} = \binom{|\alpha \cap \beta|}{t},
$$

and consequently

$$
W_{t,k}^T W_{t,k} = \sum_r \binom{k-r}{t} A_r.
$$

We can state the result of these computations as follows.

3.6.1 Lemma. *If x is the characteristic vector of the set of k-subsets of $\{1, \ldots, n\}$ with some fixed t-subset in common, then the projection of $x x^T$ onto the Bose-Mesner algebra of $J(n, k)$ is equal to*

$$
\frac{1}{\binom{n}{t}} W_{t,k}^T W_{t,k}.
$$

$\square$

We will see these matrices again in Section 6.2, and they are used in Section 6.5 to compute the eigenvalues of the Johnson scheme.

3.7 The clique bound

One of the most important parts of the theory of association schemes concerns the linear programming bound on the size of cliques and cocliques. This theory was developed by Delsarte [51], and its importance is difficult to overstate.

For the remainder of this chapter we let $\mathcal{A} = \{A_0, A_1, \ldots, A_d\}$ be an association scheme on v vertices with d classes. We denote the basis of idempotents of this association scheme by $\{E_0, E_1, \ldots, E_d\}$. If T is a subset of $\{1, \ldots, d\}$, let X_T be the graph formed by the union of the graphs X_i for i in T. We define a T-*clique* to be a clique in X_T. The adjacency matrix of X_T is $A_T = \sum_{i \in T} A_i$.

Recall that for any graph X, a symmetric matrix M is *compatible* with X if $M_{u,v} = 0$ whenever u and v are nonadjacent vertices in X. If M is a compatible matrix for X with least eigenvalue τ, then $M - \tau I$ is positive semidefinite and compatible with X.

Let M be a positive semidefinite matrix compatible with X. Since M is positive semidefinite, it can be written as $M = N^*N$ where N is a $|V(X)| \times |V(X)|$ matrix. The rows in N corresponding to any pair of nonadjacent vertices in X must have dot product 0, since the corresponding entry in M is 0. If we assign the rows in N to their corresponding vertex in X, then we have what is known as an *orthogonal representation* of the complement of X (i.e., an assignment of vectors to the vertices in a graph such that adjacent vertices receive orthogonal vectors).

3.7.1 Theorem. *Let $\mathcal{A}$ be an association scheme with d classes and let T be a subset of $\{1, \ldots, d\}$. If C is a T-clique, then*

$$|C| \leq \max_{M \in \mathcal{M}} \frac{\operatorname{sum}(M)}{\operatorname{tr}(M)}$$

where $\mathcal{M}$ is the set of all positive semidefinite matrices in $\mathbb{C}[\mathcal{A}]$ that are compatible with X_T.

Proof. Let C be a T-clique in the association scheme with characteristic vector x. Let M be the orthogonal projection of xx^T onto $\mathbb{C}[\mathcal{A}]$. Then, from Theorem 3.5.1,

$$M = \sum_{i=0}^{d} \frac{x^T A_i x}{v v_i} A_i = \sum_{j=0}^{d} \frac{x^T E_j x}{m_j} E_j.$$

Since $tr(A_i) = 0$ for all $i > 0$, we have that

$$\mathrm{tr}(M) = \frac{xx^T}{v} \, \mathrm{tr}(A_0) = |C|.$$

Further, $\mathrm{sum}(E_j) = 0$ for all $j = 1, \ldots, d$, which implies that

$$\mathrm{sum}(M) = \frac{x^T E_0 x}{m_0} \, \mathrm{sum}(E_0) = |C|^2.$$

Hence we have the result

$$|C| = \frac{\mathrm{sum}(M)}{\mathrm{tr}(M)}. \tag{3.7.1}$$

Since M is positive semidefinite matrix in $\mathbb{C}[\mathcal{A}]$ and is compatible with X_T, our theorem follows. $\qquad\square$

This theorem is unusual in that we show that there must exist a positive semidefinite matrix in $\mathbb{C}[\mathcal{A}]$ that is compatible with X_T (specifically M, the projection of xx^T). This matrix can be used to determine the exact size of a clique, but we can also get an upper bound by considering the all positive semidefinite matrices in $\mathbb{C}[\mathcal{A}]$ that are compatible with X_T.

Determining the optimum value of the bound in Theorem 3.7.1 is a linear programming problem. The simplest case arises when T has size 1. In this case, X_T is just one of the graphs of the scheme, say X_i. Any positive semidefinite matrix compatible with X_T from the Bose-Mesner algebra of the association scheme must have one of the following two forms:

$$A_i - tI, \qquad sI - A_i.$$

In the first case, t is less than or equal to the least eigenvalue τ of A, and the bound is

$$\frac{\mathrm{sum}(A_i - tI)}{\mathrm{tr}(A_i - tI)} = \frac{vv_i - vt}{-vt} = 1 - \frac{v_i}{t};$$

this is maximized when $t = \tau$. In the second case, s is greater than or equal to the valency v_i of X_i, and

$$\frac{\mathrm{sum}(sI - A_i)}{\mathrm{tr}(sI - A_i)} = \frac{vs - vv_i}{vs} = 1 - \frac{v_i}{s}.$$

Since $0 < v_i \le s$, this number is no more than 1, and, unless X_i is the empty graph, the first case will give the maximum. Thus we have we have the following result, which is called the *ratio bound for cliques*.

3.7.2 Corollary. *Let X be a graph in an association scheme and assume that X has valency k and least eigenvalue τ. If C is a clique in X, then*

$$|C| \leq 1 - \frac{k}{\tau}.$$

If C is a clique that meets the bound with equality, then the characteristic vector x_C of C is orthogonal to the τ-eigenspace.

Proof. The bound on the size of a clique holds from the statements before the corollary. Assume that C is a clique of size $1 - k/\tau$ in X and that x_C is the characteristic vector of C. Let M be the projection of $x_C x_C^T$ into $\mathbb{C}[\mathcal{A}]$; then

$$M = A_i - \tau I.$$

Let v be a τ-eigenvector for A_i, and let $j(\tau)$ be the index on the matrix idempotent corresponding to τ (i.e., $A_i E_{j(\tau)} = \tau E_{j(\tau)}$). Then $Mv = 0$, and $E_j v = 0$ unless $j = j(\tau)$. Together these imply that

$$0 = Mv = \sum_{j=0}^{d} \frac{x_C^T E_j x_C}{m_j} E_j v = \frac{x_C^T E_{j(\tau)} x_C}{m_{j(\tau)}} v.$$

From this we can conclude that $x_C^T E_{j(\tau)} x_C = 0$ and $E_{j(\tau)} x_C = 0$. $\qquad\square$

3.8 The clique-coclique bound

We derive a clique-coclique bound for graphs in an association scheme.

3.8.1 Lemma. *Let M and N be matrices in the Bose-Mesner algebra of $\mathcal{A}$, where $\mathcal{A}$ is an association scheme on v vertices. If M and N are positive semidefinite and $M \circ N = \alpha I$ for some constant α, then*

$$\frac{\text{sum}(M)}{\text{tr}(M)} \frac{\text{sum}(N)}{\text{tr}(N)} \leq v.$$

Equality holds if and only if $MN = \beta J$ for some β.

Proof. Since $N, M \in \mathbb{C}[\mathcal{A}]$, by (3.5.6)

$$N = \sum_j \frac{\langle N, E_j \rangle}{m_j} E_j, \quad M = \sum_j \frac{\langle M, E_j \rangle}{m_j} E_j$$

and therefore

$$\text{tr}(MN) = \sum_j \frac{\langle M, E_j \rangle \langle N, E_j \rangle}{m_j^2} \text{tr}(E_j) = \sum_j \frac{\langle M, E_j \rangle \langle N, E_j \rangle}{m_j}.$$

As M and N are positive semidefinite, the terms in this sum are non-negative, and accordingly

$$\operatorname{tr}(MN) \geq \frac{\langle M, E_0 \rangle \langle N, E_0 \rangle}{m_0} = \frac{\operatorname{sum}(M)\operatorname{sum}(N)}{v^2}.$$

Now

$$\operatorname{tr}(MN) = \operatorname{sum}(M \circ N) = \operatorname{sum}(\alpha I) = \alpha v.$$

Since M and N are in $\mathbb{C}[\mathcal{A}]$, both have constant diagonal and the product of the value on the diagonal in M and the value of the diagonal in N equals α. Thus we also have that

$$\operatorname{tr}(M)\operatorname{tr}(N) = \alpha v^2$$

and so

$$\operatorname{tr}(MN) = \frac{\operatorname{tr}(M)\operatorname{tr}(N)}{v}.$$

This yields the inequality.

We note also that if equality holds, then for all $1 \leq j \leq d$

$$\langle M, E_j \rangle \langle N, E_j \rangle = 0.$$

This means that for every $j \in \{1, \ldots, d\}$ either $ME_j = 0$ or $NE_j = 0$. Since $\mathbb{C}[\mathcal{A}]$ is commutative, this implies that MN is a multiple of J. $\qquad\square$

This lemma has a number of important consequences. The first is a result similar to the bound on the size of a T-clique given in Theorem 3.7.1. Let T be a subset of classes in an association scheme. A T-*coclique* in the association scheme is a set of vertices in which no two are i-related for any $i \in T$. A T-coclique is a coclique in the graph X_T.

3.8.2 Theorem. *Let $\mathcal{A}$ be an association scheme on v vertices with d classes and suppose $T \subseteq \{1, \ldots, d\}$. If S is a T-coclique, then*

$$|S| \leq \min_{N \in \mathcal{N}} \frac{v \operatorname{tr}(N)}{\operatorname{sum}(N)}$$

where $\mathcal{N}$ is the set of all positive semidefinite matrices in $\mathbb{C}[\mathcal{A}]$ that are compatible with X_T.

Proof. Let x be the characteristic vector of S and let M be the projection of xx^T onto $\mathbb{C}[\mathcal{A}]$ (clearly M is positive semidefinite). Using the same argument used to prove (3.7.1), we have that

$$|S| = \frac{\operatorname{sum}(M)}{\operatorname{tr}(M)}.$$

For any $N \in \mathbb{C}[\mathcal{A}]$ that is compatible with X_T, the matrix $M \circ N$ is a scalar matrix. Hence the theorem follows directly from Lemma 3.8.1. $\qquad\square$

From this theorem we can derive a ratio bound for cocliques.

3.8.3 Corollary. *Let $\mathcal{A}$ be an association scheme on v vertices with d classes. Let T be a subset of $\{1, \ldots, d\}$ and assume k_T is the valency of X_T and τ_T is the least eigenvalue of X_T. If S is a T-coclique, then*

$$|S| \le \frac{v}{1 - \frac{k_T}{\tau_T}}.$$

Proof. Suppose N is a positive semidefinite matrix in $\mathbb{C}[\mathcal{A}]$ and that N is compatible with X_T. Since N is in $\mathbb{C}[\mathcal{A}]$, it has constant row sum and a constant diagonal. So we can assume that the diagonal entries of N are equal to $v \ge 0$ and that each row of N sums to $k_N + v$. Then

$$\frac{\text{sum}(N)}{\text{tr}(N)} = 1 + \frac{k_N}{v}.$$

Further, by Theorem 3.8.2,

$$|S| \le \frac{v \, \text{tr}(N)}{\text{sum}(N)}$$

for any positive semidefinite matrix N in $\mathbb{C}[\mathcal{A}]$ that is compatible with X_T.

Set N to be $A(X_T) - \tau_T I$ (this is clearly a semidefinite matrix in $\mathbb{C}[\mathcal{A}]$ that is compatible with X_T). For this matrix $v = -\tau_T$ and $k_N = k_T$, so we have that

$$|S| \le v \frac{\text{tr}(N)}{\text{sum}(N)} = \frac{v}{1 - \frac{k_T}{\tau_T}}. \qquad\square$$

We next present the clique-coclique bound. The original proof of this was given in Delsarte's thesis [51, Theorem 3.9]. We give an alternate proof to Delsarte.

3.8.4 Theorem. *Let $\mathcal{A}$ be an association scheme on v vertices and let X be the union of some of the graphs in the association scheme. If C is a clique and S is a coclique in X, then*

$$|C| \, |S| \le v.$$

Moreover, if equality holds and x and y are the characteristic vectors of C and S, respectively, then for all $i = 1, \ldots, d$,

$$x^T E_j x \; y^T E_j y = 0.$$

Proof. Suppose C is a clique and S is a coclique in X and let x and y be their respective characteristic vectors. Define M and N to be the projections of $x x^T$

and yy^T, respectively, onto $\mathbb{C}[\mathcal{A}]$. In particular,

$$M = \sum_{i=0}^{d} \frac{x^T A_i x}{v v_i} A_i, \qquad N = \sum_{i=0}^{d} \frac{y^T A_i y}{v v_i} A_i.$$

Since C is a clique, for any $i > 0$ for which A_i is not one of the graphs in X, we have that $x^T A_i x = 0$. Similarly, since S is a coclique, for any $i > 0$ for which A_i is a graph in X, we have that $y^T A_i y = 0$. These two facts imply that $M \circ N$ is a scalar matrix. Thus we can apply Lemma 3.8.1 to get that

$$\frac{\text{sum}(M)}{\text{tr}(M)} \frac{\text{sum}(N)}{\text{tr}(N)} \le v.$$

As we saw in the proof of Theorem 3.7.1,

$$|C| = \frac{\text{sum}(M)}{\text{tr}(M)}, \qquad |S| = \frac{\text{sum}(N)}{\text{tr}(N)}$$

and we have our inequality.

If the bound is tight, then MN is a multiple of J. Therefore, for any $r \in \{1, \ldots, d\}$, we have

$$MNE_r = 0.$$

Since

$$NE_r = \frac{y^T E_r y}{m_r} E_r, \qquad ME_r = \frac{x^T E_r x}{m_r} E_r,$$

this proves the second part of the theorem. □

Let $\mathcal{A}$ be an association scheme with d classes, and let S be a subset of its vertices with characteristic vector x. The *degree set of S* is the set of indices i in $\{1, \ldots, d\}$ such that $x^T A_i x \ne 0$. Its *dual degree set* is the set of indices j in $\{1, \ldots, d\}$ such that $x^T E_j x \ne 0$. Clearly a clique C and a coclique S have disjoint degree sets. The previous theorems shows that if $|C| \, |S| = v$, then the dual degree sets of C and S are disjoint.

3.8.5 Corollary. *Let $\mathcal{A}$ be an association scheme with d classes and suppose $T \subseteq \{1, \ldots, d\}$. If C is a T-clique and S is a T-coclique with $|C| \, |S| = v$, then for all $j > 0$ at most one of the vectors $E_j x$ and $E_j y$ is not zero.*

Proof. If $j > 0$, then

$$x^T E_j x \; y^T E_j y = 0.$$

Since E_j is positive semidefinite, $z^T E_j z = 0$ if and only if $E_j z = 0$. □

3.8.6 Corollary. *Let $\mathcal{A}$ be an association scheme on v vertices and let X be the union of some of the graphs in the scheme. If C is a clique and S is a coclique in X with $|C|\,|S| = v$, then $|C \cap S| = 1$.*

Proof. Assume x and y are the characteristic vectors of C and S, respectively. From Corollary 3.8.5, if $|C|\,|S| = v$, then for every $j \in \{1, \ldots, d\}$ either $E_j x = 0$ or $E_j y = 0$. This implies that the vectors $\sum_{i=1}^{d} E_i x$ and $\sum_{i=1}^{d} E_i y$ are orthogonal.

Since $E_0 x = \frac{|C|}{v}\mathbf{1}$ and $E_0 y = \frac{|S|}{v}\mathbf{1}$, using the fact that $\sum_{i=0}^{d} E_i = I$ we have

$$\sum_{i=1}^{d} E_i x = x - \frac{|C|}{v}\mathbf{1}, \qquad \sum_{i=1}^{d} E_i y = y - \frac{|S|}{v}\mathbf{1}.$$

This means that the vectors

$$x - \frac{|C|}{v}\mathbf{1}, \qquad y - \frac{|S|}{v}\mathbf{1}$$

are orthogonal. So

$$0 = \left(x - \frac{|C|}{v}\mathbf{1}\right)^T \left(y - \frac{|S|}{v}\mathbf{1}\right) = x^T y - 2\frac{|C||S|}{v} + \frac{|C||S|}{v} = x^T y - 1.$$

Therefore if the clique-coclique bound is tight, then $x^T y = 1$. This implies that $|S \cap C| = 1$.   □

The clique-coclique bound provides a simple proof of the EKR bound in some special cases. Suppose that we have a t-$(n, k, 1)$ design, that is, a Steiner system $\mathcal{D}$. Then any two distinct blocks of $\mathcal{D}$ have at most $t - 1$ points in common, and therefore the blocks of $\mathcal{D}$ form a T-clique in the Johnson scheme, where $T = \{k - t + 1, \ldots, k\}$. A T-coclique is then a t-intersecting family, and so we deduce that the size of a t-intersecting family is at most

$$\frac{\binom{n}{k}}{|\mathcal{D}|} = \frac{\binom{n}{k}}{\binom{n}{t}/\binom{k}{t}} = \binom{n - t}{k - t}.$$

This is the EKR bound.

3.9 Exercises

3.1 Let $\mathcal{A} = \{A_0, \ldots, A_d\}$ be an association scheme. Show that any graph in the Bose-Mesner algebra is a union of graphs in $\mathcal{A}$.

3.2 Prove that every Schur-closed matrix algebra has a basis of Schur orthogonal 01-matrices.

3.3 Let $\mathcal{A}$ be an association scheme with d classes and let $E_0, \ldots, E_d$ be the matrix idempotents of the scheme. Suppose $F_0, \ldots, F_d$ is a set of d

pairwise orthogonal idempotents in the Bose-Mesner algebra of $\mathcal{A}$ and $\sum_i F_i = I$. Prove that, after a reordering if needed, $F_i = E_i$.

3.4 Show that the commutant of a set of $v \times v$ permutation matrices is a Schur-closed matrix algebra.

3.5 Assume that $\mathcal{A}$ is a basis of the commutant of a set of $v \times v$ permutation matrices that is composed of Schur orthogonal 01-matrices. Show that if the commutant is commutative, then $\mathcal{A}$ is an association scheme.

3.6 Represent each permutation σ in $\mathrm{Sym}(n)$ as a permutation matrix M_σ by its action on the k-subsets of an n-set. Show that the adjacency matrices of the graphs in the Johnson scheme are a basis of the commutant of $\{M_\sigma : \sigma \in \mathrm{Sym}(n)\}$.

3.7 Each permutation σ in $\mathrm{Sym}(n)$ determines a permutation of the binary sequences of length n and hence gives rise to a permutation matrix M_σ of order $2^n \times 2^n$. Find a basis of the commutant of $\{M_\sigma : \sigma \in \mathrm{Sym}(n)\}$.

3.8 Let G be a group and C_i and C_j be two conjugacy classes of G. Show that $C_i C_j = C_j C_i$.

3.9 Is the conjugacy class scheme on $\mathrm{Sym}(n)$ primitive?

3.10 A matrix M is called *Schur connected* if no diagonal entry is equal to an off-diagonal entry. Prove that an association scheme is primitive if and only if the idempotents $E_0, \ldots, E_d$ are each Schur connected.

Notes

Association schemes were first defined by Bose and others who were working in the area of design of statistical experiments. Now association schemes are widely used in algebraic combinatorics, particularly in the study of combinatorial designs and coding theory, and there are diverse approaches to the theory of association schemes.

In the literature, there is some confusion over the definition of an association scheme. For example, Zieschang in [177] does not require that an association scheme be commutative. Other sources define a set of matrices that satisfy all our conditions for an association scheme except commutativity to be either a *cellular algebra* or a *homogeneous coherent configuration*.

There are many excellent references on association schemes, such as the book by Bannai and Ito [14] or the book by Brouwer, Cohen and Neumaier [31]. Chapters 10, 11 and 12 of [78] give an accessible introduction to association schemes. Different approaches to association schemes are given by Zieschang in [177] and by Bailey in [13]. We also recommend the paper by Martin and Tanaka [123].

4

Distance-regular graphs

A critical family of association schemes arise from *distance-regular graphs*. A graph is distance regular if for any two vertices u and v at distance k in the graph, the number of vertices w that are at distance i from u and j from v depends only on k, and not on the vertices u and v.

Suppose X is a graph with diameter d. The ith *distance graph* X_i of X has the same vertex set as X, and two vertices are adjacent in X_l if and only if they are distance i in X. For any graph X, the adjacency matrices of the distance graphs of X are called the *distance matrices*. The distance matrices for any graph are linearly independent and sum to $J - I$. It has been left as an exercise to show that if X is a distance-regular graph, then the distance matrices form an association scheme. Such an association scheme is called *metric*; these are discussed in detail in Section 4.1.

Let X be a graph and S a subset of the vertices of X. Denote by S_i the set of vertices of X at distance i from S. The *distance partition* of X relative to S is the partition $\delta_S = \{S_1, \ldots, S_r\}$ of $V(X)$. We say that S is a *completely regular subset* of X if δ_S is an equitable partition. If X is a distance-regular graph, then for any $x \in V(X)$ the set $S = \{x\}$ is completely regular. Moreover, the quotient graphs of X with respect to each $\delta_{\{x\}}$ are isomorphic.

For any distance partition $\delta_S = \{S_1, \ldots, S_r\}$ of X, vertices in S_i can only be adjacent to vertices in S_{i-1}, S_i and S_{i+1}. Consequently, if X is a distance-regular graph, then in the quotient graph $X/\delta_{\{x\}}$, the vertex S_i can only have edges to S_{i-1}, S_i and S_{i+1}. If we arrange the cells of the partition in order of distance, then the adjacency matrix of the quotient graph is a tridiagonal matrix.

If X is a distance-regular graph, then quotient matrix $X/\delta_{\{x\}}$ has the form

$$\begin{pmatrix} 0 & b_0 & & & & \\ c_1 & a_1 & b_1 & & & \\ & c_2 & a_2 & b_2 & & \\ & & & \ddots & & \\ & & c_{d-1} & a_{d-1} & b_{d-1} \\ & & & c_d & a_d \end{pmatrix}.$$

The parameter a_i is the (i, i)-entry of the quotient matrix; it represents the number of vertices in S_i that are adjacent to a specific vertex in S_i. The $(i, i + 1)$-entry of this matrix is denoted by b_i and is the number of vertices in S_{i+1} that are adjacent to a fixed vertex in S_i. Finally, c_i is the entry in the $(i - 1, i)$-position and is the number of vertices in S_{i-1} adjacent to a vertex in S_i. Note that $c_1 = 1$, b_0 is the valency k and each row sums to k. In some circumstances it is useful to assume $c_0 = b_d = 0$. The numbers a_i, b_i and c_i are called the *parameters* of the distance-regular graph.

In the next four chapters we see several examples of distance-regular graphs. For now we point out that the Petersen graph defined in Section 2.2 is distance regular. The cells in the distance partition with respect to the vertex $\{1, 2\}$ are $S_1 = \{\{1, 2\}\}$, $S_2 = \{\{3, 4\}, \{3, 5\}, \{4, 5\}\}$ and $S_3 = \{\{1, 3\}, \{1, 4\}, \{1, 5\}, \{2, 3\}, \{2, 4\}, \{2, 5\}\}$, and the quotient matrix is

$$\begin{pmatrix} 0 & 3 & 0 \\ 1 & 0 & 2 \\ 0 & 1 & 2 \end{pmatrix}.$$

4.1 Metric schemes

If $\mathcal{A} = \{A_0, \ldots, A_d\}$ is an association scheme, then the Bose-Mesner algebra of $\mathcal{A}$ will have the minimal possible dimension, namely $d + 1$. This implies a bound on the diameter of the graphs in the algebra.

4.1.1 Lemma. *If $\mathcal{A}$ is an association scheme with d classes and X is a graph in $\mathbb{C}[\mathcal{A}]$, then the diameter of X is at most d.*

Proof. Assume that the diameter of X is r and let A denote the adjacency matrix of X. Then

$$((I + A)^r)_{u,v}$$

is positive if $\text{dist}(u, v) \leq r$, and zero otherwise. Therefore the powers $(I + A)^i$ for $i = 0, \ldots, r$ are linearly independent. Since the dimension of $\mathbb{C}[\mathcal{A}]$ is $d + 1$, this implies that $r \leq d$. $\qquad\square$

In the Exercises you are asked to show that the distance matrices of a distance-regular graph form an association scheme. The next result is a sort of converse of this. It gives a condition on when the graphs of an association scheme are distance graphs.

4.1.2 Lemma. *Suppose $\mathcal{A}$ is an association scheme with d classes. If X is a graph in $\mathbb{C}[\mathcal{A}]$ with diameter d, then the graphs in $\mathcal{A}$ are the distance graphs of X.*

Proof. Suppose $\mathcal{A} = \{A_0, \ldots, A_d\}$ and let $A = A(X)$. Provided that $r \le d$, the (u, v)-entry of the matrix $(I + A)^{r+1}$ is greater than or equal to the (u, v)-entry in $(I + A)^r$ and the matrix $(I + A)^{r+1}$ will have strictly fewer zero entries than $(I + A)^r$. (In fact, the (u, v)-entry of $(I + A)^r$ is zero and $(I + A)^{r+1}$ is not equal to zero exactly when u and v are at distance $r + 1$ in X.)

For each r define

$$F_r = \{i : A_i \circ (I + A)^r \ne 0\}.$$

If $1 \le r \le d$, then

$$\sum_{j \in F_r \backslash F_{r-1}} A_j$$

is the r-th distance matrix of X. Since X has $d + 1$ distance matrices and the Schur product of any two of these matrices is zero, we see that each distance matrix of X must be equal to A_j for some j. $\qquad\square$

Note that this lemma implies that if X is a graph in an association scheme $\{A_0, \ldots, A_d\}$ and X has diameter d, then $A(X) = A_i$ for some i.

Let $\mathcal{A}$ be an association scheme with d classes. If there is a graph X in the scheme with diameter d, we say that the scheme is *metric* relative to X (equivalently, if $A = A(X)$, then we say the scheme is metric relative to A). Metric association schemes are also known as *P-polynomial* association schemes. Examples of metric schemes include the Johnson scheme (defined in the introduction to Chapter 3) and the Grassmann and the Hamming schemes, which are discussed in detail in Chapters 9 and 10. Association schemes can be metric with respect to more than one matrix; for example, the Johnson scheme $J(2k + 1, k)$ is metric relative to the graph $J(2k + 1, k, 1)$ and to the Kneser graph $K(2k + 1, k)$ (this is the graph $J(2k + 1, k, k)$). It is easy to check that a primitive strongly regular graph is metric relative to both non-identity graphs in the association scheme (see Section 5.1 for more details). Further, it is not too hard to prove that an association scheme with d classes can be metric relative to at most two graphs X_1 and X_d (where each graph is the dth distance graph of the other) – this is given (with hints) as Exercise 12 in Chapter 12 of [78].

In the following results we give some examples of the extra structure possessed by metric association schemes. We leave the proof of the next result as an exercise. (The previous proof contains the ideas required.)

4.1.3 Corollary. *If the d-class scheme $\mathcal{A}$ is metric relative to X_1, then the matrices $A_0, \ldots, A_d$ can be ordered so that A_i is a polynomial in degree i in A_1.* $\square$

4.2 A three-term recurrence

Suppose $\mathcal{A} = \{A_0, \ldots, A_d\}$ is an association scheme that is metric relative to X_1 and $A_1 = A(X_1)$. Then, by Corollary 4.1.3, we can assume that the matrices A_i are ordered so A_i is a polynomial of degree i in A_1. This implies that $A_1 A_r$ is a linear combination of $A_0, \ldots, A_{r+1}$. Since $(A_1 A_r)_{u,v}$ is equal to the number of vertices in X_1 adjacent to u and at distance r from v, it follows that $(A_1 A_r)_{u,v} = 0$ if $\mathrm{dist}(u, v)$ is not $r - 1$, r or $r + 1$. Accordingly there are scalars b_r, a_r, c_r such that

$$A_1 A_r = b_{r-1} A_{r-1} + a_r A_r + c_{r+1} A_{r+1}.$$

In other words the matrices $A_0, \ldots, A_d$ satisfy a *three-term recurrence*.

The scalars b_r, a_r and c_r in the three-term recurrence have simple interpretations: given two vertices u and v at distance r in X_1, these numbers are the number of vertices w adjacent to y and at distance $r + 1$, r and $r - 1$ respectively from x. In fact, we have already seen these scalars, they are the parameters for the graph X_1. These numbers are also called the *intersection numbers* for the distance-regular graph.

Thus for $r = 0, \ldots, d$,

$$b_r + a_r + c_r = v_1$$

where v_1 is the degree of X_1. If v_r denotes the number of vertices at distance r from a fixed vertex, then

$$v_r b_r = v_{r+1} c_{r+1}.$$

The next result also follows from the fact that any metric association scheme satisfies a three-term recurrence relation.

4.2.1 Corollary. *Suppose the d-class scheme $\mathcal{A}$ is metric relative to A_1, and the matrices $A_0, \ldots, A_d$ are ordered so that A_i is a polynomial in degree i in A_1. Then there are polynomials $q_0, \ldots, q_d$ such that $A_{d-i} = q_i(A_1)A_d$.* $\square$

This implies that $q_d(A_1)A_d = I$, and therefore A_d is an invertible matrix.

4.3 Cometric schemes

We have seen that the Bose-Mesner algebra of an association scheme has two bases, the matrices of the association scheme and the idempotents. Any property of an association scheme that is defined by some property on the matrices of the association scheme has a dual version formulated by defining a related property on the idempotents. In this section we define the dual version of a metric association scheme.

Although it may seem contrived, we could define the diameter of a graph X to be the minimum degree of a polynomial p such that $p(A(X))$ has a Schur inverse (since this implies that no entry in $p(A(x))$ is equal to zero). Using this definition, we can define a dual notion of diameter. If p is a polynomial and M is a matrix, then we define $p \circ M$ to be the matrix we get by applying p to each entry of M (this is called the *Schur polynomial*). So, in symbols,

$$(p \circ M)_{i,j} = p(M_{i,j}).$$

In the case where $p(x) = x^r$, the matrix $p \circ M$ is called the r-*Schur power* of M and is also denoted by $M^{\circ(r)}$. We define the *Schur diameter* of a matrix M to be the least degree of a polynomial p such that $p \circ M$ is invertible.

4.3.1 Lemma. *If E is a matrix with Schur diameter s, then the Schur powers*

$$J, E, \ldots, E^{\circ(s)}$$

are linearly independent.

Proof. Let U_r be the span of $\{J, E, \ldots, E^{\circ(r)}\}$; we claim that $U_0, \ldots, U_s$ form a strictly increasing sequence of subspaces. Indeed, if $E^{\circ(r+1)} \in U_r$, then U_r is invariant under Schur multiplication by E. This implies that all Schur polynomials in E are in U_r. But this is a contradiction with the fact that the Schur diameter of E is s. □

Now we have the dual result to Lemma 4.1.1.

4.3.2 Lemma. *The Schur diameter of any idempotent in a primitive association scheme with d classes is no more than d.* □

For an association scheme $\mathcal{A}$ with d classes, each matrix M in $\mathbb{C}[\mathcal{A}]$ has constant diagonal and the off-diagonal entries can take at most d different values. If no off-diagonal entry of M is equal to the diagonal entries, then there is a polynomial p of degree at most d such that $p \circ M = I$. In Chapter 3 we defined a matrix to be *Schur connected* if no diagonal entry of the matrix is equal to an off-diagonal entry; it was left as an exercise to prove that an association scheme is primitive if and only the idempotents $E_1, \ldots, E_d$ are each Schur connected (see Exercise 3.10 of Chapter 3).

We can now define the dual notion of a metric scheme. An association scheme is *cometric* (or *Q-polynomial*) relative to the idempotent E if the Schur diameter of E is d. Suppose $\mathcal{A}$ is cometric relative to E. Then the Schur powers

$$J, E, \ldots, E^{\circ(d)}$$

are linearly independent and consequently they form a basis for $\mathbb{C}[\mathcal{A}]$. Therefore each primitive idempotent E_r is a Schur polynomial in E; in fact we have a dual version of Corollary 4.1.3.

4.3.3 Lemma. *If the association scheme $\mathcal{A}$ is cometric relative to the matrix idempotent E, then we can reorder the matrix idempotents so that $E_1 = E$ and each E_r is a Schur polynomial of degree r in E_1.*

Proof. Each Schur power $E^{\circ(r)}$ is in $\mathbb{C}[\mathcal{A}]$ and therefore is a linear combination of $E_0, \ldots, E_d$. Define

$$S_r = \{ j \ : \ E_j E^{\circ(s)} \neq 0 \text{ for some } s \leq r \}.$$

Clearly $S_0 = \{0\}$, $S_d = \{0, \ldots, d\}$, and $S_r \subseteq S_{r+1}$. If $r < d$ and $S_{r+1} = S_r$, then a straightforward induction argument shows that

$$S_r = S_{r+1} = \cdots = S_d.$$

But then $S_r = \{0, \ldots, d\}$ and the first r Schur powers of E will span the algebra. Thus we conclude for $r < d$ that $|S_r| < |S_{r+1}|$ and $|S_{r+1} \setminus S_r| = 1$. The unique idempotent in $S_{r+1} \setminus S_r$ is a Schur polynomial in E with degree $r + 1$. $\qquad \square$

If $\mathcal{A}$ is cometric relative to an idempotent E, we normally assume that $E_1 = E$ and E_r is a Schur polynomial of degree r in E_1. It is not difficult to show that $E_0, \ldots, E_d$ satisfy a three-term recurrence, so we leave it as an exercise.

The association scheme consisting of the distance matrices of any primitive strongly regular graph is cometric. In subsequent chapters, we realize more examples of cometric association schemes; specifically, we see that the Johnson scheme, the Grassmann scheme and the Hamming scheme are all cometric.

4.4 Intersection numbers and Krein parameters

It follows from the axioms for an association scheme that there are non-negative integers $p_{i,j}(r)$ such that

$$A_i A_j = \sum_{r=0}^{d} p_{i,j}(r) A_r.$$

These are called the *intersection numbers* of the scheme.

There is another viewpoint on the intersection numbers of an association scheme. The Bose-Mesner algebra $\mathbb{C}[\mathcal{A}]$ of a scheme $\mathcal{A}$ is not just a vector space, but a matrix algebra. Therefore if $M \in \mathbb{C}[\mathcal{A}]$, then multiplication by M is a linear mapping of $\mathbb{C}[\mathcal{A}]$ to itself. If $\mathcal{A}$ has d classes, the action of each matrix in its Bose-Mesner algebra can be represented by a matrix of order $(d + 1) \times (d + 1)$. The nonzero entries in the matrix representing A_i are the intersection numbers $p_{i,j}(k)$. If the scheme is metric relative to X_1 and the indices are ordered so that X_r is the rth distance graph of X_1, then the matrix that represents A_1 is the adjacency matrix of a weighted path. The converse is also true, but we leave both parts as an exercise.

An association scheme also has parameters that are the duals to the intersection numbers. Assume the association scheme $\mathcal{A}$ has n vertices. Since the Bose-Mesner algebra is closed under Schur multiplication, there are numbers $q_{i,j}(r)$ such that

$$E_i \circ E_j = \frac{1}{n} \sum_{r=0}^{d} q_{i,j}(r) E_r.$$

These numbers are known as the *Krein parameters* of the scheme. Note that they are eigenvalues for $E_i \circ E_j$. If the scheme is cometric relative to E_1, and we use the corresponding natural order of the idempotents, then the matrix that represents Schur multiplication by E_1 is again the adjacency matrix of a weighted path.

The Krein parameters need not be integers. Since the idempotents E_r are Hermitian, it follows that the Krein parameters are real.

4.4.1 Lemma. *The Krein parameters of an association scheme are non-negative real numbers.*

Proof. Observe that $E_i \circ E_j$ is a principal submatrix of the Kronecker product $E_i \otimes E_j$. This Kronecker product is Hermitian and idempotent; hence it is positive semidefinite, and consequently $E_i \circ E_j$ is positive semidefinite. This implies that the eigenvalues of $E_i \circ E_j$ are non-negative. $\qquad\square$

4.5 Coding theory

For any subset C of the vertices from a metric association scheme, there are a number of useful parameters we will consider. Many of these are useful in coding theory (hence the title of this section).

Assume $\mathcal{A}$ is an association scheme with d classes. Let C be a subset of the vertices of $\mathcal{A}$ and denote the characteristic vector of C by x. The *degree* of C is the number of indices $i \in \{1, \ldots, d\}$ such that $x^T A_i x \neq 0$. The *dual degree* of C is the number of indices $i \in \{1, \ldots, d\}$ such that $x^T E_i x \neq 0$. We normally denote the degree of C by s and its dual degree by s^*. In the context of coding theory, if C is a linear code, then the dual degree of C is the degree of its dual code. In general there is no combinatorial interpretation of the dual degree, but we will see that it is useful.

If $\mathcal{A}$ is metric, we define the *width* w of C to be the greatest index i such that $x^T A_i x \neq 0$. (This is just the maximum distance between two vertices of C.) A set in X_1 has width 1 if and only if it is a clique. If $\mathcal{A}$ is cometric relative to E_1, then the *dual width*, w^*, of C is the maximum integer i such that $x^T E_i x \neq 0$. Clearly $s \leq w$ and $s^* \leq w^*$.

If $\mathcal{A}$ is metric and C_i is the set of vertices of the scheme at distance i from C (so $C_0 = C$), then the maximum index r such that $C_r \neq \emptyset$ is the *covering radius* of C. It is shown in Lemma 4.5.1 that $r \leq s^*$. The vertex partition formed by the sets C_i is the *distance partition* relative to C. Recall that a set C is completely regular if its distance partition is equitable relative to X_1.

If C is a subset of the vertices of a d-class association scheme with v vertices and x is the characteristic vector of C, we define the *outer distribution matrix* of C to be the $v \times (d + 1)$ matrix with columns $A_0 x, A_1 x, \ldots, A_d x$.

To see why these concepts are important to us, consider the Johnson graph $J(n, k, 1)$. Let C be the set of all k-sets that contain a specified set of t points. The width of C is $k - t$ and the covering radius is t. The EKR Theorem is the assertion that a set of width $k - t$ has size at most $\binom{n-t}{k-t}$, and a set of width $k - t$ whose size meets this bound must consist of all k-subsets that contain a given set of t points.

With this motivation we can show relations between these parameters.

4.5.1 Lemma. *Suppose C is a subset of the vertices of a d-class association scheme $\mathcal{A}$ with v vertices, and let B be the outer distribution matrix of C. Then:*

(a) the dual degree of C is $\mathrm{rk}(B) - 1$;

(b) if $\mathcal{A}$ is metric and the covering radius of C is r, then $r + 1 \leq \mathrm{rk}(B)$.

Proof. Let x be the characteristic vector of C. The vectors $E_j x$ span the same subspace of $\mathbb{R}^v$ as the columns of B, and more precisely the nonzero vectors

$E_j x$ form an orthogonal basis for the column space of B. The number of such nonzero vectors is $s^* + 1$.

Now suppose $\mathcal{A}$ is metric and that the covering radius of C is r. Then for $i = 0, \ldots, r$, the size of the support of $(A_1 + I)^i x$ is a strictly increasing function of i. Hence these $r + 1$ vectors are linearly independent. Since the span of the powers $(A_1 + I)^i$ for $i = 0, \ldots, d$ is equal to the span of the matrices $A_0, \ldots, A_r$, we conclude that the first $r + 1$ columns of B are linearly independent.   □

If C is a single vertex u, then its covering radius is the diameter of X_1 and $e_u^T E_i e_u \neq 0$ for all i. So the inequality $r \leq s^*$ yields (an unnecessarily complex) proof of the fact that the number of distinct eigenvalues of A_1 is at least $d + 1$.

Next we offer a characterization of completely regular subsets.

4.5.2 Lemma. *Let C be a subset of the vertices of the metric scheme $\mathcal{A}$ with covering radius r. If the number of distinct rows in the outer distribution matrix of C is equal to $r + 1$, then the rank of B equals $r + 1$ and C is completely regular.*

Proof. Let B be the outer distribution matrix of C. If dist$(u, C) = i$ for some vertex u in the scheme, then the first nonzero entry in the u row of B is in position i. Therefore the number of distinct rows of B is at least $r + 1$, and since B has exactly $r + 1$ distinct rows, the u-row of B is determined by the distance from u to C. Consequently rk$(B) \leq r + 1$, and from the previous lemma we conclude that rk$(B) = r + 1$.

We also see that, viewed as a function on vertices, each column of B is constant on the cells of the distance partition of C. If P denotes the characteristic matrix of this partition, we have that col(B) is a subspace of col(P). Since

$$\mathrm{rk}(B) = r + 1 = \mathrm{rk}(P)$$

we have col$(B) = $ col(P). Since col(B) is A_1-invariant, so is col(P) and hence the distance partition of C is equitable.   □

If C is a subset of the vertices of the d-class association scheme $\mathcal{A}$, define B^* to be the matrix with columns $E_0 x, \ldots, E_d x$. We used B^* implicitly in Lemma 4.5.1, making use of the fact that B and B^* have the same column space. We will use the following reformulation of this fact; the proof is an easy consequence of the identities in Section 3.5.

4.5.3 Lemma. *Let C be a subset of the vertices of the d-class association scheme $\mathcal{A}$ and let B be its outer distribution matrix. Let P be the matrix of eigenvalues of $\mathcal{A}$ and Q the matrix of dual eigenvalues. Then $B = B^* P$ and $BQ = vB^*$.*   □

4.6 Sturm sequences

In Section 2.3 we defined *interlacing* for the eigenvalues of a matrix; now we define interlacing for the roots of two polynomials. Let p be a real polynomial of degree n with real zeros $\theta_1 \geq \cdots \geq \theta_n$. If q is a real polynomial of degree $n - 1$, we say that the zeros of q *interlace* the zeros of p if each interval $[\theta_{r-1}, \theta_r]$ contains a zero of q. Note that if $\theta_{r-1} = \theta_r$, then θ_r must also be a zero of q.

We now state two results about interlacing for the roots of two polynomials, which the reader might easily verify.

4.6.1 Lemma. *If p and q are real polynomials with positive leading terms and* $\deg(q) = \deg(p) - 1$, *then the zeros of q interlace the zeros of p if and only if all of the numerators in the partial fraction expansion of q/p are positive.* $\square$

4.6.2 Lemma. *Suppose p and q are coprime real polynomials with* $\deg(q) = \deg(p) - 1$, *and let r be the remainder when we divide p by q. If the zeros of q interlace the zeros of p, then r and q are coprime and the zeros of r interlace the zeros of q.* $\square$

The number of sign changes in a sequence $a_0, \ldots, a_n$ of nonzero real numbers is the number of indices i such that $a_{i-1} a_i < 0$. The number of sign changes in a sequence of real numbers, where some of the terms are equal to zero, is the number of sign changes of the sequence obtained by removing the zero terms.

Let p be a polynomial of degree n and q a polynomial of degree $n - 1$ whose roots interlace the roots of p. If the zeros of p are distinct and the leading terms of p and q are positive, then using Lemma 4.6.1 and Lemma 4.6.2, we can obtain a sequence of polynomials $p_0, \ldots, p_n$ satisfying:

(a) $p_n = p$, and $p_{n-1} = q$;
(b) $\deg(p_r) = r$ for $r = 0, \ldots, n$, and its leading term is positive;
(c) for $i \leq n - 2$ the polynomial $-p_i$ is the remainder when p_{i+2} is divided by p_{i+1}.

Such a sequence is called a *Sturm sequence*. For any polynomial p, it is possible to define a Sturm sequence by setting p_{n-1} to be the derivative of p.

Sturm sequences can be used to count the number of real roots of a polynomial and to determine the number of roots in an interval. For a Sturm sequence $p_n, \ldots, p_0$, define $S(x)$ to be the number of sign changes in the sequence $p_n(x), \ldots, p_0(x)$. In 1829, Jacques Sturm proved the following.

4.6.3 Theorem. *The number of roots of $p(x)$ in the interval $[a, b]$ is $S(a) - S(b)$.* $\square$

A convenient reference for this is [93, Theorem 6.3a]. One simple corollary of this result is the following.

4.6.4 Corollary. *If $p_n, \ldots, p_0$ is a Sturm sequence where the leading coefficients of p_n and p_{n-1} are positive, then the number of sign changes in the sequence*

$$p_n(\lambda), \, p_{n-1}(\lambda), \, \ldots, \, p_0(\lambda)$$

is equal to the number of zeros of p_n that are greater than λ.   □

Let B be a tridiagonal matrix of order $(d+1) \times (d+1)$, given in the form

$$B = \begin{pmatrix} a_0 & b_0 & & & & \\ c_1 & a_1 & b_1 & & & \\ & c_2 & a_2 & b_2 & & \\ & & & \ddots & & \\ & & & c_{d-1} & a_{d-1} & b_{d-1} \\ & & & & c_d & a_d \end{pmatrix}. \tag{4.6.1}$$

We say that B is *indecomposable* if $b_{i-1}c_i \neq 0$ for $i = 1, \ldots, d$. If $b_{i-1}c_i \geq 0$ and we set $\sigma_i = \sqrt{b_{i-1}c_i}$, then there is an invertible diagonal matrix D such that

$$D^{-1}BD = \begin{pmatrix} a_0 & \sigma_1 & & & & \\ \sigma_1 & a_1 & \sigma_2 & & & \\ & \sigma_2 & a_2 & \sigma_3 & & \\ & & & \ddots & & \\ & & & \sigma_{d-1} & a_{d-1} & \sigma_d \\ & & & & \sigma_d & a_d \end{pmatrix}.$$

It follows that if $b_{i-1}c_i \geq 0$ for all i (as it will be in all cases of interest to us), then B is similar to a symmetric real matrix and therefore its eigenvalues are all real. Further, if we assume that $b_{i-1}c_i > 0$ for all i, then the submatrix of $B - tI$ formed by deleting the first row and last column is upper triangular with nonzero diagonal entries and hence this submatrix has rank d. Thus for any eigenvalue t, the rank of the matrix $B - tI$ is d and it follows that, if B is indecomposable, all its eigenvalues are simple.

If w is an eigenvector for B with eigenvalue λ, we have the recurrence

$$\lambda w_r = c_r w_{r-1} + a_r w_r + b_r w_{r+1},$$

for $r = 0, \ldots, d$ (with the understanding that $c_0 = b_d = 0$). It follows that there are polynomials $\psi_0, \ldots, \psi_d$ such that $w_r = \psi_r(\lambda)$. These polynomials satisfy the recurrence

$$t\psi_r(t) = c_r\psi_{r-1}(t) + a_r\psi_r(t) + b_r\psi_{r+1}(t)$$

for $r = 0, \ldots, d$. Using this we deduce that the polynomials ψ_r form a Sturm sequence. Further, with this recurrence we can inductively determine a formula for ψ_r.

4.6.5 Lemma. *For any $(d + 1) \times (d + 1)$ tridiagonal matrix B, we have $\psi_0 = 1$ and if $r \geq 1$, then ψ_r is a constant multiple of the characteristic polynomial of the leading $(r + 1) \times (r + 1)$ principal submatrix of B.* □

For the matrix B given at (4.6.1), if we let $p_{r+1}(t)$ be the characteristic polynomial of the leading $(r + 1) \times (r + 1)$ principal submatrix of B, then

$$\psi_r(t) = \frac{1}{b_{r-1}b_{r-2}\cdots b_0} \, p_{r+1}(t).$$

We can use the polynomials ψ_i to determine some facts about the entries of the eigenvector w.

4.6.6 Lemma. *Suppose B is a $(d + 1) \times (d + 1)$ tridiagonal indecomposable matrix. Further, assume that its off-diagonal entries are non-negative, and that its eigenvalues are $\theta_0 > \cdots > \theta_d$, in decreasing order. Then the following hold:*

(a) *the eigenvector with eigenvalue θ_i has exactly i sign changes;*
(b) *if all row sums of B are equal, then the entries of the θ_1-eigenvector form a strictly monotone sequence.*

Proof. Let $w = (w_0, w_1, \ldots, w_d)$ be an eigenvector with eigenvalue θ_i. By Lemma 4.6.5, $w_r = \psi_r(\theta_i)$. By Corollary 4.6.4, the number of sign changes in the sequence $\psi_0(\theta_i), \psi_1(\theta_i), \ldots, \psi_d(\theta_i)$ is the number of zeros of ψ_d that are greater than θ_i. This number is exactly i.

To prove the second statement, consider the matrix B, defined in (4.6.1), and the following matrix:

$$S = \begin{pmatrix} 1 & -1 & & & & \\ & 1 & 1 & & & \\ & & 1 & -1 & & \\ & & & \ddots & & \\ & & & & 1 & -1 \\ & & & & & 1 \end{pmatrix}.$$

The matrix S is invertible and

$$SBS^{-1} = \begin{pmatrix} k - b_0 - c_1 & b_1 & & & & \\ c_1 & k - b_1 - c_2 & b_2 & & & \\ & c_2 & k - b_2 - c_3 & b_3 & & \\ & & & \ddots & & \\ & & & c_{d-1} & k - b_{d-1} - c_d & 0 \\ & & & & c_d & k \end{pmatrix}.$$

The vector $Sw = (w_0 - w_1, w_1 - w_2, \ldots, w_{d-1} - w_d, w_d)$ is a θ_i eigenvector for SBS^{-1}, and, provided that $i \neq 0$, it is also an eigenvector for the tridiagonal matrix M formed from the first d rows and columns of SBS^{-1}. It is clear from the structure of SBS^{-1} that the eigenvalues of M are exactly $\theta_1, \theta_2, \ldots, \theta_d$. In particular, if w is an eigenvalue for θ_1, the vector $(w_0 - w_1, w_1 - w_2, \ldots, w_{d-1} - w_d)$ is an eigenvector for M corresponding to the largest eigenvalue. Thus, by the first part of this theorem, it has no sign changes and the second statement holds. □

We illustrate this lemma with an example on 4×4 matrices. The intersection array for the Johnson graph $J(7, 3, 1)$ is the following:

$$B = \begin{pmatrix} 0 & 12 & 0 & 0 \\ 1 & 5 & 6 & 0 \\ 0 & 4 & 6 & 2 \\ 0 & 0 & 9 & 3 \end{pmatrix}.$$

The eigenvalues are $\{12, 5, 0, -3\}$ (note that all have multiplicity 1) with corresponding eigenvectors

$$w(0) = \begin{pmatrix} 1 \\ 1 \\ 1 \\ 1 \end{pmatrix}, \quad w(1) = \begin{pmatrix} -12 \\ -5 \\ 2 \\ 9 \end{pmatrix}, \quad w(2) = \begin{pmatrix} 6 \\ 0 \\ -1 \\ 3 \end{pmatrix}, \quad w(3) = \begin{pmatrix} -12 \\ 3 \\ -2 \\ 3 \end{pmatrix}.$$

It is clear that the number of sign changes in $w(i)$ is exactly i for each eigenvector. Further, for each nonconstant vector we can consider the vector $w' = (w_0 - w_1, w_1 - w_2, w_2 - w_3)$:

$$w'(1) = \begin{pmatrix} -7 \\ -7 \\ -7 \end{pmatrix}, \quad w'(2) = \begin{pmatrix} 6 \\ 1 \\ -4 \end{pmatrix}, \quad w'(3) = \begin{pmatrix} -15 \\ 5 \\ -5 \end{pmatrix}.$$

Note that for each eigenvector w, the vector w' will have one less sign change.

4.7 Classical parameters

There is a family of distance-regular graphs that is defined by an additional restriction on their parameters. This family is called the distance-regular graphs with *classical parameters*, and it includes many of the most interesting distance-regular graphs.

Recall from Section 4.2 that each distance-regular graph of valency k and diameter d has a set of $3(d + 1)$ parameters

$$b_i, \ a_i, \ c_i$$

with $i = 0, \ldots, d$. Further, since $b_i + a_i + c_i = k$, it is sufficient to specify the values of b_i and c_i for $i = 0, \ldots, d$.

We define

$$[j]_q = \frac{q^j - 1}{q - 1},$$

with the understanding that if $q = 1$, then $[j]_q = j$. Then we say that a distance-regular graph has *classical parameters* if in addition to the diameter d, there are three parameters q, α, β such that

$$b_i = \left([d]_q - [i]_q\right)\left(\beta - \alpha[i]_q\right)$$
$$c_i = [i]_q\left(1 + \alpha[i - 1]_q\right).$$

We leave it as an exercise to confirm that the Johnson graph $J(n, k, 1)$ is an example of a distance-regular graph with classical parameters using

$$q = 1, \quad \alpha = 1, \quad \beta = n - k.$$

Our next example is the *Grassmann graph* (this graph is discussed in detail in Chapter 9). The Grassmann graph $J_q(n, k)$ is defined to be the graph with the k-dimensional subspaces of an n-dimensional vector space over $GF(q)$ as its vertices, where two vertices are adjacent if their intersection is a subspace of dimension $k - 1$. This gives rise to a metric association scheme where two subspaces are at distance i if their intersection has dimension $k - i$. The Grassmann graph has classical parameters:

$$q = q, \quad \alpha = q, \quad \beta = [n - k + 1]_q - 1.$$

If $q = 1$, these are simply the parameters of the Johnson graph.

Our third example is the *bilinear forms graph*. This graph has the $m \times n$ matrices over a field of order q as its vertices, where the distance between two matrices M and N is $\mathrm{rk}(M - N)$. In particular, M and N are adjacent exactly when $\mathrm{rk}(M - N) = 1$. These graphs are distance regular with classical parameters:

$$q = q, \quad \alpha = q - 1, \quad \beta = q^n - 1.$$

The bilinear forms graph is also discussed in Section 9.11.

Our final example is a family of graphs called the *dual polar graphs*. These graphs are distance regular with classical parameters:

$$q = q, \quad \alpha = 0, \quad \beta = q^e$$

where

$$e \in \left\{ 0, \frac{1}{2}, 1, \frac{3}{2}, 2 \right\}$$

and q is a square if e is not an integer. These graphs are induced subgraphs of a Grassmann graph with a large automorphism group. They arise in connection with forms – bilinear, Hermitian or quadratic – on vector spaces. For a detailed discussion of the dual polar graphs see [31, Section 9.4]; here we will give only one family of examples of the dual polar graphs.

Let H be an invertible $d \times d$ matrix over $GF(q)$. Further, if q is odd we assume $H^T = -H$; if q is even we assume $H = H^T$ and $H_{i,i} = 0$ for all i. (These conditions also force us to assume that d is even.) Then the map

$$(x, y) \mapsto x^T H y$$

is an alternating form on the vector space $V = GF(q)^d$. Using this form we can provide a generalization of the orthogonal complement of a vector space. If U is a subspace of V, define

$$U^\perp = \{ v \in V : \forall u \in U, u^T H v = 0 \}.$$

With this definition, $\dim(U^\perp) = d - \dim(U)$ and $(U^\perp)^\perp = U$. A subspace U is *isotropic* if $U \subseteq U^\perp$. For $d = 2e$, the *symplectic dual polar graph* has the isotropic subspace of dimension e as its vertices, and two such subspaces are adjacent if their intersection has dimension $e - 1$.

4.8 Exercises

4.1 Prove that the distance matrices of a distance-regular graph form an association scheme.

4.2 Let S be the collection of all k-subsets of $\{1, \ldots, n\}$ that contain the element 1. Show that S is a completely regular subset of the Kneser graph $K(n, k)$.

4.3 Prove that for all k the Kneser graph $K(2k + 1, k)$ is distance regular (these graphs are known as the *odd graphs*). Calculate the intersection numbers for these graphs.

4.4 Prove that if an association scheme is metric relative to a graph X, then X must be distance regular.

4.5 Let $\mathcal{A}$ be a d-class scheme that is metric relative to X_1. Show that the matrices $A_0, \ldots, A_d$ can be ordered so that A_i is a polynomial of degree i in A_1. (This is Corollary 4.1.3.)

4.6 Show that the idempotents in any cometric association scheme satisfy a three-term recurrence relation. Further, prove that in a cometric association scheme the Krein parameters $q_{i,j}(r) = 0$ for $0 \le r < |i - j|$.

4.7 Give the distance partition of the set C in the Johnson graph $J(n, k, 1)$ when C consists of one vertex. Show that this C is completely regular. Finally, show that the Johnson scheme is metric.

4.8 Calculate the intersection numbers for the Johnson graph $J(n, k, 1)$.

4.9 Show that the Johnson graph $J(n, k, 1)$ is a distance-regular graph with classical parameters.

4.10 A *perfect e-code* in a graph X is a set of vertices C with the property that for each vertex $x \in V(X)$ there is a unique vertex in C at distance at most e from it. Let X be a k-regular graph and C a perfect 1-code in X. Show that the partition $\pi = \{C, V(X) \setminus C\}$ is equitable and find the eigenvalues of the quotient graph.

4.11 Let $J(7, 3, 1)$ be the Johnson graph. Consider the following subset of vertices of $J(7, 3, 1)$:

$$C = \{124, 235, 346, 457, 561, 672, 713\}.$$

Find the degree, the width and the covering radius of this set. Next, show that C is a perfect 1-code. Finally, find the quotient graph of $J(7, 3, 1)$ with this partition $\pi = \{C, V(J(7, 3, 1)) \setminus C\}$ and give the eigenvalues of the quotient graph.

4.12 Let X be a distance-regular graph and C a subset of $V(X)$. Define the *outer distribution module* to be the column space of the outer distribution matrix and let v_C be the characteristic vector of C. Prove that the following are equivalent:

(a) C is a complete regular code;

(b) the outer distribution module is closed under Schur multiplication and the covering radius equals s^*;

(c) the distance partition of X with respect to C is an equitable partition.

4.13 Let X be a distance-regular graph and $C \subseteq V(X)$. Assume that the dual degree of C is s^*, the covering radius is r and $d(C)$ is the largest distance between any two vertices in C. Show that $r \leq s^*$ and $d(C) \leq 2s^* - 1$. Also show if $d(C) = 2s^* + 1$, then C is completely regular.

4.14 Prove that the Grassmann graphs have classical parameters.

4.15 Prove that the bilinear forms graphs have classical parameters.

4.16 Prove that the symplectic dual polar graphs have classical parameters.

Notes

For more reading on distance-regular graphs see [31, 36, 78]. Both distance-regular graphs with classical parameters and dual polar graphs are treated in some detail in [31].

Exercise 4.12 is based on Neumaier [137] and Exercise 4.13 comes from Delsarte [51]. Martin's paper [122] contains both results along with an excellent introduction to completely regular codes and some of their history. More discussion of codes can be found in [78, Chapter 11].

5

Strongly regular graphs

A graph on v vertices is a *strongly regular* graph with parameters $(v, k; a, c)$ if:

(a) it is k-regular;

(b) each pair of adjacent vertices in the graph have exactly a common neighbors;

(c) each pair of distinct nonadjacent vertices in the graph have exactly c common neighbors.

If X is a strongly regular graph, then its complement is also a strongly regular graph. A strongly regular graph X is *primitive* if both X and its complement are connected. If X is not primitive, we call it *imprimitive*. The imprimitive strongly regular graphs are exactly the disjoint unions of complete graphs and their complements, the complete multipartite graph. It is not difficult to show that a strongly regular graph is primitive if and only if $0 < c < k$. It is customary to declare that the complete and empty graphs are not strongly regular, and we follow this custom.

Two examples of strongly regular graphs are the line graphs of complete graphs and the line graphs of complete bipartite graphs. We will meet other large classes of examples as we go through this chapter – most of which arise from well-known combinatorial objects. In this chapter we provide detailed information about the cliques and cocliques for many of these graphs. We see that many interesting objects from design theory and finite geometry occur encoded as cliques and cocliques in these graphs, and this leads to interesting variants of the EKR Theorem.

5.1 An association scheme

As stated in the previous chapter, association schemes with two classes are equivalent to strongly regular graphs. Throughout this section we denote the adjacency matrix of a graph X by A, rather than $A(X)$, and the adjacency matrix of the complement of X by $\overline{A}$, rather than $A(\overline{X})$.

5.1.1 Lemma. *Let X be a graph. Then X is strongly regular if and only if $\mathcal{A} = \{I, A, \overline{A}\}$ is an association scheme.*

Proof. Assume that X is a strongly regular graph with parameters $(v, k; a, c)$. It is clear that the set of matrices $\mathcal{A} = \{I, A, \overline{A}\}$ satisfies properties (a) through (c) in the definition of an association scheme given in Section 3.1. We only need to confirm that the product of any two matrices in $\mathcal{A}$ is a linear combination of other matrices in $\mathcal{A}$ (in doing this, we also show that the matrices commute).

Since the (x, y)-entry in A^2 gives the number of length-2 paths between the vertices x and y, we have that

$$A^2 = kI + aA + c\overline{A}.$$

Thus A^2 is in the span of the other matrices, and we can conclude the same for $(\overline{A})^2$, since it is also an adjacency matrix for a strongly regular graph. Finally, we note that both $\overline{A}A$ and $A\overline{A}$ are equal to

$$(k - a)A + (k - c)\overline{A};$$

this also proves that the matrices A and $\overline{A}$ commute.

Conversely, assume that the matrices $\{I, A, \overline{A}\}$ are an association scheme. Then,

$$A^2 = \alpha I + \beta A + \gamma \overline{A}.$$

Let X be the graph that corresponds to A. From the equation, it is clear that any vertex in X is adjacent to exactly α vertices. Further, if two vertices of X are adjacent, then they share β common neighbors and nonadjacent vertices share γ common neighbors. Thus X is strongly regular with parameters $(v, \alpha; \beta, \gamma)$. $\qquad\square$

5.1.2 Corollary. *In any symmetric association scheme with two classes, the classes correspond to strongly regular graphs.* $\qquad\square$

5.2 Eigenvalues

A primitive strongly regular graph has exactly three distinct eigenvalues, and the converse of this is also true – a connected regular graph with exactly three eigenvalues must be strongly regular. We denote the eigenvalues of a strongly regular graph by k (its valency), θ and τ. We will always choose θ and τ so that

$\theta > \tau$ and denote their multiplicities by m_θ and m_τ. Since a strongly regular graph is connected, the multiplicity of k will always be 1 and the all-ones vector, $\mathbf{1}$, is an eigenvector for this eigenvalue. We call the following matrix the *modified matrix of eigenvalues* of the strongly regular graph. The first column gives the dimensions of the eigenspaces (the multiplicities of the eigenvalues), the second column contains the eigenvalues of the graph, and the third the eigenvalues of its complement.

$$\begin{pmatrix} 1 & k & v - 1 - k \\ m_\theta & \theta & -1 - \theta \\ m_\tau & \tau & -1 - \tau \end{pmatrix}$$

It is possible to express the eigenvalues and their multiplicities in terms of the parameters of the graph. We do not include a proof of this result; rather we refer the reader to [87, Section 10.2].

5.2.1 Theorem. *If X is a primitive strongly regular graph with parameters $(v, k; a, c)$ and define*

$$\Delta = \sqrt{(a - c)^2 + 4(k - c)},$$

then the three eigenvalues of X are

$$k, \quad \theta = \frac{1}{2}(a - c + \Delta), \quad \tau = \frac{1}{2}(a - c - \Delta),$$

with respective multiplicities

$$m_k = 1, \quad m_\theta = -\frac{(v - 1)\tau + k}{\theta - \tau}, \quad m_\tau = \frac{(v - 1)\theta + k}{\theta - \tau}. \qquad \square$$

As $k > c$, it follows that $\theta > 0$ and $\tau < 0$; thus the maximum and minimum eigenvalues are

$$k, \quad \tau = \frac{(a - c) - \sqrt{(a - c)^2 + 4(k - c)}}{2}. \tag{5.2.1}$$

These values can be used in the ratio bound for cliques, Corollary 3.7.2, and the ratio bound for cocliques, Theorem 2.4.1. We can also apply the inertia bound, Theorem 2.9.1, to any strongly regular graph X to get that $\alpha(X) \le m_\tau$.

Either of the ratio bound on cocliques or the inertia bound can be stronger. For example, by work of Brouwer and Haemers [33], we know that there is a unique strongly regular graph with parameters $(81, 20; 1, 6)$. Its modified matrix of eigenvalues is

$$
\begin{pmatrix}
1 & 20 & 60 \\
60 & 2 & -3 \\
20 & -7 & 6
\end{pmatrix}.
$$

The value given by the ratio bound for cocliques in this graph is 21, while the inertia bound shows that the size of any coclique is no more than 20. Further, the ratio bound for cliques gives that a clique in this graph is no larger than $27/7$. Clearly, neither of the ratio bounds can be tight for this graph.

The remainder of this chapter is devoted to examples of strongly regular graphs in which the ratio bound for cliques holds with equality and can be interpreted as an EKR-type result.

5.3 Designs

A 2-$(n, m, 1)$ *design* is a collection of m-sets of an n-set with the property that every pair from the n-set is in exactly one set. A specific 2-$(n, m, 1)$ design is denoted by $(V, \mathcal{B})$, where V is the n-set (which we call the *base set*) and $\mathcal{B}$ is the collection of m-sets – these are called the *blocks* of the design. A 2-$(n, m, 1)$ design may also be called a 2-design. A simple counting argument shows that the number of blocks in a 2-$(n, m, 1)$ design is $\frac{n(n-1)}{m(m-1)}$ and each element of V occurs in exactly $\frac{n-1}{m-1}$ blocks (this is usually called the *replication number*). There are many references for more information on 2-designs; we simply recommend the *Handbook of Combinatorial Designs* [46] and the references within.

The blocks of a 2-design are a set system, and every pair from the base set occurs in exactly one block. Thus two distinct blocks of a 2-design must have intersection size 0 or 1. An intersecting set system from a 2-design is a set of blocks from the design in which any two have intersection of size exactly 1. The question we now ask is, what is the largest possible such set? Clearly if we take the collection of all blocks that contain a fixed element, we will have a system of size $\frac{n-1}{m-1}$. An EKR-type theorem for 2-designs would state that this is the largest possible set of intersecting blocks and determine the conditions when the only intersecting sets of blocks that has this size is the set of all blocks that contain a fixed element. (The first result would be the bound in the EKR Theorem, and the second would be the characterization.) In this section we show that the bound always holds and that the uniqueness holds in some cases. To do this, we define a graph so that the cliques in the graph are exactly the intersecting set systems from the design and then we determine the size and structure of the cliques.

The *block graph of a* 2-$(n, m, 1)$ *design* $(V, \mathcal{B})$ is the graph with the blocks of the design as the vertices in which two blocks are adjacent if and only if they intersect. In a 2-design, any two blocks that intersect meet in exactly one point. The block graph of a design $(V, \mathcal{B})$ is denoted by $X_{(V,\mathcal{B})}$. Alternatively, we could define a graph on the same vertex set in which two vertices are adjacent if and only if the blocks do not intersect – this graph is simply the complement of the block graph. A clique in the block graph $X_{(V,\mathcal{B})}$ (or a coclique in its complement) is an intersecting set system from $(V, \mathcal{B})$.

Fisher's inequality implies that the number of blocks in a 2-design is at least n; if equality holds, the design is said to be *symmetric* and the block graph of a symmetric 2-design is the complete graph K_n. To avoid this trivial case, we assume that our designs are not symmetric.

The block graph of a 2-$(n, m, 1)$ design is strongly regular – this can be seen by simply calculating the parameters. From these parameters the eigenvalues of the association scheme can also be calculated (this is not a pleasant task, so we leave it as an exercise!).

5.3.1 Theorem. *The block graph of a* 2-$(n, m, 1)$ *design (that is not symmetric) is strongly regular with parameters*

$$\left(\frac{n(n-1)}{m(m-1)}, \quad \frac{m(n-m)}{(m-1)}; \quad (m-1)^2 + \frac{n-1}{m-1} - 2, \quad m^2 \right).$$

The modified matrix of eigenvalues is

$$\begin{pmatrix} 1 & \left| \begin{array}{cc} \frac{m(n-m)}{m-1} & \frac{(n-1)(n-m^2)}{m(m-1)} + m - 1 \end{array} \right. \\ n-1 & \left. \begin{array}{cc} \frac{n-m^2}{m-1} & -1 - \frac{n-m^2}{m-1} \end{array} \right. \\ \frac{n(n-1)}{m(m-1)} - n & \left| \begin{array}{cc} -m & m-1 \end{array} \right. \end{pmatrix}. \qquad \square$$

This association scheme is similar to the Johnson scheme in that relations are defined on sets by the size of their intersection, but since the sets are from a 2-design, there are only two possible sizes of intersections and hence only two classes.

For any nonsymmetric 2-design $(V, \mathcal{B})$ with block graph $X_{(V,\mathcal{B})}$, by the ratio bound for cliques, Corollary 3.7.2,

$$\omega(X_{(V,\mathcal{B})}) \leq 1 - \frac{k}{\tau} = 1 - \frac{\frac{m(n-m)}{(m-1)}}{-m} = \frac{n-1}{m-1}.$$

It is not difficult to construct a clique of this size: for any $i \in \{1, \ldots, n\}$ let S_i be the collection of all blocks in the design that contain i. We call the cliques S_i the *canonical cliques* of the block graph.

5.3.2 Theorem. *If $X_{(V, \mathcal{B})}$ is the block graph of* 2-$(n, m, 1)$ *design, then*

$$\omega(X) = \frac{n - 1}{m - 1}. \qquad \square$$

From this theorem, we know that a set of intersecting blocks in a 2-design is no larger than the set of all blocks that contain a common point – this is the bound for an EKR-type theorem for the blocks in a design.

5.3.3 Theorem. *The largest set of intersecting blocks from a* 2-$(n, m, 1)$ *design has size $\frac{n-1}{m-1}$.* $\qquad \square$

It is not known for which designs these are the only maximal intersecting sets. We can offer a partial result.

5.3.4 Theorem. *If a clique in the block graph of a* 2-$(n, m, 1)$ *design does not consist of all the blocks that contain a given point, then its size is at most $m^2 - m + 1$.*

Proof. Assume that C is a non-canonical clique and that the set $\{1, \ldots, m\}$ is in C. Divide the other vertices in the clique into m groups, labeled G_i such that each vertex in group G_i contains the element i.

Assume that G_1 is the largest group. Since the clique is non-canonical, there is a vertex in G_i for some $i > 1$. All the vertices in G_1 must intersect this vertex so each vertex of G_1 must contain one of the $m - 1$ elements in this vertex (but not the element i, as 1 and i are both in the set $\{1, \ldots, m\}$). Since no two vertices of G_1 can contain the same element from the vertex in G_i, the size of G_1 can be no more than $(m - 1)$. Since G_1 is the largest group the size of the clique is no more than $m(m - 1) + 1$. $\qquad \square$

A corollary of this is an analog of the EKR Theorem, with the characterization of maximal families, for intersecting sets of blocks in a 2-$(n, m, 1)$ design.

5.3.5 Corollary. *The only cliques of size $\frac{n-1}{m-1}$ in the block graph $X_{(V, \mathcal{B})}$ of a* 2-$(n, m, 1)$ *design with $n > m^3 - 2m^2 + 2m$ are the sets of blocks that contain a given point i in $\{1, \ldots, n\}$.* $\qquad \square$

The characterization in this corollary may fail if $n \leq m^3 - 2m^2 + 2m$. For example, consider the projective geometry $PG(3, 2)$. The points of this geometry can be identified with the 15 nonzero vectors in a 4-dimensional vector space V over $GF(2)$, and the lines with the 35 subspaces of dimension 2. This gives us a design with parameters 2-(15, 3, 1), where each block consists of the three nonzero vectors in a 2-dimensional subspace. There are exactly 15 subspaces of V with dimension 3, and each such subspace contains exactly seven points and exactly seven lines and so provides a copy of the projective plane of order two. In the block graph, the seven lines in any one of these projective planes forms a clique of size 7. In addition, each point of the design lies on exactly seven lines, and this provides a second family of 15 cliques of size 7.

We will offer a few comments on how it might be possible to characterize all the cliques in the block graph of a 2-design. Since equality holds in the ratio bound for cliques in the block graph for any 2-$(n, m, 1)$ design, by Corollary 3.7.2, the characteristic vectors of the sets S_i are orthogonal to the τ-eigenspace. Thus these vectors lie in the sum of the eigenspace with dimension $n - 1$ and the eigenspace of dimension 1. The following theorem shows that characteristic vectors of the sets S_i actually span the sum of these eigenspaces. With this fact, the characteristic vector of any clique of maximum size must be a linear combination of characteristic vectors of the sets S_i. A possible way to characterize the maximum cliques is to characterize all linear combinations of the characteristic vectors of S_i that gives a 01-vector with weight $\frac{n-1}{m-1}$. It is left as an exercise to show that this method works for the block graph of the 2-(15, 3, 1) design described earlier.

Let H be the matrix whose columns are the characteristic vectors of S_i – thus the rows of H are the characteristic vectors of the blocks in the design. The next result implies that H has rank n by proving a more general result that we will use later. A t-(n, k, λ) *design* is a collection of k-subsets (called the *blocks*) from an n-set with the property that any t-subset from the n-set is contained in exactly λ blocks.

5.3.6 Lemma. *For any t-(n, k, λ) design with $t \geq 2$ let H be the matrix whose rows are the characteristic vectors of the blocks of the design. If $n \neq 2k - 1$ then H has full rank.*

Proof. By counting the number of blocks that contain a single element from the point set, and counting the number of blocks that contain a pair of elements from the point set, we get that

$$H^T H = \lambda \frac{\binom{n-1}{t-1}}{\binom{k-1}{t-1}} I_n + \lambda \frac{\binom{n-2}{t-2}}{\binom{k-2}{t-2}} (J_n - I_n).$$

This matrix has zero as an eigenvalue if and only if $n = 2k - 1$. Thus $\text{rk}(H) = \text{rk}(H^T H) = n$, provided that $n \neq 2k - 1$. □

Finally, we consider the cocliques in the block graph. Since $X_{(V,\mathcal{B})}$ is regular, the ratio bound for cocliques can be applied. Using the eigenvalues given in Theorem 5.3.1, this bound gives that $\alpha(X) \leq n/m$. Since this number is not always an integer, it is not surprising that it is not always possible to find a coclique of this size. A coclique in the block graph with size n/m is a set of blocks that partition the point set. Such a set in a 2-$(n, m, 1)$ design is called a *parallel class*. The ratio bound for cocliques holds with equality if and only if the design has a parallel class. If a design is *resolvable* then its block set can be partitioned into parallel classes, and clearly equality holds in this bound. The standard examples of resolvable designs are the affine planes, which can be defined as designs with parameters 2-$(m^2, m, 1)$. There are many other examples. In fact, Ray-Chaudhuri amd Wilson [145] proved that whenever n is large enough and divisible by m, there is a resolvable 2-$(n, m, 1)$ design.

Finally, since we have a bound on the size of the largest coclique, we also have the following bound on the chromatic number.

5.3.7 Lemma. *If $X_{(V,\mathcal{B})}$ is the block graph of a 2-$(n, m, 1)$ design then*

$$\chi(X) \geq \frac{n-1}{m-1}$$

and equality holds if and only if the design is resolvable.   □

5.4 The Witt graph

We now consider a specific design where the method described in the previous section can be used to completely determine the cliques in the block graph of a design. The design that we use is a 3-$(22, 6, 1)$ design – such a design can be constructed by taking the derived design of the Witt design with parameters 4-$(23, 7, 1)$. This design has 77 blocks and each element is in exactly 21 blocks. Throughout this section we denote the block graph of this design by X. The graph X is strongly regular with parameters $(77, 60; 47, 45)$ and its eigenvalues are $\{60, -3, 5\}$ with multiplicities $\{1, 55, 21\}$, respectively. The complement of X is known as the *Witt graph*.

The ratio bound on the cliques of X is

$$1 - \frac{60}{-3} = 21.$$

The set of all blocks that contain a fixed point forms a clique of this size. We call such cliques the *canonical cliques* of X; we will show that the canonical cliques are the only maximum cliques in X.

Define a matrix H to have the characteristic vectors of the canonical cliques as its columns. By Lemma 5.3.6, H has full rank and we conclude that the columns of H span the orthogonal complement of the (-3)-eigenspace. Since the ratio bound holds with equality, the characteristic vector of any maximum clique is in this orthogonal complement and hence is a linear combination of the columns of H. Assume that C is a maximum clique in this graph and denote its characteristic vector by v_C. Then there exists a vector y such that $Hy = v_C$. We will show that y must be the vector with all entries equal to 0 except one entry that is equal to 1.

Let B be a block in the design that is also in the clique. Order the rows of H as follows: first is the block B; then all blocks that are disjoint from B (in any order); finally all remaining blocks (in any order). Further, order the columns of H so that the characteristic vectors of the canonical clique corresponding to elements $i \in B$ occur before those that correspond to elements $i \notin B$. With this ordering we can write the matrix H and the vector y in blocks so that

$$Hy = \left(\begin{array}{c|c} \mathbf{1}_{1\times 6} & \mathbf{0}_{1\times 16} \\ \hline \mathbf{0}_{6\times 16} & M \\ \hline H_1 & H_2 \end{array} \right) \left(\begin{array}{c} y_1 \\ \hline y_2 \end{array} \right) = v_C.$$

Since none of the blocks disjoint from B can be in the clique, it must be that

$$\left(\mathbf{0}_{6\times 16} \middle| M \right) \left(\begin{array}{c} y_1 \\ \hline y_2 \end{array} \right) = \mathbf{0}_{6\times 1}.$$

At this point we need to consider the design that we used to construct the graph. This particular design has the property that the set of all blocks that do not contain any of the points from B forms a 2-$(16, 6, 2)$ design (this is not hard to see and a proof is given in [84]). Lemma 5.3.6 shows that M is full rank, so we conclude that y_2 must be the zero vector.

Considering the multiplication of the first row of H with y, we see that the sum of the first six entries of y must be 1. Finally, the multiplication of H_1 and y_1 must produce a 01-vector. For any two columns in H_1 there is a row that has a one in each of these columns and zeros everywhere else. Considering this, some short calculations show that exactly one entry in y_1 is one and all other entries are zero. So we conclude that y must be a 01-vector with exactly one entry equal to 1 and that v_C is one of the columns of H.

The Witt graph is the graph induced by the vertices in the Higman-Sims graph that are not adjacent to a given vertex. So each clique in our graph corresponds to a coclique of size 22 in the Higman-Sims graph. Thus we have characterized all cocliques of size 22 in the Higman-Sims graph.

5.5 Orthogonal arrays

An *orthogonal array* $OA(m, n)$ is an $m \times n^2$ array with entries from $\{1, \ldots, n\}$ with the property that the columns of any $2 \times n^2$ subarray consist of all n^2 possible pairs. In particular, between any two rows each pair from $\{1, \ldots, n\}$ occurs in exactly one column. This implies that in each row, each element occurs in exactly n columns. There are many applications of orthogonal arrays, particularly to test design, and orthogonal arrays are related to many other interesting combinatorial designs (for example, they are equivalent to a set of mutually orthogonal Latin squares). See [46, Part III] for more details.

The *block graph of an orthogonal array* $OA(m, n)$ is defined to be the graph whose vertices are columns of the orthogonal array, where two columns are adjacent if there exists a row where they have the same entry. We denote the block graph for an orthogonal array $OA(m, n)$ by $X_{OA(m,n)}$.

It is well known for any $OA(m, n)$ that $m \leq n + 1$ and equality holds if and only if there exists a projective plane of order n (see [46, Part III, Section 3]). It is left as an exercise to show that for any n, the block graph of $OA(n + 1, n)$ is isomorphic to the complete graph on n^2 vertices. So for the remainder of this section we assume that $m < n + 1$.

Any two columns in an orthogonal array can have at most one row in which they have the same entry. Thus the complement of $X_{OA(m,n)}$ is the graph in which columns are adjacent if and only if there are no rows in which they have the same entry. This, as in the case of the block graph for a 2-design, produces a two-class association scheme. To prove this, we show that the block graph is a strongly regular graph. The parameters can be calculated directly (in fact, the calculations are much easier than for block graphs of 2-designs, so they are not left as an exercise).

5.5.1 Theorem. *If $OA(m, n)$ is an orthogonal array where $m < n + 1$, then its block graph $X_{OA(m,n)}$ is strongly regular, with parameters*

$$\left(n^2, \quad m(n - 1); \quad (m - 1)(m - 2) + n - 2, \quad m(m - 1) \right),$$

and modified matrix of eigenvalues

$$\begin{pmatrix} 1 & \begin{array}{|cc} m(n - 1) & (n - 1)(n + 1 - m) \\ \end{array} \\ m(n - 1) & \begin{array}{|cc} n - m & m - n - 1 \\ \end{array} \\ (n - 1)(n + 1 - m) & \begin{array}{|cc} -m & m - 1 \\ \end{array} \end{pmatrix}. \qquad \square$$

We can apply the ratio bound for cliques to the block graph of an orthogonal array to see that

$$\omega(X_{OA(m,n)}) \leq 1 - \frac{m(n-1)}{-m} = n.$$

It is straightforward to construct cliques that meet this bound. If $i \in \{1, \ldots, n\}$ let $S_{r,i}$ be the set of columns of $OA(m, n)$ that have the entry i in row r. Clearly these sets are cliques, and since each element of $\{1, \ldots, n\}$ occurs exactly n times in each row, the size of $S_{r,i}$ is n for all i and r. These cliques are called the *canonical cliques* in the block graph of the orthogonal array.

If we view columns of an orthogonal array that have the same entry in the same row as *intersecting columns*, then we can view this bound as the bound in the EKR Theorem for intersecting columns of an orthogonal array. The question we ask is, under what conditions will all cliques of size n in the graph $X_{OA(m,n)}$ be canonical? The following answer can be viewed as the uniqueness part of the EKR Theorem – it tells us that if $n > (m-1)^2$, then any clique of maximal size is a canonical clique. This is equivalent to saying that the largest set of intersecting columns in an orthogonal array is the set of all columns that have the same entry in the some row, and these sets are the only maximum intersecting sets.

5.5.2 Theorem. *Let $OA(m, n)$ be an orthogonal array. If S is a* non-canonical *clique in the block graph $X_{OA(m,n)}$ then $|S| \leq (m-1)^2$.*

Proof. Assume the column of all zeros is in S (this can be done without loss of generality by swapping numbers in the array). Then every other vertex in the clique is a column with exactly one zero. Split these vertices into m classes according to which row this unique zero is in. Denote the class that has the unique zero in row i by C_i. Assume that C_1 is the largest such class.

Since S is not a canonical clique, there is an $i > 1$ such that the class C_i is non-empty; thus there is a column c that is in the clique, with a zero in the ith position. Every vertex in the first class, C_1, must be intersecting with c. There are only $m - 2$ rows where this intersection can occur (since they cannot intersect where the columns have zeros) and no two distinct columns in C_1 can intersect c in the same place (since no pair can be repeated in a $2 \times n^2$ subarray).

This means that C_1 has size no bigger than $m - 2$. Since there are m groups, each of size no more than $m - 2$, plus the column of all zeros, the total size of clique is no more than $m(m-2) + 1 = (m-1)^2$. $\qquad\square$

5.5.3 Corollary. *If $OA(m, n)$ is an orthogonal array with $n > (m-1)^2$, then the only cliques of size n in $X_{OA(m,n)}$ are canonical cliques.* $\qquad\square$

Without the bound on n, it is possible to construct orthogonal arrays for which there are maximum cliques in the block graph that are not canonical cliques. Let $m - 1$ be a prime power; then there exists an $OA(m, m - 1)$ and, using MacNeish's construction [158, Section 6.4.2], it is possible to construct an $OA(m, (m - 1)^2)$ from this array. This larger orthogonal array has $OA(m, m - 1)$ as a subarray, and thus the graph $X_{OA(m,(m-1)^2)}$ has the graph $X_{OA(m,m-1)}$ as an induced subgraph. Since this subgraph is isomorphic to $K_{(m-1)^2}$, it is a clique of size $(m - 1)^2$ in $X_{OA(m,(m-1)^2)}$ that is not canonical.

For example, the following $OA(3, 2)$

$$\begin{bmatrix} 0 & 0 & 1 & 1 \\ 0 & 1 & 0 & 1 \\ 0 & 1 & 1 & 0 \end{bmatrix}$$

can be used to construct this $OA(3, 4)$:

$$\begin{bmatrix} 0011 & 0011 & 2233 & 2233 \\ 0101 & 2323 & 0101 & 2323 \\ 0110 & 2332 & 2332 & 0110 \end{bmatrix}.$$

The first four columns form a maximal clique that is not a canonical clique (there are several other non-canonical maximum cliques).

In general, if the array $OA(m, n)$ has a subarray with m rows that is an $OA(m, m - 1)$, then the columns of this subarray form a clique in $X_{OA(m,n)}$ of size $(m - 1)^2$. By the ratio bound, it must be that $(m - 1)^2 \leq n$ (and if $(m - 1)^2 = n$, then this would be a example of an orthogonal array whose block graph has maximum cliques that are not canonical cliques). This can be interpreted as the following result for orthogonal arrays.

5.5.4 Lemma. *Let m and n be integers with $m - 1 < n < (m - 1)^2$. An orthogonal array $OA(m, n)$ does not have a subarray that is an $OA(m, m - 1)$.* $\square$

For orthogonal arrays $O(m, n)$ with $n \leq (m - 1)^2$ it is not known when the only maximal cliques are canonical cliques, but we now outline a method that may be useful in determining when this is the case.

Since equality holds in the clique bound, the characteristic vector of any clique of maximal size is orthogonal to the $(-m)$-eigenspace; this means that it is in the sum of the $m(n - 1)$-eigenspace and the $(n - m)$-eigenspace. We will show that the characteristic vectors of the sets $S_{r,i}$ span this vector space. Thus the characteristic vector of any maximal clique is a linear combination of the characteristic vectors of $S_{r,i}$. If we can prove, for the particular orthogonal array, that the only linear combination of these vectors that gives a 01-vector with weight n is a trivial combination, then the only maximal cliques are canonical cliques. This result would be the uniqueness part of the EKR Theorem.

To show that the characteristic vectors of the sets $S_{r,i}$ span this vector space, we show that a subset of them are a basis for the $(n-m)$-eigenspace and simply note that the sum of all the characteristic vectors is a multiple of $\mathbf{1}$.

5.5.5 Theorem. *Let $OA(m, n)$ be an orthogonal array. Let N be the incidence matrix for the canonical cliques of the orthogonal array. The columns of N^T indexed by the points*

$$(r, i) \in \{1, 2, \ldots, m\} \times \{1, 2, \ldots, n-1\}$$

form a basis for the $(n-m)$-eigenspace.

Proof. The columns of N are the characteristic vectors of the sets $S_{r,i}$, so we can index the columns by (r, i). Define H be to the submatrix of N formed by the columns (r, i) with $i \leq n-1$. Order the columns so that the columns with the same value of i are together in a block, and within the blocks the columns are arranged so the value of r is increasing. Then

$$H^T H = (J_{n-1} \otimes (J_m - I_m)) + n I_{m(n-1)}.$$

The eigenvalues of this matrix are

$$\{(n-1)(m-1)+n, \quad n, \quad 1\}.$$

All the eigenvalues are nonzero and the rank of $H^T H$ is $m(n-1)$. Since the rank of H equals the rank of $H^T H$, the rank of H is $m(n-1)$, and the columns of H are a basis of the $(n-m)$-eigenspace. $\quad\square$

The ratio bound for cocliques also gives the bound that

$$\alpha(X_{OA(m,n)}) \leq n.$$

Unlike the ratio bound for cliques, this bound cannot be met for all orthogonal arrays. A coclique of size c in the block graph of an orthogonal array corresponds to a set of c columns that are pairwise nonintersecting (such a set of columns is called *disjoint*). Not all covering arrays have a set of n disjoint columns. It is an active area of research to determine orthogonal arrays that do, since orthogonal arrays that have a large number of disjoint columns can be used to construct covering arrays with few columns – see [157] for details.

Finally we consider the case when $m = 3$. An $OA(3, n)$ is equivalent to an $n \times n$ Latin square (the first two rows of the array describe a position in the square, and the final row gives the entry in that position). For this reason the block graph on an orthogonal array with three rows is also known as a *Latin square graph*. If the Latin square has a *transversal* (a set of n positions, each in different rows and different columns, and each containing a different entry) then the block graph of the array has a coclique set of size n. Although it is conjectured by Ryser that every Latin square of odd order has a transversal, there are large families of Latin squares that do not have a transversal [167] (the multiplication tables for abelian groups with cyclic Sylow 2-subgroups are examples). Thus there are many orthogonal arrays whose block graph does not have a ratio tight coclique.

5.6 Partial geometries

The previous two examples can be generalized in one design. A *partial geometry*, denoted $(\mathcal{P}, \mathcal{L})$, with parameters (s, t, v) is an incidence structure of points (denoted $\mathcal{P}$) and lines (denoted $\mathcal{L}$) such that:

(a) each pair of distinct points lies on at most one line (and so any two distinct lines have at most one point in common);

(b) each line contains exactly $s + 1$ points, and each point is on $t + 1$ lines;

(c) if x is a point and ℓ is a line not on x, then exactly v of the lines on x contain a point on ℓ.

The first of these conditions says that our incidence structure is a *partial linear space*. The second condition says that the structure is both point-regular and line-regular. The last is equivalent to the condition that exactly v of the points on a line ℓ are collinear with a point x not on ℓ. It follows that if $(\mathcal{P}, \mathcal{L})$ are the points and lines of a partial geometry with parameters (s, t, v), then $(\mathcal{L}, \mathcal{P})$ is also a partial geometry with parameters (t, s, v). The geometry $(\mathcal{L}, \mathcal{P})$ is called the *dual geometry*.

We can determine the number of points in a partial geometry with parameters (s, t, v) by a simple counting argument. Fix a line ℓ and consider the ordered pairs (x, y) where $x \notin \ell$ and y is a point on ℓ collinear with x. Counting these pairs, we find that

$$(s + 1)ts = (|\mathcal{P}| - s - 1)v.$$

This gives that the number of points is

$$|\mathcal{P}| = \frac{(ts + v)(s + 1)}{v}.$$

A similar argument counting pairs of intersecting lines (one of which contains a fixed point x) yields that the number of lines is

$$|\mathcal{L}| = \frac{(ts + v)(t + 1)}{v}.$$

We have already met two important families of partial geometries. A design with parameters 2-$(n, m, 1)$ is a partial geometry with parameters

$$s = m - 1, \quad t = \frac{n - m}{m - 1}, \quad v = m.$$

Further, it can be shown that any partial geometry with $v = s + 1$ must come from a 2-$(n, m, 1)$ design (see [28]).

Our second class of partial geometries comes from orthogonal arrays. Given an $OA(m, n)$ we construct a geometry with point set

$$\{1, \ldots, m\} \times \{1, \ldots, n\}.$$

The lines are the n^2 columns of the array, and a point (i, j) is incident with a given column if the ith entry of the column is equal to j. The parameters of this geometry are

$$s = m - 1, \quad t = n - 1, \quad v = m - 1.$$

Further, any partial geometry with $v = s$ must come from this construction. The corresponding incidence structures are known as *transversal designs*.

There are two graphs that can be defined on a partial geometry. The *point graph* of a partial geometry is the graph on the points of the geometry, where two distinct points are adjacent if they are collinear. The *line graph* has the lines of the partial geometry as its vertices, and two lines are adjacent if they intersect; the line graph is the point graph of the dual geometry. Because of this we focus our attention mainly on the point graphs of partial geometries.

The point graph of the geometry arising from a 2-design is the complete graph, and the line graph is its block graph. The point graph of the partial geometry associated to an $OA(m, n)$ is $\overline{m K_n}$ and the line graph is the block graph of the array. We have seen that the block graphs for 2-designs and orthogonal arrays are strongly regular graphs. In this section, we see that the point and line graphs of a partial geometry are strongly regular and we determine the spectrums of these graphs. Our approach in this section is to work directly with the geometries. In the next section we re-derive this information using linear algebra.

5.6.1 Theorem. *The point graph of a partial geometry with parameters (s, t, v) is a strongly regular graph with parameters*

$$((s + 1)(st + v)/v, \quad (t + 1)s; \quad -1 + t(v - 1), \quad (t + 1)v).$$

The modified matrix of eigenvalues of this graph is

$$\begin{pmatrix} 1 & (t + 1)s & \dfrac{st}{v}(s - v + 1) \\ m_\theta & s - v & -1 - s + v \\ m_\tau & -1 - t & t \end{pmatrix}$$

where

$$m_\theta = \frac{(t + 1)((s + 1)(st + v) - v - sv)}{v(s + t + 1 - v)},$$

$$m_\tau = \frac{((s + 1)(st + v) - v)(s - v) + vs(t + 1)}{v(s + t + 1 - v)}.$$

Proof. We have determined that the number of vertices in the point graph is $(st + v)(s + 1)/v$. To determine the degree of the graph, let x be a point in the geometry. Then x lies on $t + 1$ lines, any two of which have only x in common. Hence x is collinear with exactly $(t + 1)s$ points.

To compute the parameter c, consider two noncollinear points x and y. There are $t + 1$ lines on x and each of these lines contains exactly v points collinear with y. Therefore $c = (t + 1)v$.

To complete the list of parameters, take two distinct collinear points x and y, and let ℓ be the unique line that contains both of them. Each of the $s - 1$ points on ℓ distinct from x and y is collinear with both. Also, on each of the t lines on y distinct from ℓ there are $v - 1$ points distinct from y and collinear with x. This gives our stated value for a.

The eigenvalues and their multiplicities can now be determined from the parameters. We leave this as a straightforward but somewhat unpleasant exercise. $\qquad\square$

Applying the ratio bound for cliques, Corollary 3.7.2, to the point graph X of a partial geometry with parameters (s, t, v) shows that

$$\omega(X) \leq 1 + s.$$

For any line of the geometry, the vertices on the line form a clique of size $s + 1$ in the point graph. Moreover, the set of all lines define a set of edge-disjoint cliques of size $s + 1$. We define these cliques to be the *canonical cliques*, and the ratio bound proves that no clique in the graph is larger than a canonical clique. Since the line graph is the point graph for the dual geometry, these bounds also hold for the line graph. Specifically, the size of the maximal clique in a line graph of a partial geometry with parameters (t, s, v) is $s + 1$. These bounds can be considered to be EKR-type theorems; we only state this as a result on line graphs, where the concept of intersection is more natural.

5.6.2 Theorem. *In a partial geometry, $(\mathcal{P}, \mathcal{L})$, the set of all lines through a fixed point of $\mathcal{P}$ is a maximum subset of $\mathcal{L}$ such that any two lines in the subset intersect.* □

Again, the next question to ask is, when are the canonical cliques the only maximal cliques in the point (or line) graph of a partial geometry? We consider this question in the next section. Our approach is similar to, but more generalized than, the approach to taken for the block graphs of 2-designs in Section 5.3, and for the block graphs of orthogonal arrays in Section 5.5.

5.7 Eigenspaces of point and line graphs

Let $(\mathcal{P}, \mathcal{L})$ be a partial geometry with parameters (s, t, v) and let N denote its point-block incidence matrix where the blocks are the lines of the geometry. Then it is easy to see that

$$NJ = (t + 1)J, \qquad N^T J = (s + 1)J. \tag{5.7.1}$$

Let A be the adjacency matrix of the point graph of the geometry and let B be the adjacency matrix of the line graph. Then

$$NN^T = (t + 1)I + A, \qquad N^T N = (s + 1)I + B. \tag{5.7.2}$$

Now let x be a point and ℓ a line of the geometry. If $x \in \ell$, there are exactly s points on ℓ distinct from and collinear with x. If $x \notin \ell$, there are exactly v points on ℓ distinct from and collinear with x. Thus

$$AN = sN + v(J - N) = (s - v)N + vJ. \tag{5.7.3}$$

If we multiply the last equation on the right by N^T, we find that

$$ANN^T = (s - v)NN^T + v(t + 1)J;$$

using (5.7.2) to replace A and rearranging yields

$$(NN^T)^2 - (s - v + t + 1)NN^T = v(t + 1)J.$$

This, with (5.7.1), implies that the eigenvalues of NN^T are

$$0, \quad s - v + t + 1, \quad (s + 1)(t + 1).$$

Accordingly the eigenvalues of $A = NN^T - (t + 1)I$ are

$$-t - 1, \quad s - v, \quad s(t + 1).$$

Now we have the eigenvalues of the adjacency matrix. We can also determine the dimensions of the eigenspaces. The only eigenvectors belonging to the eigenvalue $(t + 1)s$ are the constant vectors. Thus the eigenspace is spanned by the all-ones vector and has dimension 1. If $z \in \ker(N^T)$, then

$$(A + (t + 1)I)z = NN^T z = 0.$$

So if $z \neq 0$, it is an eigenvector of A with eigenvalue $-t - 1$. We see that $-t - 1$ has multiplicity

$$\dim(\ker(N^T)) = |\mathcal{P}| - \mathrm{rk}(N).$$

Another consequence of (5.7.3) is that the column space of N is A-invariant, and so it is spanned by eigenvectors of A. We can determine which eigenvectors are in this span. From (5.7.3), we see that if z is a balanced vector (i.e., $z^T \mathbf{1} = 0$) then Nz is an eigenvector with eigenvalue $s - v$. If z is balanced, then Nz is also balanced because $\mathbf{1}^T N = (s + 1)\mathbf{1}^T$. Therefore the set of balanced vectors Nz is equal to the subspace of balanced vectors in $\mathrm{col}(N)$, which has codimension 1 in $\mathrm{col}(N)$. Hence we deduce that $s - v$ is an eigenvalue of A with multiplicity at least $\mathrm{rk}(N) - 1$.

The theory just presented could be used to confirm Theorem 5.6.1 (and Theorems 5.3.1 and 5.5.1), but we leave this for the reader. However, it is worth explicitly considering the incidence matrices.

The columns of N are the characteristic vectors of the canonical cliques in the point graph. Since equality holds in the clique bound, Corollary 3.7.2, each column of N is orthogonal to the $(-1 - t)$-eigenspace. Thus each column is in the sum of the $(s - v)$-eigenspace and the $s(t + 1)$-eigenspace. The dimension of this vector space is exactly $\mathrm{rk}(N)$, and we conclude that the columns of N span this space. Since equality holds in the clique bound, any clique in the line graph is a linear combination of the columns of N. To show that the canonical cliques are the only maximum cliques, we would need to show that the only 01-vectors of weight $s + 1$ in $\mathrm{col}(N)$ are the columns of N.

We now shift our focus to the cocliques of the point graph of a partial geometry. Each point in a coclique S lies on $t + 1$ lines, and these sets of lines are pairwise disjoint. Thus the number of lines that intersect a point in S is bounded by the total number of lines, so

$$|S|(t + 1) \leq \frac{1}{v}(t + 1)(st + v),$$

and therefore

$$\alpha(X) \le \frac{st + v}{v}.$$

A quick calculation shows that this is the same bound given by the ratio bound for cocliques. When $v = 1$ a coclique of this size is called an *ovoid*. We use the term ovoid for any coclique of size $(st + v)/v$ in the point graph of any partial geometry.

If an ovoid exists in the point graph of a partial geometry, then the clique-coclique bound holds with equality. This implies that the balanced characteristic vectors of an ovoid and a clique are orthogonal (see the comments following Corollary 2.1.3). We have seen that the balanced characteristic vectors of the canonical cliques always span the the θ-eigenspace. It is interesting to consider, for a partial geometry that has ovoids of size $(st + v)/v$, whether the balanced characteristic vectors of the ovoids span the τ-eigenspace.

Finally, the chromatic number of the point graph is at least $s + 1$. If equality holds then the point set of the geometry can be partitioned into $s + 1$ ovoids.

5.8 Paley graphs

The final family of strongly regular graphs that we consider is the *Paley graphs*. Let $\mathbb{F}$ be a finite field of order q. The vertices of the Paley graph, $P(q)$, are the elements of $\mathbb{F}$, and vertices are adjacent if and only if their difference is a square in $\mathbb{F}$. We will assume $q \equiv 1 \pmod 4$, since this is the only case where $P(q)$ is an undirected graph. The following theorem lists the basic properties of the Paley graphs.

5.8.1 Theorem. *Let $P(q)$ be a Paley graph with $q \equiv 1 \pmod 4$. Then*

   (a) $P(q)$ is self-complementary and arc transitive;

   (b) $P(q)$ is a strongly regular graph with parameters

$$(q, \quad (q-1)/2; \quad (q-5)/4, \quad (q-1)/4);$$

   (c) the modified matrix of eigenvalues for $P(q)$ is

$$\begin{pmatrix} 1 & (q-1)/2 & (q-1)/2 \\ (q-1)/2 & (-1+\sqrt{q})/2 & (-1-\sqrt{q})/2 \\ (q-1)/2 & (-1-\sqrt{q})/2 & (-1+\sqrt{q})/2 \end{pmatrix}.$$

Proof. Let S denote the set of nonzero squares in $\mathbb{F}$. Then $|S| = (q-1)/2$ and $P(q)$ is the Cayley graph for the additive group of $\mathbb{F}$ relative to the connection set S; hence it is regular with valency $(q - 1)/2$. Since -1 is a square, S is closed under multiplication by -1 and the graph is not directed.

If x is not a square in $\mathbb{F}$, then xS is the set of all non-squares in $\mathbb{F}$ and is the complement of S in $\mathbb{F} \setminus 0$. Therefore the Cayley graph for $\mathbb{F}$ with connection set xS is the complement of $P(q)$. Thus the map that sends a vertex v of $P(q)$ to xv is an isomorphism from $P(q)$ to its complement, and $P(q)$ is self-complementary.

If z is a nonzero square in $\mathbb{F}$, then $zS = S$. It follows that multiplication by z is an automorphism of $P(q)$ that fixes 0 and maps 1 to z. Therefore there is an automorphism that maps any arc adjacent to 0 to any other arc adjacent to 0. From this we can conclude that $P(q)$ is arc transitive.

Similarly, if x and y are two non-squares and $z = y/x$, then z is a square. So multiplication by z is an automorphism of $P(q)$ that fixes 0 and maps x to y. It follows that the stabilizer of 0 in $\mathrm{Aut}(P(q))$ acts transitively on the vertices adjacent to 0 and on the vertices not adjacent to 0. Therefore the parameters a and c are well defined, and so $P(q)$ is strongly regular.

It remains to determine the parameters of $P(q)$. We know that $v = q$ and $k = (q - 1)/2$; it remains to calculate the parameters a and c. First, we count in two different ways the number of edges that join a neighbor of 0 to a vertex at distance 2 from 0. Thus

$$k(k - a - 1) = kc,$$

whence $a + c = k - 1$. If X is strongly regular with parameters $(v, k; a, c)$, then the number of triangles on an edge is a. So the number of triangles on an edge in the complement $\overline{X}$ is $v - 2 - 2k + c$. As $P(q)$ is self-complementary, we deduce that

$$a = v - 2 - 2k + c.$$

Since $v = 2k + 1$ we find that $a - c = -1$ and consequently $c = k/2$ and $a = (k - 2)/2$. This yields our expressions for the parameters of $P(q)$. The formulas for the eigenvalues and their multiplicities follow from Theorem 5.2.1. $\square$

Any strongly regular graph with the parameters

$$(v, (q - 1)/2; (q - 5)/4, (q - 1)/4)$$

is called a *conference graph*. These graphs are characterized by the condition that the multiplicities of θ and τ are equal. There are conference graphs whose order is not a prime power and hence are not Paley graphs. For further background on conference graphs see [31, Section 1.3].

5.9 Paley graphs of square order

Since the Paley graphs are self-complementary graphs, a bound on the size of a coclique is also a bound on the size of a clique. The ratio bound on cocliques gives

$$\alpha(P(q)) \leq \frac{q}{1 - \frac{q-1}{-(1+\sqrt{q})}} = \sqrt{q}$$

and this also means that $\omega(P(q)) \leq \sqrt{q}$. Clearly this cannot be tight if q is not a square. In fact if q is a prime, the experimental evidence is that $\alpha(P(q))$ is of order $c \log(q)$ (see [45] for more details). In this section, we show that the ratio bound $\alpha(P(q^2)) \leq q$ is realized for all odd prime powers q.

5.9.1 Lemma. *If q is a prime power and 2 does not divide q, then*

$$\omega(P(q^2)) = q.$$

Proof. Let $\mathbb{F}$ be a finite field of order q^2 and $\mathbb{E}$ be the subfield of order q. The nonzero squares in $\mathbb{F}$ form a multiplicative subgroup of $\mathbb{F}^*$ with order

$$\frac{q^2 - 1}{2} = (q - 1)\frac{q + 1}{2}.$$

The group $\mathbb{F}^*$ is cyclic with order $q^2 - 1$, so for every divisor of $q^2 - 1$ there is a unique subgroup of that order.

Let S be the set of nonzero squares in $\mathbb{F}^*$ and define a homomorphism

$$\phi : \mathbb{F}^* \to S$$

by $\phi(x) = x^2$. Then the image of ϕ is S, which is the unique subgroup of $\mathbb{F}^*$ with order $(q - 1)(q + 1)/2$. Further, S is also cyclic, so S must contain a unique subgroup of order $q - 1$. This subgroup is the unique subgroup of order $q - 1$ in $\mathbb{F}^*$. Thus this subgroup is exactly nonzero elements in $\mathbb{E}$.

Since each element of $\mathbb{E}$ is in S, every element in $\mathbb{E}$ is a square in $\mathbb{F}$. Thus the elements of $\mathbb{E}$ induce a clique in $P(q^2)$, and the ratio bound for cliques is tight. $\square$

Since the Paley graphs are self-complementary, this lemma implies that $\alpha(P(q^2)) = q$. Therefore the ratio bound for cocliques and the clique-coclique bound are tight for these graphs.

We can determine the chromatic number of these graphs (in fact, we will do more than just find the chromatic number). Let S denote the set of nonzero squares in $\mathbb{F}$. Since $\mathbb{E}^*$ is a subgroup of index $(q + 1)/2$ in S, its multiplicative cosets partition S into $(q + 1)/2$ subsets each of size $q - 1$. Since the elements of $\mathbb{E}^*$ form a clique in $P(q^2)$, each of these subsets is also a clique in $P(q^2)$.

Adding 0 to each of these cliques produces $(q + 1)/2$ cliques of size q. Further, each of these cliques forms an additive subgroup of $\mathbb{F}$.

The q additive cosets of any one of these subgroups partition $\mathbb{F}$ into q pairwise disjoint cliques. Hence, from any one of these cosets we obtain a proper q-coloring of the vertices of the complement of $P(q^2)$. Since the Paley graph is self-complementary, we conclude that

$$\chi(P(q^2)) = q.$$

We could reach the same conclusion by simply taking the clique $\mathbb{E}$ and the cosets of it in $\mathbb{F}$. But using this method we see that the cliques on 0 together with their additive cosets provide a set of $q(q + 1)/2$ cliques.

We know that the size of the maximum cliques in $P(q^2)$ is q and we know that the elements of the field of order q form a clique of maximum size. But what are the other maximum cocliques in the graph? We have seen that if S denotes the set of nonzero squares in $\mathbb{F}$, then for any a in S and any b in $\mathbb{F}$, the set

$$S(a, b) = \{ax + b : x \in \mathbb{E}\}$$

is a clique in $P(q^2)$. We say that these cliques are the *square translates* of $\mathbb{E}$ (the square refers to the fact that a is a square, since otherwise $S(a, b)$ will be a coclique). The square translates of $\mathbb{E}$ can be considered to the *canonical* cliques in the Paley graphs. An analog of the EKR Theorem for Paley graphs would be that the only cliques of size q are the canonical cliques. Blokhuis [20] gave a very interesting proof of this using polynomials. We state this result now.

5.9.2 Theorem. *A clique in $P(q^2)$ of size q is a translate of the set of squares in $GF(q^2)$.* $\square$

There is another possible approach to this theorem. Any clique of size q in $P(q^2)$ meets the ratio bound, and so its balanced characteristic vector lies in the θ-eigenspace of the Paley graph. If we could prove the following two results, we would have a second proof of Blokhuis's result.

(a) The balanced characteristic vectors of the canonical cliques span the θ-eigenspace of $P(q^2)$.

(b) The only balanced characteristic vectors of sets of size q in the eigenspace belonging to θ of $P(q^2)$ are the balanced characteristic vectors of the canonical cliques.

5.10 Exercises

5.1 Let X be a strongly regular graph with parameters $(v, k; a, c)$. What are the parameters of $\overline{X}$?

5.2 Prove that a strongly regular graph is primitive if and only if $0 < c < k$.

5.3 Prove that the line graphs of the complete graph K_n and the complete bipartite graph $K_{n,n}$ are strongly regular and find the parameters of these strongly regular graphs.

5.4 Prove that a strongly regular graph is imprimitive if and only if 0 or -1 is an eigenvalue.

5.5 A strongly regular graph is also a distance-regular graph. Find the parameters a_r, b_r and c_r for the association scheme generated by a strongly regular graph with parameters $(v, k; a, c)$.

5.6 Confirm that the eigenvalues given in Theorem 5.3.1 for the block graph of a 2-$(n, m, 1)$ design are correct. The first step of this is to confirm that for this strongly regular graph

$$\Delta = \frac{n - 2m + 1}{m - 1} + 1.$$

5.7 Let $(V, \mathcal{B})$ be a 2-$(n, m, 1)$ design with $n = m^3 - 2m^2 + 2m$. Show that any non-canonical clique of size $(n - 1)/(m - 1)$ in the block graph of $(V, \mathcal{B})$ forms a $(m^2 - m, m, 1)$ subdesign (which is a projective plane of order $m - 1$).

5.8 Let X be the block graph of the design with parameters 2-$(15, 3, 1)$ described after Corollary 5.3.5. Assume that S is a maximum clique in X and that b is a block in S. Let H be the matrix whose rows are the blocks in the design that do not intersect with b and whose columns are the characteristic vectors of S_i (where $i \notin b$). Find a basis for the kernel of H. Using this basis, show that any maximum clique in X is either S_i or the set of all 2-dimensional subspaces of a 3-dimensional space.

5.9 Prove that the block graph for an $OA(n + 1, n)$ is isomorphic to the complete graph on n^2 vertices.

5.10 Prove that the parameters for the block graph $X_{OA(m,n)}$ are

$$(n^2, \quad m(n - 1); \quad (m - 1)(m - 2) + n - 2, \quad m^2).$$

5.11 Show that the vertices of block graph $X_{OA(m,n)}$ can be partitioned into n disjoint cliques of size n.

5.12 Suppose X is the block graph of an orthogonal array $OA(m, n)$ and let f be a function from $V(X)$ to the integers $1, \ldots, n$. Extend the orthogonal array by adding an extra row, such that the ith entry of the row is the value of f on the ith vertex (or, equivalently, the ith column). Show that the new array is orthogonal if and only if f is a proper n-coloring.

5.13 Let $M = OA(m, n)$ be an orthogonal array. Form another orthogonal array M' by removing a single row from M. Show that the chromatic number of the block graph of M' is equal to n.

5.14 Calculate the eigenvalues of the point graph of a partial geometry using the parameters of the graph.

5.15 A *maximal arc* with parameters (n, m) in a projective plane of order q is a set $\mathcal{M}$ of n points such that any line in the plane is either disjoint from $\mathcal{M}$ or meets it in exactly m points. Given a maximal arc $\mathcal{M}$, we define an incidence structure whose points are the points of the plane not in the arc and whose lines are the lines of the plane that contain a point of the arc. Prove that this is a partial geometry with parameters

$$\left(q-m, \ \frac{n}{m}-1, \ \frac{n}{m}-m\right) = \left(q-m, \ q+1-\frac{q}{m}, \ \frac{1}{m}(q-m)(m-1)\right).$$

5.16 Prove that the only maximum cliques in the point graph of a partial geometry with parameters $(s, t, 1)$ are the canonical cliques. Using this, state a version of the EKR Theorem for generalized quadrangles.

5.17 A *linear code over* $GF(q)$ is a subspace of $GF(q)^m$, for some m. We represent a linear code with the rows of a $(d \times m)$ matrix M over $GF(q)$ (the code is the row space of M). Elements of the code are called words and the weight of a code word is the number of nonzero entries in it. A code is projective if the columns of M are distinct.

Suppose C is a projective code, with exactly two nonzero weights, w_1 and w_2. Define a graph $X(C)$ whose vertices are the words in C, where two words are adjacent if their difference has weight w_1. Prove that $X(C)$ is strongly regular by determining its parameters.

Notes

There are many additional references for strongly regular graphs. We recommend the first chapter of the classic book by Brouwer, Cohen and Neumaier [31], Chapter 9 of Brouwer and Haemers's *Spectra of Graphs* [36] and Chapter 10 of Godsil and Royle [87].

At the end of Section 5.2 we discussed a strongly regular graph with parameters $(81, 20; 1, 6)$. This graph can be constructed from a projective 2-weight code of length 10 and dimension 4 over $GF(3)$, with weights 6 and 9. For more details see [33].

A version of Corollary 5.3.5 (this is the EKR Theorem for t-designs) was proved by Rands [144]. Rands considers the more general case of s-intersecting sets of blocks from t-(n, m, λ)-designs. Provided that n is sufficiently large relative to m, t, s then the largest set of intersecting blocks is the collection of all blocks that contain a fixed s-set (Rands's proof uses a counting argument). The bound given by Rands for the case when $t = 2$ and $\lambda = 1$ is slightly weaker than the bound in Corollary 5.3.5.

Partial geometries with $v = 1$ are known as *generalized quadrangles*. Examples can be constructed using a nondegenerate symplectic form on $PG(3, q)$, details are given in [87, Section 5.5]. Many more examples of generalized quadrangles will be found in the book by Payne and Thas [139].

A graph is called *core complete* if the graph is itself a core or its core is a complete graph. Cameron and Kazanidis [39] asked whether all strongly regular graphs are core complete. Godsil and Royle [88] determined that many of the strongly regular graphs that we consider in this chapter are indeed core complete.

6

The Johnson scheme

The EKR Theorem can be expressed very naturally as a statement about cocliques of maximal size in the Kneser graph $K(n, k)$. We saw in the introduction to Chapter 2 that the EKR Theorem, as stated in Theorem 1.0.1, asserts that

$$\alpha(K(n, k)) = \binom{n-1}{k-1}.$$

Further, a coclique of this size consists of the k-subsets that contain a given point i from the underlying n-set. In particular all the cocliques of maximal size are equivalent under the action of the automorphism group of $K(n, k)$, and there are exactly n of them.

6.1 Graphs in the Johnson scheme

In Section 3.2, we described an association scheme that is known as the *Johnson scheme* $\mathcal{J}(n, k)$. This is an association scheme whose vertices are the k-subsets of a fixed n-set, and two k-subsets α and β are i-related if $|\alpha \cap \beta| = k - i$. For each $i \in \{0, \ldots, k\}$ define $J(n, k, i)$ to be the graph of the ith relation (so this graph has the k-subsets of a set of size n as its vertices, and two k-subsets α and β are adjacent if and only if $|\alpha \cap \beta| = k - i$). The graph $J(n, k, 1)$ is the *Johnson graph*, $J(n, k, k)$ is the *Kneser graph*, and the other graphs in this scheme are called the *generalized Johnson graphs*. Throughout this chapter, we denote the adjacency matrix of $J(n, k, i)$ by A_i; thus $\{A_0, A_1, \ldots, A_k\}$ are the matrices in the Johnson scheme.

The map that sends a k-set to its complement gives an isomorphism from $J(n, k, i)$ to $J(n, n - k, i)$. This map is an isomorphism between the Johnshon schemes $\mathcal{J}(n, k)$ and $\mathcal{J}(n, n - k)$. So we may conveniently assume that $n \geq 2k$.

Of all the graphs in the Johnson scheme, the Johnson graph and the Kneser graph are the most frequently studied. Before considering the entire Johnson scheme, we consider these two graphs. When no ambiguity results, we use $J(n, k)$ to denote the Johnson graph. If $k = 1$, the Johnson graph is the complete graph K_n. The Johnson graph $J(n, 2)$ is better known as $L(K_n)$, the line graph of K_n.

In Section 4.7, it was left as an exercise to prove that the Johnson graph $J(n, k)$ is distance regular with classical parameters. The degree of this graph can be easily calculated to be $k(n - k)$. Two vertices are i-related in the Johnson scheme if and only if they are at distance i in $J(n, k)$. Thus the graphs of the Johnson scheme are the distance graphs of $J(n, k)$; this implies that the Johnson scheme is metric.

If α is a fixed vertex in the Johnson graph, then the number of vertices at distance r from α is

$$\binom{k}{k - r}\binom{n - k}{r}.$$

We may view the vertices at distance r from α as ordered pairs (β, γ) where β is a $(k - r)$-subset of α and γ is an r-subset of the complement of α (in the underlying n-set). Two such pairs (β, γ) and (β', γ') are adjacent in the Johnson graph if and only if either

$$|\beta \cap \beta'| = k - r - 1, \quad \gamma = \gamma';$$

or

$$\beta = \beta', \quad |\gamma \cap \gamma'| = r - 1.$$

Thus the subgraph induced by the vertices at distance r from α is isomorphic to the Cartesian product of the Johnson graphs

$$J(k, k - r) \,\square\, J(n - k, r).$$

When $r = 1$, this is the Cartesian product of the complete graphs K_k and K_{n-k}.

There are two obvious families of cliques in $J(n, k)$. For the first, take all $n - k + 1$ of the k-subsets that contain a fixed $(k - 1)$-subset; for the second, take the k-subsets of a fixed set of size $k + 1$. When $n = 2k$ the cliques in these two families have the same size, but the cliques in the first set are bigger if $n > 2k$. By the complete Erdős–Ko–Rado Theorem 1.3.1 with $t = k - 1$, if $n > 2k$, these are the largest cliques in $J(n, k)$.

Applying the clique-coclique bound shows that the size of a maximal coclique can be no more than

$$\frac{1}{n-k+1}\binom{n}{k}.$$

In general $n - k + 1$ does not divide $\binom{n}{k}$, and therefore the clique-coclique bound cannot be tight for all values of n and k. Not much is known about the cocliques in the Johnson graph, nor do we know much about its chromatic number. In the exercises you are invited to determine the chromatic number of $J(n, 2)$, and to find an upper bound for all Johnson graphs. We refer the reader to [60] for more details.

We also briefly consider the Kneser graph. If $2k > n$, then $K(n, k)$ is the empty graph, and if $n = 2k$ it is the disconnected graph with $n/2$ disjoint edges. As these cases are not particularly interesting, we consider only Kneser graphs with $n > 2k$.

If S_i denotes the set of all k-subsets of $\{1, \ldots, n\}$ that contain i, then S_i is a coclique in $K(n, k)$ with size $\binom{n-1}{k-1}$. The EKR Theorem for intersecting k-sets is the assertion that the maximum size of a coclique in $K(n, k)$ is $\binom{n-1}{k-1}$, and any coclique of this size is equal to an S_i for some i. These cocliques can be used to determine a proper coloring of the Kneser graph. For $i = 1, \ldots, n - 2k + 1$ set all the vertices in S_i to be color i (if a vertex is assigned more than one color, set it to be any one of the assigned colors). The remaining vertices induce a subgraph isomorphic to $K(2k - 1, k)$, but this is an empty graph, so all the remaining vertices can be assigned a new color. Thus, the chromatic number of $K(n, k)$ is at most $n - 2k + 2$. This is in fact the actual chromatic number, but this is a major result due to Lovász [116].

There is a very easy proof of the EKR bound when k divides n, which we record here. It is straightforward to see that if k divides n, then

$$\omega(K(n, k)) = \frac{n}{k};$$

and hence

$$\frac{n}{k} = \omega(K(n, k)) \leq \omega_f(K(n, k)) = \frac{\binom{n}{k}}{\alpha(K(n, k))}.$$

Therefore $\alpha(K(n, k)) \leq \binom{n-1}{k-1}$.

6.2 A third basis

In Section 3.4, we saw that if $\mathcal{A} = \{A_0, \ldots, A_d\}$ is an association scheme, then the matrix idempotents $\{E_0, \ldots, E_d\}$ give a second basis for the Bose-Mesner algebra (in addition to the matrices A_i). In this section we show that the Johnson scheme has a third basis that can be used to determine the eigenvalues of the Johnson graph.

Define $W_{t,k}$ to be the 01-matrix with rows indexed by the t-subsets of $\{1, \ldots, n\}$, columns by the k-subsets, and with its (α, β)-entry equal to 1 if $\alpha \subseteq \beta$ (these matrices were also defined in Section 3.6). Define matrices $C_0, \ldots, C_k$ by

$$C_i = W_{i,k}^T W_{i,k}.$$

If α and β are k-sets with $\alpha \cap \beta = r$, we have $(C_i)_{\alpha,\beta} = \binom{r}{i}$. We use $A(n, k, i)$ to denote the adjacency matrix of $J(n, k, i)$, and if the values of n and k are clear we simply use A_i. For $i = 0, \ldots, k$,

$$C_i = \sum_{r \geq i} \binom{r}{i} A_{k-r}, \tag{6.2.1}$$

and therefore C_i lies in the Bose-Mesner algebra of $J(n, k)$. We can also express the matrices in the Johnson scheme in terms of C_i, but first we need a technical lemma.

6.2.1 Lemma. *If $\{M_0, \ldots, M_k\}$ and $\{N_0, \ldots, N_k\}$ are matrices with*

$$M_i = \sum_{j=i}^{k} \binom{j}{i} N_j,$$

then

$$N_i = \sum_{j=i}^{k} (-1)^{j-i} \binom{j}{i} M_j.$$

Proof. Assume that

$$M_i = \sum_{j=i}^{k} \binom{j}{i} N_j,$$

and hence

$$\sum_{i=0}^{k} t^i M_i = \sum_{\substack{i,j \\ i \leq j}} t^i \binom{j}{i} N_j = \sum_{j=0}^{k} (1+t)^j N_j.$$

If we replace t by $-t - 1$, we find that

$$\sum_{i=0}^{k} t^i (-1)^i N_i = \sum_{j=0}^{k} (1+t)^j (-1)^j M_j,$$

and therefore

$$(-1)^i N_i = \sum_{j=i}^{k} \binom{j}{i}(-1)^j M_j. \qquad \square$$

From (6.2.1) and Lemma 6.2.1 we find that

$$A_{k-r} = \sum_{j=r}^{k}(-1)^{j-r}\binom{j}{r}C_j, \qquad (6.2.2)$$

where $r = 0, \ldots, k$. This shows that the matrices C_i form a basis for for the Bose-Mesner algebra of the Johnson scheme.

6.2.2 Lemma. *The column space of $W_{j,k}^T$ is invariant under the Bose-Mesner algebra of $J(n,k)$.*

Proof. The matrices A_i and C_j commute; hence $\text{col}(C_j)$ is A_i-invariant. As

$$\text{col}(C_j) = \text{col}(W_{j,k}^T),$$

the column space of the matrix $W_{j,k}^T$ is invariant under the Bose-Mesner algebra of $J(n,k)$. $\qquad \square$

6.2.3 Lemma. *If $s \le t \le k$ then*

$$W_{s,t}W_{t,k} = \binom{k-s}{t-s}W_{s,k}.$$

Proof. If α is an s-subset and β a k-subset of $\{1, \ldots, n\}$, then the (α, β)-entry of $W_{s,t}W_{t,k}$ is the number of t-subsets of $\{1, \ldots, n\}$ that contain α and lie in β. This equals $\binom{k-s}{t-s}$. $\qquad \square$

Define U_t to be the column space of $W_{t,k}^T$. The previous lemma implies that U_t is contained in U_{t+1}. It follows that $\mathbb{R}^{\binom{n}{k}}$ is the sum of the orthogonal subspaces

$$U_t \cap U_{t-1}^{\perp},$$

where $t = 0, \ldots, k$. Further, each of these subspaces is invariant under the Bose-Mesner algebra of $J(n,k)$.

6.3 Eigenvalues of the Johnson graph

Next we demonstrate a simple approach to finding all the eigenvalues of the Johnson graph that uses the basis defined in the previous section. This approach relies on the following very useful result.

6.3.1 Lemma. *Let M be an $m \times n$ matrix and N an $n \times m$ matrix. Then MN and NM have the same nonzero eigenvalues, with the same multiplicities.*

Proof. Suppose z is an eigenvector for MN with eigenvalue λ. Then

$$NM(Nz) = N(MNz) = N(\lambda z) = \lambda Nz,$$

so, if Nz is not zero, then it is an eigenvector for NM with eigenvalue λ.

Further, N maps $\ker(MN - \lambda I)$ into $\ker(NM - \lambda I)$ and the same argument shows that M maps $\ker(NM - \lambda I)$ into $\ker(MN - I)$. Thus the composite map MN sends $\ker(MN - \lambda I)$ into itself and similarly NM sends $\ker(NM - \lambda I)$ to itself.

If $MNy = 0$ for some $y \in \ker(MN - \lambda I)$, then $MNy = \lambda y = 0$. For $\lambda \neq 0$, this implies that $y = 0$, and that the composite map is injective on $\ker(MN - \lambda I)$. It follows that N is an isomorphism from $\ker(MN - \lambda I)$ to $\ker(NM - \lambda I)$, and this proves the lemma. $\qquad\Box$

6.3.2 Theorem. *The eigenvalues of the Johnson graph $J(n, k)$ are*

$$k(n - k) - i(n + 1 - i),$$

for $i = 0, \ldots, k$, with respective multiplicities

$$\binom{n}{i} - \binom{n}{i - 1}$$

(with the convention that $\binom{n}{-1} = 0$).

Proof. In this proof, we use Johnson graphs $J(n, k)$ with different values of k. We use the notation $A(n, k, 1)$ for the Johnson graph $J(n, k)$. It is easy to calculate that

$$W_{k,k+1} W_{k,k+1}^T = A(n, k, 1) + (n - k)I,$$

$$W_{k,k+1}^T W_{k,k+1} = A(n, k + 1, 1) + (k + 1)I.$$

By Lemma 6.3.1, $W_{k,k+1} W_{k,k+1}^T$ and $W_{k,k+1}^T W_{k,k+1}$ have the same nonzero eigenvalues, with the same multiplicities.

Since the difference in the sizes of the matrices is $\binom{n}{k+1} - \binom{n}{k}$, this is a lower bound on the multiplicity of zero as an eigenvector of the second matrix. It follows that $-k - 1$ is an eigenvalue of $A(n, k + 1, 1)$ with multiplicity at least $\binom{n}{k+1} - \binom{n}{k}$.

If $\theta_0 > \theta_1 > \cdots > \theta_k$ are the eigenvalues of $A(n, k, 1)$, then the eigenvalues of $A(n, k + 1, 1)$ are $\theta_i + n - 2k - 1$ for $i = 0, \ldots, k$. Our expression for the eigenvalues follows by a simple induction argument starting with the base case $k = 1$ where the Johnson graph is the complete graph. $\qquad\Box$

Table 6.1 *Eigenvalues of $A(n, k, 1)$ for $k = 1, \ldots, 5$*

$k = 1$:	$n - 1$	-1			
$k = 2$:	$2n - 4$	$n - 4$	-2		
$k = 3$:	$3n - 9$	$2n - 9$	$n - 7$	-3	
$k = 4$:	$4n - 16$	$3n - 16$	$2n - 14$	$n - 10$	-4
$k = 5$:	$5n - 25$	$4n - 25$	$3n - 23$	$2n - 19$	$n - 13$	-5

We can summarize the induction in Table 6.1. In row $k + 1$, the entry in each column is either $\theta_i(k) + n - 2k - 1$, where $\theta_i(k)$ is the entry above it, or $-(k + 1)$.

6.3.3 Theorem. *The ith eigenspace of $J(n, k)$ is the orthogonal complement to $\mathrm{col}(W_{i-1,k}^T)$ in $\mathrm{col}(W_{i,k}^T)$ and has dimension $\binom{n}{i} - \binom{n}{i-1}$.*

Proof. By Theorem 6.3.2, we see that $A(n, k, 1) + (n - k)I$ is invertible and, from the proof of Lemma 6.3.1, if U is one of the eigenspaces of $A(n, k, 1)$, then $W_{k,k+1}^T U$ is an eigenspace of $A(n, k + 1, 1)$ with the same dimension. Using induction, we can assume that the ith eigenspace of $A(n, k, 1)$ is the orthogonal complement to $\mathrm{col}(W_{i-1,k}^T)$ in $\mathrm{col}(W_{i,k}^T)$ and has dimension $\binom{n}{i} - \binom{n}{i-1}$. From Lemma 6.2.3,

$$W_{k,k+1}^T W_{j,k}^T = \binom{k + 1 - j}{k - j} W_{j,k+1}^T,$$

and it follows that for $i = 0, \ldots, k$, the ith eigenspace of $A(n, k + 1, 1)$ is the orthogonal complement to $\mathrm{col}(W_{i-1,k+1}^T)$ in $\mathrm{col}(W_{i,k+1}^T)$ (and the dimension is $\binom{n}{i} - \binom{n}{i-1}$). The final eigenspace (corresponding to $i = k + 1$) is the kernel of $W_{k,k+1}$. $\qquad\square$

6.3.4 Corollary. *If $E_0, \ldots, E_k$ are the matrix idempotents in $J(n, k)$, then $E_0 + \cdots + E_t$ represents orthogonal projection onto the column space of $W_{t,k}^T$.* $\qquad\square$

6.4 A fourth basis

We introduce a class of matrices which form a fourth basis for the Bose-Mesner algebra of the Johnson scheme. Let $\overline{W}_{t,k}$ be the 01-matrix with rows indexed by the t-subsets of $\{1, \ldots, n\}$, columns by the k-subsets, and with its (α, β)-entry equal to 1 if $\alpha \cap \beta = \emptyset$. Define square matrices D_i by

$$D_i = W_{i,k}^T \overline{W}_{i,k}.$$

If α and β are k-subsets, then

$$(D_i)_{\alpha,\beta} = \binom{|\alpha \setminus \beta|}{i} = \binom{k - |\alpha \cap \beta|}{i};$$

so, despite appearances, D_i is a symmetric matrix. We note that $D_0 = J$ and

$$(D_1)_{\alpha,\beta} = k - |\alpha \cap \beta|.$$

Further

$$D_k = \overline{W}_{k,k} = A_k,$$

and thus D_k is the adjacency matrix of the Kneser graph.

The next lemma (which is used in the following section to derive the eigenvalues of D_i) shows that D_k can also be expressed as an alternating sum of the product of matrices $W_{i,j}$.

6.4.1 Lemma. *For integers k and t*

$$\overline{W}_{t,k} = \sum_{i=0}^{\min\{t,k\}} (-1)^i W_{i,t}^T W_{i,k}.$$

Proof. Suppose α is a t-subset and β a k-subset of our underlying set. Since

$$(W_{i,t}^T W_{i,k})_{\alpha,\beta} = \binom{|\alpha \cap \beta|}{i},$$

the (α, β)-entry of the sum in the statement of the lemma is

$$\sum_{i=0}^{\min\{t,k\}} (-1)^i \binom{|\alpha \cap \beta|}{i} = \begin{cases} 1, & \text{if } \alpha \cap \beta = \emptyset; \\ 0, & \text{otherwise.} \end{cases}$$

By the definition of $(\overline{W}_{t,k})_{\alpha,\beta}$ the lemma follows. $\qquad\square$

6.4.2 Lemma. *The matrices $D_0, \dots, D_k$ form a basis for the Bose-Mesner algebra of the Johnson scheme.*

Proof. To prove this, we only need to show that in the Johnson scheme $\mathcal{A} = \{A_0, \dots, A_d\}$ each A_i can be expressed as a linear combination of the matrices D_i.

The expression just given for the entries of D_i implies that

$$D_i = \sum_{j=i}^{k} \binom{j}{i} A_j,$$

and by Lemma 6.2.1,

$$(-1)^i A_i = \sum_{j=i}^{k} \binom{j}{i} (-1)^j D_j. \qquad \square$$

6.5 Eigenvalues of the Johnson scheme

In this section we determine the eigenvalues of the matrices D_r in the Bose-Mesner algebra of $\mathcal{J}(n, k)$ defined in the previous section. From this we can also determine the eigenvalues of all the graphs in the Johnson scheme $\mathcal{J}(n, k)$.

6.5.1 Theorem. *The nonzero eigenvalues of D_r are*

$$(-1)^j \binom{k-j}{r-j} \binom{n-r-j}{k-j}$$

for $j = 0, \ldots, r$, with respective multiplicities $\binom{n}{j} - \binom{n}{j-1}$.

Proof. We consider the matrices D_r in the Bose-Mesner algebra of $\mathcal{J}(n, k)$ for multiple values of k. So we denote

$$D_i(k) = W_{i,k}^T \overline{W}_{i,k}$$

to make it clear which k we are using. Also, we denote the ith idempotent matrix from $J(n, k)$ as $E_i(k)$.

We start by finding the eigenvalues for $D_k(k)$. Since $D_k(k) = \overline{W}_{k,k}$ it is easy to calculate that

$$D_k(k)W_{t,k}^T = \overline{W}_{k,k}W_{t,k}^T = \binom{n-k-t}{k-t} \overline{W}_{k,t}. \qquad (6.5.1)$$

From the fact that $E_t(k)W_{i,k}^T = 0$ if $i < t$, it follows from Lemma 6.4.1 that

$$E_t(t)\overline{W}_{t,k} = (-1)^t E_t(t)W_{t,t}^T W_{t,k} = (-1)^t E_t(t)W_{t,k}.$$

By taking the transpose,

$$\overline{W}_{k,t}E_t(t) = (-1)^t W_{t,k}^T E_t(t).$$

Combining this with (6.5.1) we have that

$$D_k(k)W_{t,k}^T E_t(t) = \binom{n-k-t}{k-t} \overline{W}_{k,t}E_t(t) = (-1)^t \binom{n-k-t}{k-t} W_{t,k}^T E_t(t)$$

and we see that $(-1)^t \binom{n-k-t}{k-t}$ is an eigenvalue of $D_k(k)$ with multiplicity $\binom{n}{t} - \binom{n}{t-1}$ (the multiplicity follows from Theorem 6.3.3). This can be done for any

$t \in \{0, \ldots, k\}$, so the eigenvalues of $D_k(k)$ are

$$(-1)^j \binom{n-k-j}{k-j},$$

where $j \in \{0, \ldots, k\}$.

To get the eigenvalues for $D_r(k)$, note that since

$$D_r(k) = W_{r,k}^T \overline{W}_{r,k}$$

and

$$\overline{W}_{r,k} W_{r,k}^T = \binom{n-2r}{k-r} \overline{W}_{r,r} = \binom{n-2r}{k-r} D_r(r),$$

the matrices

$$D_r(k), \qquad \binom{n-2r}{k-r} D_r(r)$$

have the same nonzero eigenvalues, with the same multiplicities (this follows from Lemma 6.3.1).

Since both matrices have rank $\binom{n}{r}$ we deduce that the nonzero eigenvalues of $D_r(k)$ are

$$\binom{n-2r}{k-r} \left((-1)^j \binom{n-r-j}{r-j} \right) = (-1)^j \binom{k-j}{r-j} \binom{n-r-j}{k-j} \qquad (6.5.2)$$

for $j = 0, \ldots, r$. $\qquad\square$

Since

$$A_i = \sum_{r=i}^{k} (-1)^{r-i} \binom{r}{i} D_r,$$

we can use (6.5.2) in the proof of Theorem 6.5.1 to find an expresssion for the eigenvalues of A_i.

6.5.2 Theorem. *The eigenvalue of A_i from $\mathcal{J}(n, k)$ on the jth eigenspace is*

$$p_i(j) = \sum_{r=i}^{k} (-1)^{r-i+j} \binom{r}{i} \binom{n-2r}{k-r} \binom{n-r-j}{r-j}. \qquad\square$$

We can also use the matrices D_r (and what we know about their eigenvalues) to show that the Johnson scheme is cometric.

6.5.3 Lemma. *The Johnson scheme is cometric relative to the first idempotent E_1.*

Proof. Since the matrices D_r form a basis for the Bose-Mesner algebra of the Johnson scheme, we can express each idempotent E_j as a linear combination of the matrices D_r (the coefficients can be determined by taking the inner products of the matrices). From Theorem 6.5.1 we know the eigenvalues of each D_r, and since E_j is the projection to the eigenspaces we also have that $D_r E_j = 0$ if $j > r$ and $D_r E_r \neq 0$. So E_j is a linear combination of $D_0, \dots, D_j$, where the coefficient of D_j is not zero. Now

$$(D_r)_{\alpha,\beta} = \binom{|\alpha \setminus \beta|}{r}$$

and since $\binom{x}{r}$ is a polynomial of degree r in x, it follows that D_r is a Schur polynomial of degree r in D_1. As D_1 is a linear combination of E_1 and E_0, we conclude that E_j is a Schur polynomial in E_1 with degree j. $\qquad\square$

6.6 Kneser graphs

In this section, we show how the characterization of the sets that meet the bound in the EKR Theorem can be found by considering the eigenspaces of the Johnson scheme. We restrict ourselves to the non-empty Kneser graphs, so throughout this section we require that $n \geq 2k$.

To start, set $i = k$ in Theorem 6.5.2 to get the eigenvalues of the Kneser graph.

6.6.1 Corollary. *The eigenvalues of $K(n, k)$ are the integers*

$$(-1)^i \binom{n - k - i}{k - i},$$

where $i = 0, \dots, k$, with corresponding multiplicities

$$\binom{n}{i} - \binom{n}{i - 1}. \qquad\square$$

Thus the least eigenvalue of the Kneser graph $K(n, k)$ is

$$\tau = -\binom{n - k - 1}{k - 1},$$

with multiplicity $n - 1$. Since the valency of $K(n, k)$ is $\binom{n-k}{k}$, the ratio bound for cocliques (Theorem 2.4.1) is

$$\alpha(K(n, k)) \leq \frac{\binom{n}{k}}{1 + \frac{n-k}{k}} = \binom{n - 1}{k - 1}$$

(as we derived in Section 2.5). The fact that equality holds in the ratio bound provides the additional information that the characteristic vector of a coclique of this size is a linear combination of **1**, and an eigenvector for $K(n, k)$ with eigenvalue τ.

The cocliques of the Kneser graph that are formed from all the k-subsets that contain a given point are cocliques of maximum size. The characteristic vectors of these cocliques are the rows of $W_{1,k}$. Our goal is to show that there are no other cocliques of maximum size.

From Theorem 6.3.3 we know that $\mathrm{col}(W_{1,k}^T)$ is the sum of the first two eigenspaces for $J(n, k)$; this means that it is the subspace spanned the constant vectors and the τ-eigenvectors of $K(n, k)$. It follows that if S is a coclique in the Kneser graph of maximum size, then its characteristic vector z lies in $\mathrm{col}(W_{1,k}^T)$; so there is a vector h such that

$$W_{1,k}^T h = z.$$

Let w_β denote the row of $W_{1,k}^T$ indexed by the k-subset β. We see that if $\beta \notin S$, then h must be orthogonal to w_β. There is no loss of generality in assuming that S contains the k-subset

$$\alpha = \{1, \ldots, k\}.$$

In this case we see that h must be orthogonal to each vector w_β, where $\beta \cap \alpha = \emptyset$. This gives $\binom{n-k}{k}$ constraints on h. If we let $M_{1,k}^T$ denote the $\binom{n-k}{k} \times (n - k)$ submatrix of $W_{1,k}$ formed by taking the rows indexed by subsets that do not intersect with α and the columns indexed by the elements not in α, we can express these constraints in matrix form as

$$\begin{pmatrix} 0 & M_{1,k}^T \end{pmatrix} h = 0$$

(here the zero-block has order $\binom{n-k}{k} \times k$).

If $n > 2k$, then the rank of $M_{1,k}$ is $n - k$ (simply consider $M_{1,k} M_{1,k}^T$ to see this fact) and we conclude from this that if $i \notin \alpha$ then $h_i = 0$. Equivalently we have shown that

$$\mathrm{supp}(h) \subseteq \alpha.$$

Since this must hold true for all k-subsets α in the coclique S, the intersection of the k-subsets contains $\mathrm{supp}(h)$. Since $\mathrm{supp}(h)$ is not empty, the intersection of all sets in S is also not empty. Thus we have proved the EKR Theorem for intersecting families of k-subsets.

6.7 Exercises

6.1 Prove that two vertices are adjacent in the ith graph of the Johnson scheme if and only if they are at distance i in $J(n, k)$.

6.2 Determine all parameters n, k, i for which the generalized Johnson graph $J(n, k, i)$ is connected.

6.3 The Johnson graph $J(n, 2)$ is also known as the *triangle graph*. Find the size of the maximum coclique in $J(n, 2)$, and the chromatic number of $J(n, 2)$.

6.4 For all n and k, show that $\chi(J(n, k)) \le n$.

6.5 Determine the values of k for which the Kneser graph $K(n, k)$ is strongly regular.

6.6 Show that each of the Johnson graphs $J(n, 2, 1)$, $J(7, 3, 1)$ and $J(10, 3, 1)$ is strongly regular. Show that there are no other strongly regular Johnson graphs.

6.7 Show that the two families of cliques in the Johnson graph described in Section 6.1 are all the maximal cliques in the graph.

6.8 Using the maximal cliques in the previous question, determine the automorphism group of $J(n, k)$ for $n \ne 2k$.

6.9 Show that $W_{t,k}$ and $\overline{W}_{t,k}$ have the same row space. Deduce from this that $W_{k,n-k}$ is invertible.

6.10 Using Exercise 6.9 and the identity

$$W_{t,h} W_{h,n-t} = \binom{n - 2t}{h - t} W_{t,n-t},$$

prove that if $t \le \min\{h, n - h\}$ then the rows of $W_{t,h}$ are linearly independent.

6.11 Prove that

$$W_{s,k} W_{t,k}^T = \sum_{i=0}^{\min\{s,t\}} \binom{n - s - t}{n - k - i} W_{i,s}^T W_{i,t}.$$

Consider the case $s = t$. Prove that $W_{t,k} W_{t,k}^T$ is the sum of positive semidefinite matrices, at least one of which is invertible, and deduce that the rows of $W_{t,k}$ are linearly independent (when $t \le \min\{k, n - k\}$).

6.12 Show directly that the matrices C_i commute (find an expression for $C_i C_j$ as a linear combination of matrices C_r where $r \le \min\{i, j\}$).

Notes

A coclique in the Johnson graph $J(n, k)$ is a constant weight code with word length n, weight k, and minimum Hamming distance 4 (take the characteristic vectors of the sets in the coclique). Clearly the size of a maximum coclique is the largest possible code. A proper coloring of the graph $J(n, k)$ gives a partition of the k-subsets from $\{1, 2, \ldots, n\}$ into such constant weight codes, and thus the chromatic number gives the minimum number of parts in such a partition. The goal of Exercise 6.4 is to show that $\chi(J(n, k)) \leq n$. This was originally done by Graham and Sloane [89], who gave the construction of a very simple coloring of this size. There are many cases where the inequality is strict; see [60] for examples. For tables of constant weight codes and some constructions see [34].

The proof of the chromatic number of the Kneser graph is, for many graph theorists, an unsatisfying one. The first proof was given by Lovász in 1978 and used topological methods [116]. Bárány [15] soon afterward gave a simplified proof that relied on the Borsuk–Ulam theorem (a topological result). In 2002, Greene [90] gave another, very interesting and short topological proof of this result. A purely combinatorial proof was given by Matoušek in 2004 [125]. Matoušek's proof uses a combinatorial analog of the Borsuk–Ulam theorem called Tucker's lemma. It is both surprising and frustrating that there still is no proof of the chromatic number of the Kneser graph that uses only standard graph theory!

A proof of Exercise 6.2 is given by Jones in [100].

Exercises 6.7 and 6.8 are based on [143]. The automorphism groups for each of the graphs in the Johnson scheme are given in [100].

From Exercise 6.6 we know that the triangle graphs $J(n, 2)$ and the two Johnson graphs $J(7, 3)$ and $J(10, 3)$ are strongly regular. Further, it is not hard to see that the Kneser graphs $K(2n, n)$ and $K(n, 2)$ are also strongly regular graphs. In [41] it is shown that there are no other strongly regular graphs among the generalized Johnson graphs.

7

Polytopes

Suppose U is an $n \times d$ matrix whose columns form an orthonormal basis for an eigenspace of a graph X. Then the rows of U also specify n points in $\mathbb{R}^d$, and these points generate a convex polytope. These polytopes are surprisingly useful. For example, in the previous chapter we proved that the only characteristic vectors of maximum cocliques of the Kneser graph $K(n, k)$ that lie in $\mathrm{col}(W_{1,k}^T)$ are the columns of $W_{1,k}^T$ (see Section 6.6). With this fact, we were able to give a proof of the characterization of the maximum intersecting sets in the EKR Theorem.

In this chapter we use polytopes to prove various forms of the EKR Theorem for intersecting families and show how this approach can be used in other situations. In Chapter 8, we use the methods developed in this chapter to characterize the t-intersecting families of k-sets of maximum size.

7.1 Convex polytopes

A *convex polytope* is the convex hull of a finite set of points in a real vector space or, equivalently, a bounded subset of a real vector space formed by the intersection of a finite number of closed half-spaces. Since we consider only polytopes that are convex, we use the term *polytope* rather than convex polytope. The *dimension* of a polytope is the smallest dimension of an affine space that contains it. If $\mathcal{P}$ is a polytope, then an affine hyperplane H is called a *supporting hyperplane* if $H \cap \mathcal{P} \neq \emptyset$ and H does not separate two points of $\mathcal{P}$. A *face* of a polytope P is the intersection of P with a supporting hyperplane (note that a face is itself necessarily a polytope). The maximal proper faces of $\mathcal{P}$ are its *facets*. The faces of dimension 0 and 1 are the *vertices* and *edges*, respectively. (This is where the graph theory terms come from.) If $\mathcal{P}$ is the convex hull of a finite set of points, then the vertices of $\mathcal{P}$ are a subset of these points.

The graph formed by the vertices and edges of a polytope is its *1-skeleton*. A classical result of Steinitz [155] asserts that any 3-connected planar graph arises as the 1-skeleton of a polytope in $\mathbb{R}^3$. The 1-skeleton of a polytope does not determine the polytope. For example if $n \geq 4$, then there are many distinct polytopes whose 1-skeleton is a complete graph.

The ordered pair (h, γ), where $h \in \mathbb{R}^d$ and $\gamma \in \mathbb{R}$, defines the affine hyperplane

$$\{x : h^T x = \gamma\}$$

(and any affine hyperplane can be similarly be defined with a vector h and a scalar γ). If $h^T x \leq \gamma$ (or $h^T x \geq \gamma$) for all x in a polytope $\mathcal{P}$ and there is some $x \in \mathcal{P}$ with $h^T x = \gamma$, then (h, γ) is a *supporting* hyperplane for $\mathcal{P}$. Since we are free to replace h by $-h$, we usually assume that h is chosen so that $h^T x \leq \gamma$. Note that each nonzero vector h determines two distinct faces of a polytope (one is the supporting hyperplane defined by (h, γ) where γ is the maximum value of $h^T x$ over all $x \in \mathcal{P}$, and the other is where γ is the minimum). These two faces span subspaces perpendicular to the line spanned by h.

The polytopes we consider will often arise as follows. Let U be a real $n \times d$ matrix, with rank d for convenience. Then the rows of U provide n points in $\mathbb{R}^d$, and we denote their convex hull by $\mathcal{P}(U)$. For a matrix U we denote the ith row of U by U_i.

7.1.1 Lemma. *Let U be an $n \times d$ matrix and $\mathcal{P} = \mathcal{P}(U)$ the polytope generated by the rows of U. Then for each face F of $\mathcal{P}$, there is vector h such that F is the convex hull of the set of points $u \in \mathcal{P}$ on which $h^T u$ is maximal.*

Proof. Suppose F is a face of $\mathcal{P}$. Then there is a supporting hyperplane that defines it, assume this is defined by (h, γ). Then either $h^T x \leq \gamma$ for all $x \in \mathcal{P}$, or $h^T x \geq \gamma$ (for all x in $\mathcal{P}$). By replacing h with $-h$ if necessary, we have that $\gamma = \max\{h^T x : x \in \mathcal{P}\}$ and the theorem follows. $\square$

For a matrix U, the vertices of $\mathcal{P}(U)$ are a subset of the rows of U. If every row of U has the same norm, then every row of U is a vertex of $\mathcal{P}(U)$. To see this, consider $h = (U_i)^T$ and note that $U_j h = \langle U_j, U_i \rangle$. By the Cauchy-Schwartz inequality, if U_i and U_j are linearly independent, then

$$\langle U_j, U_i \rangle < \|U_j\| \, \|U_i\| = \|U_i\|^2 = U_i h.$$

Since every row has the same norm, the only way U_j and U_i are not linearly independent is if $U_i = \pm U_j$. So for all j with $U_j \neq U_i$ it is the case that $U_j h < U_i h$. Thus each row of U is a vertex of $\mathcal{P}$.

In our applications, the rows of U will all have the same norm and the column space of U will be a union of eigenspaces of an association scheme. Consider vectors h such that $Uh = x$ is a 01-vector. In this case, the convex hull of the vectors in the support of x forms a face of the polytope $\mathcal{P}(U)$ (just consider the affine hyperplane defined by $(h, 1)$). Similarly, replacing h with $-h$ shows that convex hull of the complement of the support of x forms another face of the polytope.

But lest the reader gain a sense of assurance, the theory of polytopes has its simple aspects, but it also has traps for even the moderately wary. For example, it is not true that there is an edge between any pair of vertices in a polytope that are as close as possible (for example, consider a parallelogram in $\mathbb{R}^2$).

7.2 A polytope from the Johnson graph

Assume k and n are integers with $n > 2k$ and consider the polytope $\mathcal{P}$ generated by the rows of $W = W_{1,k}^T$. The row of W corresponding to the k-subset α will be denoted by w_α. So for any k-set β, the β-entry of $W w_\alpha$ is $|\alpha \cap \beta|$; in particular, the α-entry is k and all other entries are at most $k - 1$. Hence the vertices of $\mathcal{P}$ are exactly the rows of W. So each vertex of the polytope is the characteristic vector of a unique k-set.

Next, we determine the edges of $\mathcal{P}$. If α and β are k-sets with $|\alpha \cap \beta| = k - 1$, consider

$$h = \frac{1}{2}(w_\alpha + w_\beta).$$

The α-entry and the β-entry of Wh both equal $(2k - 1)/2$, whereas any other entry is at most $k - 1$. Hence w_α and w_β generate an edge of $\mathcal{P}$. It is left as an exercise to show that there is no edge between α and β if $|\alpha \cap \beta| < k - 1$. So we conclude that $J(n, k)$ is isomorphic to the 1-skeleton of $\mathcal{P}$.

Next we describe all faces of $\mathcal{P}$. These were first listed in [79].

7.2.1 Theorem. *For any $A \subseteq B \subseteq \{1, \ldots, n\}$, let*

$$S_{A,B} = \{S : A \subseteq S \subseteq B, |S| = k\}.$$

The sets $S_{A,B}$ for all $A \subseteq B \subseteq \{1, \ldots, n\}$ are exactly the vertex sets of the faces of $\mathcal{P}$.

Proof. We prove first that the given families are faces of $\mathcal{P}$. Suppose $A \subseteq B \subseteq \{1, \ldots, n\}$ with $|A| \leq k \leq |B|$. Let $h \in \mathbb{R}^n$ be given by

$$h_i = \begin{cases} 2, & i \in A; \\ 1, & i \in B \setminus A; \\ 0, & \text{otherwise.} \end{cases}$$

Since $A \subseteq B$, for any set α the value of $(Wh)_\alpha$ is

$$2|\alpha \cap A| + |\alpha \cap (B \setminus A)| = |\alpha \cap A| + |\alpha \cap B|$$

and this attains its maximum value of $|A| + k$ if and only if $A \subseteq \alpha \subseteq B$.

Now suppose F is a face determined by a vector $h = (h_1, \dots, h_n)$. Assume for the moment that

$$h_1 \geq \cdots \geq h_n.$$

Let j be the greatest integer such that $h_j > h_k$ and let ℓ be the greatest integer such that $h_\ell = h_k$. Then the k-set α lies in F if and only if

$$\{1, \dots, j\} \subseteq \alpha \subseteq \{1, \dots, \ell\}.$$

Since $\mathcal{P}$ is invariant under the action of $\mathrm{Sym}(n)$, the theorem follows. $\qquad \square$

We can now use the polytope $\mathcal{P}$ to give the characterization in the EKR Theorem. Suppose S is a coclique in $K(n, k)$ with size $\binom{n-1}{k-1}$ and characteristic vector z. As in Section 6.6, the vector z lies in $\mathrm{col}(W_{1,k}^T)$. In particular, there is a vector h such that $Wh = z$. Since z is the characteristic vector of S, the rows on which Wh achieves its maximum value are exactly the rows that correspond to sets in S. Thus by Lemma 7.1.1, the k-subsets α such that $\alpha \subseteq S$ form a face of our polytope $\mathcal{P}$. We deduce from Theorem 7.2.1 that there are subsets A and B of $\{1, \dots, n\}$ such that

$$S = \{\alpha : A \subseteq \alpha \subseteq B\}.$$

If $A \neq \emptyset$, then the size of S implies that $|A| = 1$ and we are done. If $A = \emptyset$, then S is the set of all k-subsets of B. Since any two elements of S have a vertex in common, $|B| \leq 2k - 1$, and therefore

$$|S| = \binom{2k-1}{k} = \binom{2k-1}{k-1}.$$

But since $n \geq 2k + 1$ this is a contradiction with the size of S; indeed,

$$|S| = \binom{n-1}{k-1} \geq \binom{2k}{k-1} = \binom{2k-1}{k-1} + \binom{2k-1}{k-2} > \binom{2k-1}{k-1}.$$

We conclude that $A \neq \emptyset$, and this completes our second proof of the characterization of the maximum intersecting sets in the EKR Theorem.

7.3 Perfect matching polytopes

For a given graph, a *matching* is a set of vertex-disjoint edges and a matching is *perfect* if it includes all the vertices in the graph. Two matchings in a graph *intersect* if they have an edge in common. Given this setup (and what we are about) it is natural to ask for the maximum size of an intersecting family of matchings and for a characterization of the intersecting families of maximum size. We use polytopes to help study these questions.

We view a matching as a subset of the edges of a graph, and hence we may represent matchings by their characteristic vectors (these are vectors in $\mathbb{R}^{|E(X)|}$). The convex hull of the characteristic vectors of the perfect matchings of a graph X is called the *perfect matching polytope* of X. Further, the *incidence matrix* for the matchings of a graph X is the matrix U whose rows are the characteristic vectors of the matchings of X. The columns of U are indexed by the edges of X and the rows by the matchings. From Lemma 7.1.1, a face of the perfect matching polytope is the convex hull of the rows where Uh achieves its maximum for some vector h.

We introduce some terminology and notation to allow us to provide information on the structure of the faces of the perfect matching polytope. If X is a graph and S is a subset of $V(X)$, we define its *boundary* $\partial(S)$ to be the set of edges that join a vertex in S to a vertex not in S. The boundary of S is also known as an *edge cut*. If S consists of a single vertex u, then ∂S is just the set of edges incident with u. If S is a subset of $V(X)$ of odd size, then any perfect matching in X must contain at least one edge from $\partial(S)$. Clearly if S is a single vertex, any perfect matching contains exactly one element of ∂S. It is an amazing classical result of Edmonds that these two constraints characterize the perfect matching polytope for any graph. For a short proof we refer the reader to Schrijver [150].

7.3.1 Theorem. *Let X be a graph. A vector x in $\mathbb{R}^{|E(X)|}$ lies in the perfect matching polytope of X if and only if:*

    (a) $x \geq 0$;
    (b) if $S = \{u\}$ for some $u \in V(X)$, then $\sum_{e \in \partial S} x(e) = 1$;
    (c) if S is an odd subset of $V(X)$ with $|S| \geq 3$, then $\sum_{e \in \partial S} x(e) \geq 1$.

If X is bipartite, then x lies in the perfect matching polytope if and only if the first two conditions hold.   □

The constraints in (b) define an affine subspace of $\mathbb{R}^{|E(X)|}$. The perfect matching polytope is the intersection of this subspace with affine half-spaces defined by the conditions in (a) and (c); hence the points in a proper face of the polytope must satisfy at least one of these conditions with equality.

In particular, if X is bipartite, any facet of the perfect matching polytope is the intersection of the polytope with the hyperplane given by the condition $x(e) = 0$ for some edge of X. Therefore the vertices of a facet consist of the perfect matchings that do not contain a given edge. If X is not bipartite, we still have facets of the type just given, but we also have facets defined by equations of the form

$$\sum_{e \in \partial S} x(e) = 1$$

for some odd subset of $V(X)$. If X is not bipartite, then the vertices of a facet are either the perfect matchings that miss a given edge, or the perfect matchings that contain exactly one edge from ∂S for some odd subset S.

The only graphs of interest to us are $K_{n,n}$ and K_n; these are considered in the following two sections.

7.4 Perfect matchings in complete graphs

If $n = 2m$, then the number of perfect matchings in the complete graph K_n is

$$\frac{(2m)!}{2^m m!} = (2m - 1)(2m - 3) \cdots 1 = (2m - 1)!!. \tag{7.4.1}$$

Consequently if $n = 2m + 1$, the number of matchings in K_n of size m is

$$(2m + 1)(2m - 1)!! = (2m + 1)!!.$$

It follows from Theorem 7.3.1 that every perfect matching in K_{2m} is a vertex in the perfect matching polytope. But we can also determine the vertices of every facet in this polytope.

7.4.1 Lemma. *In the matching polytope of K_{2m}, the vertices of a facet of maximum size are the perfect matchings that do not contain a given edge.*

Proof. Let F be a facet of the polytope of maximum size. From the comments in the previous section, equality holds in at least one of equations

$$\sum_{e \in \partial S} x(e) \geq 1$$

for all $x \in F$. Suppose S is the subset that defines such an equation; then S is an odd subset of the vertices in K_{2m} for which $\sum_{e \in \partial S} x(e) = 1$ for all $x \in F$.

Let s be the size of S. Each perfect matching with exactly one edge in ∂S consists of the following: a matching of size $(s - 1)/2$ covering all but one vertex of S; an edge joining this missed vertex of S to a vertex in $\overline{S}$; and a matching of size $(2m - s - 1)/2$ covering all but one vertex in $\overline{S}$. Hence there are

$$(s - 2)!! \, s(2m - s) \, (2m - s - 2)!! = s!!(n - s)!!$$

such perfect matchings. We denote this number by $N(s)$ and observe that

$$\frac{N(s-2)}{N(s)} = \frac{n-s+2}{s}.$$

Hence for all s such that $3 \le s \le m$ we see that the values $N(s)$ are strictly decreasing, so the maximum size of a set of such vertices is $N(3) = 3(2m - 3)!!$.

On the other hand, the number of perfect matchings in K_{2m} that do not contain a given edge is

$$(2m-1)!! - (2m-3)!! = (2m-2)(2m-3)!!.$$

Since this is always larger than $N(3)$, the lemma follows. $\square$

Define a graph $M(2m)$ whose vertices are the perfect matchings of K_{2m} and two perfect matchings are adjacent if and only if they are not intersecting. This graph is called the *perfect matching graph* and is discussed in detail in Section 15.2. This graph plays the same role as the Kneser graph does in the EKR Theorem for sets; a set of intersecting matchings in K_{2m} is a coclique $M(2m)$.

For any edge e in K_{2m}, define the set S_e to be the set of matchings that contain e. Clearly S_e is a coclique of size $(2k-3)!!$ in $M(2k)$. In Section 15.2 we will see that S_e is a maximum coclique, and we ask if these are all maximum cocliques in the matching graph. At this point we can show that if the characteristic vector of coclique of size $(2k-3)!!$ is in the column space of the incidence matrix for the perfect matchings, then the coclique must be of the form S_e for some edge e.

7.4.2 Lemma. *Let B be the incidence matrix for the perfect matchings of K_{2m}, and let z be the characteristic vector of a coclique S of size $(2m-3)!!$. If $z \in \mathrm{col}(B)$, then $S = S_e$ for some edge e from K_{2m}.*

Proof. The convex hull of the rows of B is the perfect matching polytope of K_{2m}. If $z \in \mathrm{col}(B)$, then $z = Bh$ for some h. If $S = \mathrm{supp}(z)$, it follows that S is the set of vertices of a face of the polytope formed by the rows of B; similarly, the matchings in the complement, $\overline{S}$, also form a face in the perfect matching polytope of K_n

By Lemma 7.4.1 the vertices of a maximum facet are the perfect matchings that miss a given edge. Thus the complement $\overline{S}$ of S is a face of maximal size. $\square$

In Section 15.6, we show that intersecting set of perfect matchings with maximum size must lie in the column space of B. We use also another method to establish an EKR-type result for perfect matchings of K_{2m} in Section 15.2.

7.5 Perfect matchings in complete bipartite graphs

Now we turn our attention to perfect matchings in complete bipartite graphs. A perfect matchings in $K_{n,n}$ corresponds to a permutation of $\{1, \ldots, n\}$. To see this, let $\{X, Y\}$ be the bipartition of $K_{n,n}$. A perfect matching pairs each $x \in X$ with a unique $y_x \in Y$; the corresponding permutation maps x to y_x.

Two matchings are intersecting if they share a common edge, so we say two such permutations σ and ρ are intersecting if and only if $\sigma(i) = \rho(i)$ for some $i \in \{1, \ldots, n\}$. Equivalently, σ and ρ are intersecting if $\rho^{-1}\sigma$ has a fixed point. Conversely, σ and ρ are not intersecting if $\rho^{-1}\sigma$ is a *derangement* (a derangement is a permutation with no fixed points). Derangements are further discussed in Section 14.1.

The Cayley graph for $\mathrm{Sym}(n)$ with the set of derangements as its connection set is the *derangement graph*. An intersecting family of perfect matchings from $K_{n,n}$ corresponds to a coclique in the derangement graph. It is easy to see that the permutations that map i to j give rise to a coclique of size $(n-1)!$. It also not too difficult to prove that this is the maximum size (you may treat this as an exercise, or refer to Section 14.1), and we refer to cocliques of this type as *canonical cocliques*.

7.5.1 Lemma. *Let B be the incidence matrix for the perfect matchings of $K_{n,n}$ and let z be the characteristic vector of a coclique S in the derangement graph of size $(n-1)!$. If $z \in \mathrm{col}(B)$, then S is canonical.*

Proof. Similar to Lemma 7.4.2, the matchings in S and in the complement $\overline{S}$ form faces in the perfect matching polytope of $K_{n,n}$.

The set of matchings that do not contain a given edge form a face of size $n! - (n-1)! = (n-1)(n-1)!$, and any face of the polytope is an intersection of these faces. It follows that $\overline{S}$ must consist of all perfect matchings that do not contain a given edge and therefore S consists of all perfect matchings that do contain a specified edge. $\qquad\square$

In Section 14.4 we see that the characteristic vector of any coclique of size $(n-1)!$ in the derangement graph must lie in the column space of B, and thus we have a proof of the EKR Theorem for intersecting permutations.

7.6 Exercises

7.1 Let $\mathcal{P}$ be the polytope generated by the rows of $W_{1,k}^T$. Show that if α and β are two sets such that $|\alpha \cap \beta| < k - 1$, then there is no edge of the polytope between the vertices corresponding to α and β in $\mathcal{P}$.

7.2 Find the vertices and the edges in the matching polytope for an even cycle.

7.3 Find the vertices and the edges in the matching polytope for an odd wheel.

Notes

Two standard references for the theory of polytopes are Brønsted [30] and Ziegler [176]. Matching polytopes are treated at length in Lovász and Plummer [118] and in many books on combinatorial optimization (e.g., Schrijver [151], Cook, Cunningham, Pulleyblank and Schrijver [47]). More information about the polytopes associated to eigenspaces of graphs appears in [77, 79, 80].

Our use of polytopes to derive EKR-type theorems is new, and is presented here in print for the first time. The results presented show that this approach provides an effective way to prove EKR Theorems for intersecting families in a range of structures. In Section 8.8 we see that polytopes are useful in characterizing the maximal t-intersecting families in the Johnson scheme.

8

The exact bound in the EKR Theorem

In this chapter we focus on the EKR Theorem for t-intersecting sets. The EKR Theorem in this case is the following.

8.0.1 Theorem. *If $n \geq (t+1)(k-t+1)$, then a t-intersecting family of k-sets has size at most $\binom{n-t}{k-t}$. Further, if $n > (t+1)(k-t+1)$, then the only family of this size consists of all k-sets that contain a fixed t-set.*

It is not difficult to prove this theorem if the condition $n \geq (t+1)(k-t+1)$ is replaced with the condition that n be sufficiently large (we do this in Section 8.2). But the real goal of this chapter is to give the details of Wilson's proof from [172] that the exact lower bound on n is $(t+1)(k-t+1)$.

8.1 A certificate for maximality

As a thought experiment, let us attempt to prove the EKR bound for t-intersecting k-sets using the clique-coclique bound. We will apply this bound to a graph X whose vertices are the k-subsets of $\{1, \ldots, n\}$ and two vertices are adjacent if and only if their intersection has size less than t. This graph is the union of graphs in the Johnson scheme, specifically

$$X = \bigcup_{i=k-t+1}^{k} J(n, k, i).$$

A t-intersecting family of k-sets is a coclique in X (and vice versa). A clique in X is a collection T of k-subsets such that any two distinct members have at most $t-1$ elements in common. For the clique-coclique bound to give a tight bound on the size of a coclique, we would need that

$$|T| = \frac{\binom{n}{k}}{\binom{n-t}{k-t}} = \frac{\binom{n}{t}}{\binom{k}{t}}.$$

Since each set in T contains exactly $\binom{k}{t}$ subsets of size t and none of these t-subsets can be contained in more than one set of T, this implies that each t-subset lies in exactly one element of T. In other words T is a t-$(n, k, 1)$ design. So we can immediately conclude that if a t-$(n, k, 1)$ design exists, then a t-intersecting family of subsets has size at most $\binom{n-t}{k-t}$. (As 1-$(n, k, 1)$ designs exist if k divides n, we obtain the simple proof of the EKR bound for when k divides n that was given in Section 6.1.)

To get further we consider the proof of the clique-coclique bound. Suppose S is a t-intersecting family of k-subsets with characteristic vector x, and let M be the projection of xx^T onto the Bose-Mesner algebra of $J(n, k)$. Then using (3.7.1) we have that

$$|S| = \frac{\mathrm{sum}(M)}{\mathrm{tr}(M)}.$$

If y is the characteristic vector of a t-$(n, k, 1)$ design and N is the projection of yy^T onto the Bose-Mesner algebra of the scheme, then $M \circ N$ is a scalar multiple of I. Lemma 3.8.1 yields that

$$|S| \le \binom{n}{k} \frac{\mathrm{tr}(N)}{\mathrm{sum}(N)}$$

with equality holding if and only if $MN = dJ$ for some d. If we use the previous inequality, rather than the clique-coclique bound, then we can derive the EKR bound without needing an actual design. If there is a positive semidefinite matrix N in the Bose-Mesner algebra such that $M \circ N = cI$ and $MN = dJ$, then S must be a t-intersecting family of maximal size. The mere existence of such an N certifies that S has maximum size. Thus we call such a matrix N a *certificate* for the Johnson scheme.

Our next result translates our question about the existence of N into a question about the existence of a suitable polynomial.

8.1.1 Theorem. *Let $\mathcal{A}$ be a metric scheme with d classes and let S be a subset of the vertices of the scheme with width $d - t$ and characteristic vector x. Then S has the maximum possible size for a subset of width $d - t$ if there is a polynomial q of degree at most $t - 1$ such that:*

(a) $q(A_1)A_d + I$ *is positive semidefinite;*
(b) *if $j \ne 0$ and $E_j x \ne 0$, then $(q(A_1)A_d + I)E_j = 0$.*

Proof. Let M be the projection of xx^T onto $\mathbb{C}[\mathcal{A}]$ and suppose

$$N = q(A_1)A_d + I.$$

Since $\mathcal{A}$ is metric, $q(A_1)A_d$ lies in the span of $A_{d-t+1}, \ldots, A_d$ and $M \circ N = \alpha I$ for some α. It follows from (b) that $NE_jM = 0$ if $j \neq 0$ and so $MN = \beta J$ for some β. Therefore by Lemma 3.8.1 we have that

$$\frac{\text{sum}(M)}{\text{tr}(M)} \frac{\text{sum}(N)}{\text{tr}(N)} = v.$$

Now suppose S_1 is an arbitrary subset of the vertices of $\mathcal{A}$ with width $d - t$ and characteristic vector y. Let M_1 be the projection of yy^T onto $\mathbb{C}[\mathcal{A}]$. Then $M_1 \circ N$ is a scalar matrix whence Lemma 3.8.1 yields that

$$\frac{\text{sum}(M_1)}{\text{tr}(M_1)} \frac{\text{sum}(N)}{\text{tr}(N)} \leq v.$$

Since

$$|S_1| = \frac{\text{sum}(M_1)}{\text{tr}(M_1)},$$

we conclude that S has maximum size for a subset of width $d - t$. $\qquad \square$

If there are t distinct indices j such that $E_jx \neq 0$, then the conditions in (b) exactly determine the polynomial q. In fact, we can build this polynomial for the Johnson scheme.

Recall that the *Lagrange interpolating polynomials* $\ell_1, \ldots, \ell_t$ on the set $\{\theta_1, \ldots, \theta_t\}$ are the polynomials of degree t such that

$$\ell_i(\theta_j) = \begin{cases} 1, & \text{if } j = i; \\ 0, & \text{otherwise.} \end{cases}$$

Thus

$$\ell_i(x) = \prod_{j \in \{1, \ldots, t\} \setminus i} \frac{x - \theta_j}{\theta_i - \theta_j},$$

and if q is a polynomial of degree less than t, then

$$q = \sum_{i=1}^{t} q(\theta_i)\ell_i.$$

8.1.2 Corollary. *Let θ_i and σ_i denote the eigenvalues of A_1 and A_k respectively on the ith eigenspace of $J(n, k)$, and let $\ell_1, \ldots, \ell_t$ be the Lagrange interpolating polynomials on $\theta_1, \ldots, \theta_t$. Then a t-intersecting family in $J(n, k)$ has size at most $\binom{n-t}{k-t}$ if*

$$\sigma_j \sum_{i=1}^{t} \sigma_i^{-1} \ell_i(\theta_j) \leq 1, \tag{8.1.1}$$

for $j = t + 1, \ldots, k$.

Proof. Let S be a canonical t-intersecting set in $J(n, k)$ and let x be its characteristic vector. By Lemma 3.6.1 we have that the projection of xx^T into the Bose-Mesner algebra is

$$M = \frac{t!(n-t)!}{n!} W_{t,k}^T W_{t,k}$$

whence $ME_j \neq 0$ if $0 \leq j \leq t$.

Assuming that (8.1.1) holds we show, using Theorem 8.1.1, that M has the maximum possible size for a subset of width $d - t$. To do this we define a polynomial q as

$$q = -\sum_{i=1}^{t} \sigma_i^{-1} \ell_i.$$

We need to prove that q satisfies the following three conditions: it has degree at most $t - 1$, $q(A_1)A_d + I$ is positive semidefinite, and $(q(A_1)A_d + I)E_j = 0$ for all $E_j x \neq 0$ with $j \neq 0$.

It is clear from the definition of q that its degree is $t - 1$. If z lies in the jth eigenspace of $J(n, k)$, then

$$q(A_1)A_k z = q(\theta_j)\sigma_j z.$$

So the eigenvalues of $q(A_1)A_d + I$ are

$$q(\theta_j)\sigma_j + 1 = -\sigma_j \sum_{i=1}^{t} \sigma_i^{-1} \ell_i(\theta_j) + 1$$

and by (8.1.1) these are non-negative. Hence $q(A_1)A_d + I$ is positive semidefinite. Further, for any $0 \leq j \leq t$,

$$(q(A_1)A_d + I)E_j = (q(\theta_j)\sigma_j + 1)E_j = \left(-\sigma_j \sum_{i=1}^{t} \sigma_i^{-1} \ell_i(\theta_j) + 1\right) E_j = 0.$$

Therefore the second condition of Theorem 8.1.1 is also met and we have the result. $\qquad\square$

We can also state Theorem 8.1.1 in terms of the existence of a matrix. We call such a matrix a *certificate* for the association scheme.

8.1.3 Corollary. *Let $\mathcal{A}$ be a metric association scheme with d classes, and let S be a subset of the vertices of the scheme with width $d - t$ and characteristic vector x. Then S has the maximum possible size for a subset of width $d - t$ if there is a matrix N such that:*

    (a) N is positive semidefinite;
    (b) if $j \neq 0$ and $E_j x \neq 0$, then $NE_j = 0$. $\qquad\square$

8.2 Special cases

In this section we consider two special cases where Corollary 8.1.2 can be used to prove the EKR Theorem. First we consider when $t = 1$, and second we consider when n is large relative to k.

If S is a subset of vertices from the Johnson scheme $J(n, k)$ with width $(d - 1)$, then S is an intersecting k-set system. By the EKR Theorem the size of the set S is maximum, so by Theorem 8.1.1 there exists a constant γ such that $\gamma A_k + I$ is positive semidefinite. We can find a suitable γ as follows. If τ is the least eigenvalue of A_k, then $-\tau^{-1} A_k + I \succcurlyeq 0$ and its least eigenvalue is zero and belongs to the E_1 eigenspace. So we have established that the collection of all k-sets that contain a given element form an intersecting family of maximum size or, more concretely, that an intersecting family has size at most $\binom{n-1}{k-1}$. The brevity of this argument gives some indication of the power of the tools at hand.

In the next result we prove the EKR bound for all t, provided that n is sufficiently large. Recall from Section 6.3 that the eigenvalues of the Johnson graph X_1 in $J(n, k)$ are

$$\theta_i = k(n - k) - i(n + 1 - i),$$

for $i = 0, \ldots, k$, and from Section 6.5 the eigenvalues of the Kneser graph are

$$\sigma_i = (-1)^i \binom{n - k - i}{k - i}$$

where $i = 0, \ldots, k$.

8.2.1 Lemma. *If n is sufficiently large, then a t-intersecting family of k-subsets has size at most*

$$\binom{n - t}{k - t}.$$

Proof. We are concerned about the size of $\ell_i(\theta_r)$ where $1 \leq i \leq t$ and $t + 1 \leq r \leq k$. Since

$$\theta_i - \theta_j = (j - i)(n + 1 - i - j)$$

we find that

$$\ell_i(\theta_r) = \prod_{j \in \{1, \ldots, t\} \setminus i} \frac{\theta_r - \theta_j}{\theta_i - \theta_j}$$

$$= \prod_{j \in \{1, \ldots, t\} \setminus i} \frac{(j - r)(n + 1 - r - j)}{(j - i)(n + 1 - i - j)}$$

$$\leq \prod_{j \in \{1, \ldots, t\} \setminus i} \frac{(j - r)}{(j - i)}.$$

From this we see that

$$|\ell_i(\theta_r)| \le \alpha r^{t-1},$$

where α does not depend on n.

Next, assuming $r > i$,

$$\frac{\sigma_r}{\sigma_i} = (-1)^{r-i} \frac{\binom{n-k-r}{k-r}}{\binom{n-k-i}{k-i}} = (-1)^{j-i} \frac{\binom{k-i}{r-i}}{\binom{n-k-i}{r-i}}.$$

Whence

$$\left| \frac{\sigma_r}{\sigma_i} \right| \le \beta(n-k-r)^{i-r} \le \beta(n-2k)^{-1}$$

where β is independent of n. Combining our two bounds we deduce that

$$\left| \frac{\sigma_r}{\sigma_i} \ell_i(\theta_r) \right| \le \alpha\beta \frac{r^{t-1}}{n-2k}$$

and, since $r \le k$, we conclude that when n is large enough,

$$\sigma_r \sum_{i=1}^{t} \sigma_i^{-1} \ell_i(\theta_r) \le 1.$$

By Corollary 8.1.2, our result follows. □

8.3 The EKR bound for 2-intersecting sets

Using Corollary 8.1.2, we can prove that the bound in the EKR Theorem holds for $t = 2$ provided that $n \ge 3(k-1)$. This is the exact lower bound needed on n for the EKR bound to hold when $t = 2$. Throughout this section, we assume that $t = 2$.

We will need the following two eigenvalues of the Johnson graph $J(n,k)$:

$$\theta_1 = k(n-k) - n, \quad \theta_2 = k(n-k) - 2(n-1),$$

and the eigenvalues of the Kneser graph $K(n,k)$:

$$\sigma_1 = -\binom{n-k-1}{k-1}, \quad \sigma_2 = \binom{n-k-2}{k-2}.$$

Note that θ_1 are σ_1 both correspond to the eigenspace of the Johnson scheme with dimension $n-1$, whereas θ_2 and σ_2 both belong to the eigenspace with dimension $\binom{n}{2} - 1$.

The Lagrange interpolating polynomials that we use in this case are

$$\ell_1 = \frac{x - \theta_2}{\theta_1 - \theta_2}, \quad \ell_2 = \frac{x - \theta_1}{\theta_2 - \theta_1}.$$

Since $\theta_1 - \theta_2 = n - 2$, we see that the EKR bound holds provided that

$$(-1)^j \binom{n-k-j}{k-j} \left(-\frac{1}{\binom{n-k-1}{k-1}} \frac{\theta_j - \theta_2}{n-2} - \frac{1}{\binom{n-k-2}{k-2}} \frac{\theta_j - \theta_1}{n-2} \right) \leq 1, \quad (8.3.1)$$

for all $1 \leq j \leq k$. This inequality clearly holds for $j = 1$ and $j = 2$. Further, since $\theta_j < \theta_2 < \theta_1$, this inequality holds if $\sigma_j < 0$ (so whenever j is odd). Therefore the first interesting case is when $j = 4$. Our inequality then becomes

$$\frac{\theta_2 - \theta_4}{\binom{n-k-1}{k-1}} + \frac{\theta_1 - \theta_4}{\binom{n-k-2}{k-2}} \leq \frac{n-2}{\binom{n-k-4}{k-4}},$$

and this holds if and only if

$$\theta_2 - \theta_4 + \frac{n-k-1}{k-1}(\theta_1 - \theta_4) \leq (n-2)\frac{\binom{n-k-1}{k-1}}{\binom{n-k-4}{k-4}}.$$

The left side is equal to

$$\theta_2 - \theta_1 + (\theta_1 - \theta_4) + \frac{n-k-1}{k-1}(\theta_1 - \theta_4) = 2 - n + \frac{n-2}{k-1}(\theta_1 - \theta_4)$$

$$= (n-2)\frac{3n-k-11}{k-1}$$

and thus we can rewrite our inequality as

$$3n - k - 11 \leq (k-1)\frac{\binom{n-k-1}{3}}{\binom{k-1}{3}}.$$

Denote the left side of this inequality by $L(n)$ and the right side by $R(n)$. It is easy to check that when $n = 3k - 3$, both $L(n)$ and $R(n)$ are equal to $4(2k - 5)$. Since

$$\frac{L(n+1)}{L(n)} = \frac{3n-k-14}{3n-k-11} = 1 - \frac{3}{3n-k-11}$$

and

$$\frac{R(n+1)}{R(n)} = \frac{n-k}{n-k-3} = 1 + \frac{3}{n-k-3},$$

we conclude that $L(n) \leq R(n)$ if $n \geq 3k - 3$ and $k \geq 3$.

We have shown that (8.3.1) holds for $j \in \{1, 2, 3, 4\}$, but we need to show that the equation holds for all $j \leq k$. If we set

$$f(j) = (-1)^j \binom{n-k-j}{k-j}((k-1)(\theta_2 - \theta_j) + (n-k-1)(\theta_1 - \theta_j)),$$

then (8.3.1) is equivalent to

$$\frac{1}{(n-k-1)(n-2)}\binom{n-k-2}{k-2}^{-1}f(j) \le \binom{n-k-2}{k-2}.$$

We have shown that this inequality holds for $j = 2, 4$ and for any odd j. If we can prove that $|f(j+1)| \le |f(j)|$ for $j \ge 4$ when $n \ge 3(k-1)$, then (8.3.1) holds for all j. The inequality $|f(j+1)| \le |f(j)|$ can be expanded to

$$(k-1)((j+1)(n-j)-2(n-1)) + (n-k-1)((j+1)(n-j)-n)$$
$$\le \frac{n-k-j}{k-j}((k-1)(j(n+1-j)-2(n-1))$$
$$+ (n-k-1)(j(n+1-j)-n)),$$

which can (with a fair amount of work!) be reduced to

$$(n-2)((k-j)n-(n-2j)j)$$
$$\le (n-2k)(j(n-j)(n-2) - n(n+k-3) + 2(k-1)).$$

If $n \ge 3(k-1)$, then for all $4 \le q \le k \le n/2$, the left-hand side of this equation is a function that is decreasing in j and the right-hand side is a function that is increasing in j. It is sufficient to show that the inequality holds for $q = 4$. This reduces to showing that the polynomial

$$3n^3 + n^2(-8k-13) + n(2k^2+46k-18) + (-4k^2-60k+64)$$

is positive for $n \ge 3(k-1)$. It is easy to see that this holds for all n if $k = 2$. Some calculation shows that the polynomial is positive for all k if $n = 3(k-1)$. Also, if $n \ge 3(k-1)$ and $k > 2$, then the derivative is positive, so the polynomial is positive for all $n > 3(k-1)$. This shows that the function $f(j)$ is decreasing for $j \ge 4$, so (8.3.1) holds for all j.

8.4 Wilson's matrix

We introduce a matrix defined by Wilson in [172] and show that it will serve as a certificate for the EKR problem for t-intersecting sets. In Section 6.5 we defined a matrix $D_i = W_{i,k}^T \overline{W}_{i,k}$ and showed that it is symmetric. Following Wilson we define

$$\Omega(n,k,t) = \sum_{i=0}^{t-1} (-1)^{t-1-i} \binom{k-1-i}{k-t} \binom{n-k-t+i}{k-t}^{-1} D_{k-i}.$$

Our goal is to prove the EKR Theorem for t-intersecting families of k-subsets using Theorem 8.1.1 and Corollary 8.1.3. We will show that if we take S in Corollary 8.1.3 to be a canonical t-intersecting set from $J(n, k)$, then the matrix $N = \Omega(n, k, t) + I$ satisfies the conditions of Corollary 8.1.3. Specifically, we need to show that N is positive semidefinite and that $NE_j = 0$ for all $j \in \{1, \ldots, t\}$. If we let $\theta_0, \ldots, \theta_k$ be the eigenvalues of $\Omega(n, k, t)$, then the following theorem implies that these two conditions hold.

8.4.1 Theorem. *If* $1 \le t \le k$ *and* $n \ge 2k$, *then:*

(a) $\theta_0 = \binom{n}{k}\binom{n-t}{k-t}^{-1} - 1$;

(b) $\theta_i = -1$ *for* $i = 1, \ldots, t$;

(c) *if* $n \ge (t + 1)(k - t + 1)$, *then* $\theta_i \ge -1$ *for all* i;

(d) *if* $t + 1 \le r \le k$, *then* $\theta_r > -1$ *unless either* $t = 1$ *and* $n = 2k$ *or* $n = (t + 1)(k - t + 1)$ *and* $r = t + 2$.

Parts (a) and (b) of the previous theorem are straightforward consequences of the following lemma.

8.4.2 Lemma. *For* $1 \le t \le k$ *and* $n \ge 2k$

$$W_{t,k}\Omega(n, k, t) = J - W_{t,k}.$$

Proof. First note that

$$W_{t,k}\overline{W}_{k-i,k}^T = \binom{n - k - t + i}{k - t}\overline{W}_{t,k-i},$$

and hence

$$W_{t,k}D_{k-i} = W_{t,k}\overline{W}_{k-i,k}^T W_{k-i,k} = \binom{n - k - t + i}{k - t}\overline{W}_{t,k-i}W_{k-i,k}.$$

By definition,

$$\binom{-a}{b} = (-1)^b\binom{a + b - 1}{b}.$$

Using this we find that

$$W_{t,k}\Omega(n, k, t) = \sum_{i=0}^{t-1}\binom{t - k - 1}{t - i - 1}\overline{W}_{t,k-i}W_{k-i,k}.$$

If α is a t-subset and β a k-subset then

$$(W_{t,k}\Omega(n, k, t))_{\alpha,\beta} = \sum_{i=0}^{t-1}\binom{t - k - 1}{t - i - 1}\binom{|\alpha \setminus \beta|}{k - i}.$$

If $\alpha \subseteq \beta$, then $|\alpha \setminus \beta| < k - i$ and the sum just given is zero. If $\alpha \not\subseteq \beta$, then the sum is equal to

$$\sum_{i=0}^{t-1} \binom{t-k-1}{t-i-1}\binom{k-|\alpha \cap \beta|}{i-|\alpha \cap \beta|} = \binom{t-1-|\alpha \cap \beta|}{t-1-|\alpha \cap \beta|} = 1. \qquad \square$$

At this point, we warn the reader that the proof of part (c) is considerably more involved (the proof of part (d) will follow from this work) and requires several pages of calculations. Hence we give the proofs of these last two parts their own section.

8.5 Eigenvalues of Wilson's matrix

From Theorem 6.5.1 we deduce that the eigenvalue θ_r of $\Omega(n, k, t)$ on the rth eigenspace of $J(n, k)$ is equal to

$$(-1)^{t-1-r} \sum_{i=0}^{t-1} (-1)^i \binom{k-1-i}{k-t}\binom{k-r}{i}\binom{n-k-r+i}{k-r}\binom{n-k-t+i}{k-t}^{-1}.$$

To complete part (c) of Theorem 8.4.1, we need to show that $\theta_r > -1$ for $r \in \{t+1, \ldots, d\}$ (the proof of part (d) will come out as we prove part (c)).

Our proof will proceed as follows. First we show that θ_r is a decreasing function in n. Next we show that if $n = (t+1)(k-t+1)$ then $\theta_{t+2} = -1$. To complete the result, we show that the absolute value of the eigenvalues θ_{t+j} strictly decrease as j increases.

We start by proving a technical lemma.

8.5.1 Lemma. *Define*

$$\delta_s(x, y; a, b) = \sum_{i=0}^{s} (-1)^i \binom{s}{i}\binom{x-i}{a}\binom{y+i}{b}^{-1}.$$

Then if $r \geq t$,

$$\theta_r = (-1)^{t-1-r} \binom{r-1}{t-1} \delta_{t-1}(k-1, n-k-t; r-1, r-t). \qquad (8.5.1)$$

Proof. We have that

$$\binom{k-1-i}{k-t}\binom{k-r}{i}\binom{n-k-r+i}{k-r}\binom{n-k-t+i}{k-t}^{-1}$$

is equal to

$$\frac{(k-1-i)!(k-r)!(n-k-r+i)!(k-t)!(n-2k+i)!}{(k-t)!(t-1-i)!i!(k-r-i)!(k-r)!(n-2k+i)!(n-k-t+i)!}$$

which can be reduced to

$$\frac{(k-1-i)!(n-k-r+i)!}{(t-1-i)!i!(k-r-i)!(n-k-t+i)!}.$$

If $r \geq t$, then

$$\frac{(k-1-i)!(n-k-r+i)!}{(t-1-i)!i!(k-r-i)!(n-k-t+i)!}$$

equals to

$$\frac{(k-1-i)!(n-k-r+i)!(r-t)!(r-1)!(t-1)!}{(r-1)!(k-r-i)!(n-k-t+i)!i!i!(t-1-i)!(r-t)!(t-1)!}$$

which can be reduced to

$$\binom{k-1-i}{r-1}\binom{n-k-t+i}{r-t}^{-1}\binom{r-1}{t-1}\binom{t-1}{i}.$$

The stated expression for θ_r follows directly from these facts. □

8.5.2 Lemma. *If $r \geq t$, then $(-1)^{t-1-r}\theta_r \geq 0$. If k, t and r are fixed and $r > t$, then $|\theta_r|$ is a strictly decreasing function of n.*

Proof. First, using Pascal's formula,

$$\delta_s(x,y;a,b) = \delta_{s-1}(x,y;a,b) - \delta_{s-1}(x-1,y+1;a,b)$$
$$= (\delta_{s-1}(x,y;a,b) - \delta_{s-1}(x-1,y;a,b))$$
$$\quad + (\delta_{s-1}(x-1,y;a,b) - \delta_{s-1}(x-1,y+1;a,b))$$
$$= \delta_{s-1}(x-1,y;a-1,b) + \frac{b}{b+1}\delta_{s-1}(x-1,y+1;a,b+1).$$

Iterating this yields

$$\delta_s(x,y;a,b) = \sum_{i=0}^{s}\binom{s}{i}\frac{b}{b+i}\delta_0(x-s,y+i;a-s+i,b+i)$$
$$= \sum_{i=0}^{s}\frac{b}{b+i}\binom{s}{i}\binom{x-s}{a-s+i}\binom{y+i}{b+i}^{-1}.$$

It follows that $\delta_s(x,y;a,b) \geq 0$, which yields the sign of θ_r. If $s \leq a \leq x$ and $0 < b \leq y$, then $\delta_s(x,y;a,b) > \delta_s(x,y+1;a,b)$, which yields that $|\theta_r|$ is a strictly decreasing function of n. □

The next result is that $\theta_{t+2} = -1$ if $n = (t+1)(k-t+1)$. If $n < (t+1)(k-t+1)$, then $\theta_{t+2} < -1$ and the EKR Theorem does not hold. We saw this result in Lemma 1.2.1. If $\mathcal{F}_1$ denotes the family of k-subsets of an n-set that meet some chosen subset of size $t + 2$ in at least $t + 1$ points, then $|\mathcal{F}_1| = \binom{n-t}{k-t}$ when $n = (t + 1)(k - t + 1)$ and if $n < (t + 1)(k - t + 1)$, then $|\mathcal{F}_1| > \binom{n-t}{k-t}$.

8.5.3 Lemma. *If $t + 2 \leq k$ and $n = (t + 1)(k - t + 1)$, then $\theta_{t+2} = -1$.*

Proof. To prove this, we construct an eigenvector that lies in the eigenspace corresponding to the idempotent E_{t+2}. This construction is motivated by the counterexample, $\mathcal{F}_1$, to the EKR Theorem for $n < (t + 1)(k - t + 1)$ given earlier. Let φ be the characteristic vector of $\mathcal{F}_1$. Although φ is not itself an eigenvector in this eigenspace, the projection $E_{t+2}\varphi$ clearly is. We need to calculate the value of $\Omega(n, k, t)E_{t+2}\varphi$. To do this, we use a related matrix.

Set

$$M = I + \Omega(n, k, t) - \binom{n-t}{k-t}^{-1} J$$

and we calculate $\varphi^T M \varphi$ in two different ways.

First, since any two sets from $\mathcal{F}_1$ intersect in at least t points, we have that $\varphi^T \Omega(n, k, t)\varphi = 0$. Since the size of $\mathcal{F}_1$ is exactly $\binom{n-t}{k-t}$, this implies that $\varphi^T M \varphi = 0$.

Next we calculate $\varphi^T M \varphi$ using the fact that $\varphi = \sum_{i=0}^{k} E_i\varphi$. We break our calculations into the following three claims.

Claim (a) Only the first $t + 2$ terms of the summation $\varphi = \sum_{i=0}^{k} E_i\varphi$ are nonzero.

Since φ is the characteristic vector of $\mathcal{F}_1$, it is equal to the sum of the rows of $W_{t+1,k}$ indexed by the $(t + 1)$-subsets of the $(t + 2)$-set used to define $\mathcal{F}_1$, minus $t + 1$ times the row of $W_{t+2,k}$ corresponding to the $(t + 2)$-subset. It follows that the characteristic vectors of the families isomorphic to $\mathcal{F}_1$ are the rows of

$$F = W_{t+1,t+2}^T W_{t+1,k} - (t + 1)W_{t+2,k}. \tag{8.5.2}$$

Therefore φ lies in the row space of $W_{t+2,k}$ (by Lemma 6.2.3 the row space of $W_{t+1,k}$ is contained in the row space of $W_{t+2,k}$), and so $E_j\varphi = 0$ if $j > t + 2$.

Claim (b) $E_{t+2}\varphi \neq 0$.

If $E_{t+2}\varphi = 0$, then φ would belong to the row space of $W_{t+1,k}$. Then from (8.5.2), a row of $W_{t+2,k}$ would belong to the row space of $W_{t+1,k}$ and therefore (by symmetry) all rows of $W_{t+2,k}$ would lie in the row space of $W_{t+1,k}$. But

$$\text{rk}(W_{t+2,k}) = \binom{n}{t+2} > \binom{n}{t+1} = \text{rk}(W_{t+1,k})$$

and so $E_{t+2}\varphi \neq 0$.

Claim (c) $E_{t+1}\varphi = 0$.

To prove this, we use (8.5.2) and note that

$$\overline{W}_{t+2,k}^T F = \overline{W}_{t+2,k}^T W_{t+1,t+2}^T W_{t+1,k} - (t+1)\overline{W}_{t+2,k}^T W_{t+2,k}$$

$$= (W_{t+1,t+2}\overline{W}_{t+2,k})^T W_{t+1,k} - (t+1)\overline{W}_{t+2,k}^T W_{t+2,k}$$

$$= ((n-k-t-1)\overline{W}_{t+1,k})^T W_{t+1,k} - (t+1)\overline{W}_{t+2,k}^T W_{t+2,k}$$

$$= (n-k-t-1)D_{t+1} - (t+1)D_{t+2}.$$

From Theorem 6.5.1, the eigenvalues of D_{t+1} and D_{t+2} on the eigenspace corresponding to E_{t+1} are

$$\lambda^{t+1} = (-1)^{t+1}\binom{n-2t-2}{k-t-1}, \quad \lambda^{t+2} = (-1)^{t+1}\binom{k-t-1}{1}\binom{n-2t-3}{k-t-1}.$$

Therefore we have that

$$\frac{\lambda^{t+2}}{\lambda^{t+1}} = \frac{(k-t-1)(n-k-t-1)}{n-2t-2}.$$

Hence the eigenvalue of $\overline{W}_{t+2,k}^T F$ on eigenspace corresponding to E_{t+1} is

$$(n-k-t-1)\lambda^{t+1}\left(1 - \frac{(t+1)(k-t-1)}{n-2t-2}\right).$$

This eigenvalue is zero (for $n > k+t+1$) if and only if

$$n = 2t+2 + (t+1)(k-t-1) = (t+1)(k-t+1).$$

We conclude that if $n = (t+1)(k-t+1)$ then $E_{t+1}\varphi = 0$.

With these three claims in hand we can now prove the lemma. Using the eigenvalues for $\Omega(n,k,t)$ given in Theorem 8.4.1 it is clear that $ME_i = 0$ for $i = 0, \ldots, t$. Therefore

$$M\varphi = M\sum_{i=0}^{k} E_i\varphi = \sum_{i=0}^{t} ME_i\varphi + ME_{t+2}\varphi = ME_{t+2}\varphi = E_{t+2}\varphi + \theta_{t+2}E_{t+2}\varphi.$$

which means that

$$\varphi^T M\varphi = (1 + \theta_{t+2})\|E_{t+2}\varphi\|^2.$$

We have already seen that $\varphi^T M\varphi = 0$, hence $\theta_{t+2} = -1$. $\square$

Since $|\theta_{t+2}|$ decreases strictly as n increases, when $n > (t+1)(k-t+1)$ we know that $|\theta_{t+2}| < 1$ and, in particular $\theta_{t+2} > -1$. Lemma 8.5.2 shows that

$\theta_{t+1} > 0$, so the final step is to prove that $\theta_{t+j} > -1$ for all $j \in \{1, \ldots, d - t - 1\}$. To do this we show that the absolute value of θ_{t+j} decreases as j increases.

8.5.4 Lemma. *For $j \in \{1, \ldots, d - t - 1\}$ we have that $|\theta_{t+j}| > |\theta_{t+j+1}|$.*

Proof. Suppose

$$\alpha_i = \frac{j}{j+i} \binom{t-1+j}{j} \binom{t-1}{i} \binom{k-t}{j+i} \binom{n-k-t+i}{j+i}^{-1}$$

and assume β_i is obtained by replacing j by $j + 1$ in this definition. From the expression for $\delta_s(x, y; a, b)$ in the proof of Lemma 8.5.2 and from (8.5.1) we have

$$|\theta_{t+j}| = \sum_{i=0}^{t-1} \alpha_i, \qquad |\theta_{t+j+1}| = \sum_{i=0}^{t-1} \beta_i.$$

Now it is sufficient to show that $\alpha_i \geq \beta_i$ for $j \geq 2$. Note that

$$\begin{aligned}
\frac{\alpha_i}{\beta_i} &= \left(\frac{j(j+i+1)}{(j+1)(j+i)} \right) \left(\frac{j+1}{t+j} \right) \left(\frac{j+i+1}{k-t-j-i} \right) \left(\frac{n-k-t-j}{j+i+1} \right) \\
&= \frac{j(j+i+1)(n-k-t-j)}{(j+t)(j+i)(k-t-j-i)} \\
&> \frac{j(n-k-t-j)}{(j+t)(k-t-j)}.
\end{aligned}$$

If $n \geq (t + 1)(k - t + 1)$, then

$$\begin{aligned}
n - k - t - j &\geq (t+1)k - t^2 + 1 - k - t - j \\
&= tk - t^2 - (t + j - 1) \\
&\geq tk - t^2 - tj \\
&= t(k - t - j).
\end{aligned}$$

Hence

$$\frac{\alpha_i}{\beta_i} > \frac{j(n-k-t-j)}{(j+t)(k-t+j)} \geq \frac{jt}{j+t} \geq 1.$$

The last inequality holds since $jt \geq j + t$ (recall that we are considering only $t \geq 2$ and $j \geq 2$). It follows that $|\theta_{t+j}| > |\theta_{t+j+1}|$. $\square$

This completes the proof of Theorem 8.4.1.

8.6 Width and dual width

Let C be a set of vertices in a metric association scheme $\mathcal{A}$. The *degree* of C is defined to be the number of nonidentity relations that occur between members of C. The *width* of C is defined to be the maximum distance in X_1 between two vertices of C. A clique in X_1 is a subset of width 1, and the set consisting of a vertex and its neighbors has width at most 2. If $\mathcal{A}$ is metric with d classes and $T = \{d - t + 1, \ldots, d\}$, then a subset with width $d - t$ in $\mathcal{A}$ is a T-coclique. The width of C is an upper bound on its degree.

If C is a set of vertices in an association scheme $\mathcal{A}$, let x be the characteristic vector of C. Then, for each i, we see that $x^T A_i x$ is the number of ordered pairs of vertices from C that are i-related. So the degree of C is the number of i such that $x^T A_i x$ is nonzero and, if the association scheme is metric, the width is the maximum i such that $x^T A_i x$ is nonzero. This motivates the definition of the dual degree and dual width.

If $\mathcal{A}$ is an association scheme and C is a subset of the vertices of $\mathcal{A}$ with characteristic vector x, we recall from Section 4.5 that its *dual degree* is the number of matrices E_j in the scheme such that $x^T E_j x \neq 0$. If our scheme is cometric, so we have an ordering of the idempotents E_j, the *dual width* is the maximum value of j such that $x^T E_j x \neq 0$. Clearly, the dual width of C is an upper bound on its dual degree.

For example, let C be the set of vertices formed by the k-subsets of $\{1, \ldots, n\}$ that contain $\{1, \ldots, t\}$ and let x denote its characteristic vector. In Lemma 3.6.1 we found that the orthogonal projection of xx^T onto the Bose-Mesner algebra of $J(n, k)$ is $\binom{n}{t}^{-1} W_{t,k}^T W_{t,k}$. Hence we have

$$\sum_i \frac{x^T A_i x}{v v_i} A_i = \sum_j \frac{x^T E_j x}{m_j} E_j = \frac{1}{\binom{n}{t}} W_{t,k}^T W_{t,k}.$$

By Corollary 6.3.4, this implies that $x^T E_j x = 0$ if $j > t$ and thus the dual width of C is t.

8.6.1 Theorem. *Let $\mathcal{A}$ be a metric scheme with d classes. If C is a subset of the vertices of the scheme with width w and dual degree s^*, then*

$$w + s^* \geq d.$$

Proof. Let x be the characteristic vector of C. By Theorem 3.5.1 we have

$$\sum_{i=0}^d \frac{x^T A_i x}{v v_i} A_i = \sum_{i=0}^d \frac{x^T E_i x}{m_i} E_i. \tag{8.6.1}$$

Since C has width w and since A_i is a polynomial in A_1 with degree i, there is a real polynomial f with degree w such that

$$f(A_1) = \sum_{i=0}^d \frac{x^T A_i x}{v v_i} A_i.$$

Since we can express A_1 as

$$A_1 = \sum_{i=0}^{d} p_1(i)E_i,$$

we also have that

$$f(A_1) = \sum_{i=0}^{d} f(p_1(i))E_i.$$

Comparing this with the right-hand side of (8.6.1), we find that

$$f(p_1(i)) = \frac{x^T E_i x}{m_i},$$

for $i = 0, \ldots, d$. Since f has degree w, the number of eigenvalues of A_1 on which f vanishes is at most w, and so $w \geq d - s^*$ and the inequality holds.   □

Now for the cometric version. (The proof of the bound is entirely analogous to the proof in the metric case; we leave it as an exercise.)

8.6.2 Theorem. *Let $\mathcal{A}$ be a cometric scheme with d classes. If C is a subset of the vertices of the scheme with dual width w^* and degree s, then*

$$w^* + s \geq d.$$ □

From our previous discussions we have the inequalities

$$s \leq w, \quad s^* \leq w^*,$$

whence we have

$$w + w^* \geq d.$$

Further, if $w + w^* = d$, then $w = s$ and $w^* = s^*$. We say that a subset of the vertices of an association scheme that is both metric and cometric is *narrow* if $w + w^* = d$. Such sets play an important role in the theory of distance-regular graphs; we pay them more attention in Section 8.10.

8.7 Equality in the width bound

We have established the EKR bound for the size of a t-intersecting family of k-sets; now our problem is to characterize the families that meet this bound. From the previous section, we have the inequality

$$w + s^* \geq d.$$

We consider what we can say when this bound is tight.

Recall from the introduction to Chapter 4 that a subset of the vertices of a metric is completely regular if its distance partition is equitable. Further, if C is a subset of vertices in a metric association scheme, then the covering radius of C is the maximum distance between C and any vertex in the association scheme. The outer distribution matrix of C records for every vertex in the association scheme the number of elements in C that are i-related to it (see Section 4.5 for these definitions).

8.7.1 Theorem. *Suppose $\mathcal{A}$ is a metric scheme, and C is a subset of its vertices with width w and dual degree s^*. If $w = d - s^*$, then C is completely regular.*

Proof. Suppose C has width w and covering radius r, and let B be the outer distribution matrix of C. Assume $|V(\mathcal{A})| = n$ and let e_r denote the rth standard basis vector in $\mathbb{R}^n$. If $\mathrm{dist}(x, C) = \ell$, the distance from x of a vertex in C is at least ℓ and at most $\ell + w$. So $e_x^T B e_i = 0$ if $i < \ell$ or $i > \ell + w$.

Since $w + r \geq d$ and $s^* \geq r$ (see Lemma 4.5.1), we see that if $w + s^* = d$, then $r = s^*$. We select vertices of the metric scheme such that the corresponding rows of B are a basis for its row space.

Choose vertices y and z in C at distance w from one another, and then choose vertices $z_0, \ldots, z_{d-w}$ such that

$$\mathrm{dist}(y, z_\ell) = w + \mathrm{dist}(z_\ell, z) = w + \ell.$$

(Thus $z_0 = z$.) Then $e_{z_\ell}^T B e_i = 0$ if $j < i$ or $i > w + \ell$, and $e_{z_\ell}^T B e_i \neq 0$ if $j \leq \ell \leq w + j$. One consequence of this is that the rows of B indexed by the vertices z_ℓ for $\ell = 0, \ldots, d - w$ are linearly independent, and since $d + 1 - w = 1 + s^* = \mathrm{rk}(B)$ they form a basis for the row space of B.

Now suppose x is a vertex at distance ℓ from C. The matrix formed by the first $d + 1 - w$ columns of the z_ℓ rows of B is upper triangular, and hence $e_x^T B$ is a linear combination of the z_ℓ-rows of B such that the support of the row is a subset of $\{\ell, \ldots, \ell + w\}$. Therefore $e_x^T B$ is a multiple of the z_ℓ-row of B. Since $e_x^T B \mathbf{1} = e_{z_\ell}^T B \mathbf{1} = |C|$, we conclude that $e_x^T B = e_{z_\ell}^T B$. Hence there are exactly $r + 1$ distinct rows in B, and so C is completely regular, by Lemma 4.5.2. $\square$

8.8 Intersecting families of maximum size

We now have the background needed to give the characterization of the maximum t-intersecting sets. Assume $n > (t + 1)(k - t + 1)$ and let C be a t-intersecting family of k-subsets of $\{1, \ldots, n\}$; equivalently C is a set of vertices in the Johnson scheme with width $k - t$. In this section we show that if $n > (t + 1)(k - t + 1)$, then C is canonical.

8.8.1 Lemma. *Let $n > (t + 1)(k - t + 1)$ and C be a set of vertices in the Johnson scheme with width $k - t$. If $|C| = \binom{n-t}{k-t}$, then the dual width of C is t.*

Proof. Recall the matrix

$$M = I + \Omega(n, k, t) - \binom{n - t}{k - t}^{-1} J$$

defined in the proof of Lemma 8.5.3. From Theorem 8.4.1, the eigenvalues of M are $\theta_i = 0$ for $i = 0, \ldots, t$ and $\theta_i > 0$ for $i = t + 1, \ldots, k$. The eigenspaces of M are the eigenspaces of the Johnson scheme.

If x is the characteristic vector of C, then $x^T M x = 0$, and since M is positive semidefinite this implies that $Mx = 0$. Hence x is in the kernel of M. Therefore x is in the span of the first t eigenspaces and C has dual width t. $\square$

For the remainder of this chapter, we assume that C has width $k - t$ and $|C| = \binom{n-t}{k-t}$. Hence $w + s^* = d$ and C is a completely regular subset. Further, if we use x to denote the characteristic vector of C, the fact that $s^* = w^*$ implies $E_1 x \neq 0$.

8.8.2 Theorem. *Suppose C is a completely regular set with characteristic vector x in a metric scheme. If $E_1 x = 0$, then C is the set of vertices of a face in the polytope associated to E_1.*

Proof. Let r be the covering radius of C. Let π be the distance partition relative to C, and let x_i be the characteristic vector of the ith cell of π. Since π is equitable, for any matrix M in the Bose-Mesner algebra of the scheme, the vector Mx is constant on the cells of π; or, equivalently, it is a linear combination of $x_0, \ldots, x_r$. In particular $E_1 x$ is constant on the cells of π and we assume that

$$E_1 x = \sum_{i=0}^{r} f_i x_i$$

for scalars $f_0, \ldots, f_r$.

We claim that the sequence $f_0, \ldots, f_r$ is strictly decreasing; note that this implies that x_0 is the characteristic vector for the set of vertices of a face in the polytope associated to E_1. To prove this claim, note first that the vector

$$f = \begin{pmatrix} f_0 \\ \vdots \\ f_r \end{pmatrix}$$

is an eigenvector of $A(X_1/\pi)$ with eigenvalue θ_1. The matrix $A(X_1/\pi)$ is tridiagonal and has k and θ_1 as eigenvalues. Since θ_1 is the second-largest eigenvalue of A, it is also the second-largest eigenvalue of $A(X/\pi)$. By Lemma 4.6.6, we conclude that the entries of f are strictly decreasing. $\quad\square$

8.8.3 Corollary. *Let $n > (t + 1)(k - t + 1)$. If C is a subset of the vertices of $J(n, k)$ with width $k - t$ and $|C| = \binom{n-t}{k-t}$, then C consists of all vertices that contain a given t-subset of $\{1, \ldots, n\}$.*

Proof. The set C is the set of vertices in a face from the polytope associated to E_1, and therefore the claim follows from Theorem 7.2.1. $\quad\square$

8.9 Equality in the dual width bound

Assume that C is a subset of the vertices in a metric association scheme with d classes. We have seen that if C has width w and dual degree s^*, then $w \geq d - s^*$. Further, if equality holds then the set C must be completely regular. In this section we consider the dual bound. We work with the inequality $w^* \geq d - s$ and characterize the cases where equality holds.

If M lies in the Bose-Mesner algebra of $\mathcal{A} = \{A_1, \ldots, A_d\}$, we say that the submatrix of M indexed by the vertices in C is the *restriction* of M to C. The restriction of A_r to C is zero if no two vertices in C are r-related. The nonzero restrictions of the matrices A_i to C form a collection of 01-matrices that sum to J.

8.9.1 Theorem. *Suppose $\mathcal{A}$ is a cometric scheme and C is a subset of its vertices with dual width w^* and degree s. If $w^* = d - s$, then the nonzero restrictions to C of the matrices A_r in $\mathcal{A}$ form a cometric scheme with s classes.*

Proof. If M lies in the Bose-Mesner algebra of $\mathcal{A}$, we use $\overline{M}$ to denote its restriction to C. Since the degree of C is s, we see that exactly s of the matrices $\overline{A}_i$ are nonzero. Therefore, the matrices $\overline{A}_i$ for $i = 0, \ldots, s$ span a vector space of dimension $s + 1$ that contains $\overline{I}$ and $\overline{J}$. With a slight abuse of notation we denote this space by $\overline{\mathcal{A}}$. The space $\overline{\mathcal{A}}$ is closed under Schur multiplication, and it follows that if we show that this space is closed under matrix multiplication, it is the Bose-Mesner algebra of an association scheme (see [31, Theorem 2.6.1]).

Since the scheme is cometric relative to E_1, the Schur product $E_1 \circ E_\ell$ is a linear combination of $E_{\ell-1}, E_\ell, E_{\ell+1}$ for any $\ell \in \{1, \ldots, d - 1\}$. This implies that $\overline{E}_1 \circ \overline{E}_\ell$ is a linear combination of $\overline{E}_{\ell-1}, \overline{E}_\ell, \overline{E}_{\ell+1}$ (so the $\overline{E}_0, \ldots, \overline{E}_d$ also satisfy a three-term recurrence relation with respect to $\overline{E}_1$). Further, each A_ℓ and each E_ℓ is a Schur polynomial in E_1. This implies that each $\overline{A}_\ell$ and each $\overline{E}_\ell$ is a Schur polynomial in $\overline{E}_1$,

The remainder of this proof is split into a number of claims.

Claim (a) If $|k - \ell| > w^*$, then $\overline{E}_k \overline{E}_\ell = 0$.

Let Δ denote the diagonal matrix with $\Delta_{u,u} = 1$ if $u \in C$ and is otherwise 0. Then

$$\overline{E}_k \overline{E}_\ell = \overline{E_k \Delta E_\ell},$$

and it is sufficient to prove that $E_k \Delta E_\ell = 0$.

Let x be the characteristic vector of C and consider

$$\langle E_k \Delta E_\ell, E_k \Delta E_\ell \rangle = \operatorname{tr}(\Delta E_k \Delta E_\ell)$$
$$= \sum_{u,v \in C} (E_k \circ E_\ell)_{u,v}$$
$$= \frac{1}{v} \sum_j q_{k,\ell}(j) x^T E_j x,$$

where $q_{k,\ell}(j)$ are the Krein parameters of the association scheme. If $|k - \ell| > w^*$, then for any $0 \le j \le w^*$ we have that $j < |k - \ell|$, which implies $q_{k,\ell}(j) = 0$ (see Exercise 4.6). Further, for all $j > w^*$ the value of $x^T E_j x$ is 0. Thus, if $|k - \ell| > w^*$ all terms in the final sum vanish, whence $E_k \Delta E_\ell = 0$.

This proves the first claim. The next five claims each give a different basis for $\overline{\mathcal{A}}$.

Claim (b) The set

$$\{\overline{E}_0, \dots, \overline{E}_s\}$$

is a basis for $\overline{\mathcal{A}}$.

Let i be the least index such that $\overline{E}_{i+1}$ is linearly dependent on $\overline{E}_0, \dots, \overline{E}_i$. Since $\overline{E}_1 \circ \overline{E}_\ell$ is a linear combination of $\overline{E}_0, \dots, \overline{E}_{\ell+1}$, it follows that the span of $\overline{E}_0, \dots, \overline{E}_i$ is closed under Schur multiplication by $\overline{E}_1$.

Since every $\overline{A}_\ell$ is a Schur polynomial in $\overline{E}_1$, every $\overline{A}_\ell$ is in the span of $\{\overline{E}_0, \dots, \overline{E}_i\}$. Therefore this span is equal to $\overline{\mathcal{A}}$ which implies $i + 1 = \dim(\overline{\mathcal{A}}) = s + 1$ and the result follows.

Claim (c) The set

$$\{\overline{E}_{d-s}, \dots, \overline{E}_d\}$$

is a basis for $\overline{\mathcal{A}}$.

Similar to the previous claim, we let i be the smallest index such that $\overline{E}_{d-i-1}$ is in the span of $\{\overline{E}_{d-i}, \dots, \overline{E}_d\}$. Since the matrices $\{\overline{E}_0, \dots, \overline{E}_d\}$ satisfy a three-term recurrence relation, the set $\{\overline{E}_{d-i}, \ldots, \overline{E}_d\}$ is closed under Schur multiplication by $\overline{E}_1$. Repeatedly using the three-term recurrence, we see that $\overline{E}_j$ is in the span of $\{\overline{E}_{d-i}, \ldots, \overline{E}_d\}$ for all $j \leq d - i - 1$. Thus the span of this set of matrices is equal to $\overline{\mathcal{A}}$; this implies that $i + 1 = \dim(\overline{\mathcal{A}}) = s + 1$ and the result follows.

Claim (d) If $1 \leq j \leq s$, the set

$$\{\overline{E}_0, \ldots, \overline{E}_{j-1}, \overline{E}_{d-s+j}, \ldots, \overline{E}_d\}$$

is a basis for $\overline{\mathcal{A}}$.

Since each of the sets

$$\{\overline{E}_0, \ldots, \overline{E}_{j-1}\}, \qquad \{\overline{E}_{d-s+j}, \ldots, \overline{E}_d\}$$

is contained in a basis, they are linearly independent. By Claim (a), the matrices in the first set are orthogonal to the matrices in the second, and so the union of the two sets is a basis.

Claim (e) The set

$$\{\overline{I}, \overline{E}_{d-s+1}, \ldots, \overline{E}_d\}$$

is a basis for $\overline{\mathcal{A}}$.

By Claim (c), it is sufficient to prove that $\overline{I}$ is not in the span of

$$\{\overline{E}_{d-s+1}, \ldots, \overline{E}_d\}.$$

If it were, we would have $\overline{I} = \sum_{\ell=d-s+1}^{d} a_\ell \overline{E}_\ell$ for some scalars a_ℓ, which would imply that

$$\overline{E}_0 = \overline{E}_0 \overline{I} = \sum_{\ell=d-s+1}^{d} a_\ell \overline{E}_0 \overline{E}_\ell.$$

But by Claim (a) each of the products on the right is zero.

Claim (f) If $1 \leq j \leq s$, the set

$$\{\overline{E}_0, \ldots, \overline{E}_{j-1}, \overline{I}, \overline{E}_{d-s+j+1}, \ldots, \overline{E}_d\}$$

is a basis for $\overline{\mathcal{A}}$. If $\overline{\mathcal{E}}_j = \mathrm{span}\{\overline{E}_0, \ldots, \overline{E}_j\}$, then $\overline{\mathcal{A}}\overline{\mathcal{E}}_j = \overline{\mathcal{E}}_j \overline{\mathcal{A}} = \overline{\mathcal{E}}_j$.

We prove both claims by induction, in parallel. We note that both claims hold when $j = 0$. Assume $j \geq 1$. We need to prove that $\overline{E}_{d-s+j}$ is in the span of

$$\{\overline{E}_0, \ldots, \overline{E}_{j-1}, \overline{I}, \overline{E}_{d-s+j+1}, \ldots, \overline{E}_d\}.$$

By Claim (d), $\bar{I}$ can be written as

$$\bar{I} = \sum_{i=0}^{j-1} a_i \bar{E}_i + \sum_{i=d-s+j}^{d} a_i \bar{E}_i. \tag{8.9.1}$$

If $a_{d-s+j} \neq 0$,

$$\bar{E}_{d-s+j} = \frac{1}{a_{d-s+j}} \bar{I} - \sum_{i=0}^{j-1} a_i \bar{E}_i + \sum_{i=d-s+j+1}^{d} a_i \bar{E}_i,$$

and $\bar{E}_{d-s+j}$ is in the span of the other matrices.

If $a_{d-s+j} = 0$, multiply both sides of (8.9.1) by $\bar{E}_j$. By induction and Claim (a), we get an expression for $\bar{E}_j$ in terms of elements of $\bar{E}_j \bar{\mathcal{E}}_{j-1}$, which, by the induction hypothesis, is contained in $\bar{\mathcal{E}}_{j-1}$. This contradicts Claim (b).

We prove the second part of the claim. Let M be in the span of $\bar{\mathcal{A}}$. By the first part of the claim

$$M = \sum_{\ell=0}^{j-1} a_\ell \bar{E}_\ell + a_j \bar{I} + \sum_{\ell=d-s+j+1}^{d} a_\ell \bar{E}_\ell.$$

So for any $k \leq j$ we have that

$$M \bar{E}_k = \sum_{\ell=0}^{j-1} a_\ell \bar{E}_\ell \bar{E}_k + a_j \bar{I} \bar{E}_k + \sum_{\ell=d-s+j+1}^{d} a_\ell \bar{E}_\ell \bar{E}_k$$

$$= \sum_{\ell=0}^{j-1} a_\ell \bar{E}_\ell \bar{E}_k + a_j \bar{E}_k.$$

This shows that the matrix $M\bar{E}_k$ is in $\bar{\mathcal{E}}_j$. It follows that for any matrix in $\bar{\mathcal{E}}_j$, so for any $N = \sum_{i=0}^{j} b_i \bar{E}_i$, we have that MN is in the span of $\bar{\mathcal{E}}_j$ (similarly, NM is also in this span). Thus $\bar{\mathcal{A}}\bar{\mathcal{E}}_j = \bar{\mathcal{E}}_j \bar{\mathcal{A}} = \bar{\mathcal{E}}_j$.

Claim (g) $\bar{\mathcal{A}}$ is a cometric scheme.

It follows from Claim (f) that $\bar{\mathcal{A}}$ is closed under matrix multiplication. Hence it is the Bose-Mesner algebra of a scheme and we need only prove that it is cometric.

Recall that the span of the set $\{\bar{E}_\ell : \ell = 0, \ldots, j\}$ is denoted by $\bar{\mathcal{E}}_j$. For each j the space $\bar{\mathcal{E}}_j$ is closed under multiplication and therefore has a basis of mutually orthogonal idempotents. Each of the j idempotents in $\bar{\mathcal{E}}_{j-1}$ is a sum of some of the $j+1$ idempotents in $\bar{\mathcal{E}}_j$. Therefore either all the minimal idempotents of $\overline{\mathcal{E}}_{j-1}$ remain minimal in $\overline{\mathcal{E}}_j$ and $\overline{\mathcal{E}}_j$ has one additional minimal idempotent, or all but one of the minimal idempotents of $\overline{\mathcal{E}}_{j-1}$ remain minimal, and the other splits into a pair of minimal idempotents for $\overline{\mathcal{E}}_j$.

We can, however, eliminate the latter possibility. For suppose F is a minimal idempotent in $\overline{\mathcal{E}}_{j-1}$ and $F = F_1 + F_2$ where F_1 and F_2 are minimal idempotents in $\overline{\mathcal{E}}_j$. Multiply F by $\overline{E}_{d-s+j}$ to get $\overline{E}_{d-s+j}F_1 = -\overline{E}_{d-s+j}F_2$, whence

$$\overline{E}_{d-s+j}F_1 = \overline{E}_{d-s+j}F_1 F_1 = -\overline{E}_{d-s+j}F_2 F_1 = 0$$

and therefore $\overline{E}_{d-s+j}\overline{E} = 0$ for any $E \in \overline{\mathcal{E}}_j$, which is impossible.

Thus $\overline{\mathcal{A}}$ has minimal idempotents $F_0, \ldots, F_s$, where $F_0, \ldots, F_j$ span $\overline{\mathcal{E}}_j$. Since the elements of $\overline{\mathcal{E}}_j \setminus \overline{\mathcal{E}}_{j-1}$ are Schur polynomials of degree j in $\overline{E}_1$ (and hence in F_1), we deduce (using Lemma 4.3.3) that our scheme is cometric. $\square$

8.10 Narrow subsets

Suppose $\mathcal{A}$ is a metric and cometric association scheme with d classes and C is a subset of its vertices with width w and dual width w^*. For any C we have seen that $w + w^* \geq d$ and if $w + w^* = d$, then C is a narrow set.

Following the notation from the previous section, for any matrix M in the Bose-Mesner algebra of $\mathcal{A}$, we denote the restriction of M to the vertices in C by $\overline{M}$.

8.10.1 Theorem. *Suppose $\mathcal{A}$ is an association scheme that is both metric and cometric. Let C be a subset of the vertices of the scheme. If C is narrow, the restriction of $\mathcal{A}$ to C is cometric and if this restriction is primitive, it is also metric.*

Proof. If C is narrow then $w^* + s = d$ and, by Theorem 8.9.1, the restricted scheme is cometric.

Let Y denote the subgraph of X_1 induced by C. Since the induced scheme is primitive, Y is connected and, since $s = w$, the distance between two vertices of Y is equal to their distance in X_1. Accordingly $\overline{A}_i$ is the matrix of the distance-i relation in Y, and therefore the restricted scheme is metric. $\square$

8.10.2 Lemma. *If C is a subset of a cometric scheme $\mathcal{A}$ with d classes and $w^* + s = d$, then the Krein parameters of the scheme obtained by restriction to C are determined by w and w^*.*

Proof. The set

$$\{\overline{A}_0, \ldots, \overline{A}_w\}$$

is the basis of adjacency matrices for the Bose-Mesner algebra of the induced scheme. By Claim (b) in the proof of Theorem 8.9.1, the set

$$\{\overline{E}_0, \ldots, \overline{E}_w\} \tag{8.10.1}$$

is a second basis. By Claim (e) the set

$$\{\overline{E}_0, \ldots, \overline{E}_{j-1}, \overline{I}, \overline{E}_{d-s+j+1}, \ldots, \overline{E}_d\} \tag{8.10.2}$$

is a third basis. Since

$$\overline{E}_r = \frac{1}{v} \sum_j q_r(j)\overline{A}_j$$

(where the coefficients $q_r(j)$ are the dual eigenvalues of $\mathcal{A}$) the change-of-basis matrix between any two of these three bases does not depend on C.

Each product $\overline{E}_i\overline{E}_j$ is a linear combination of the matrices $\overline{E}_0, \ldots, \overline{E}_w$, and we need to show that these coefficients do not depend on C. We assume that

$$\overline{E}_i\overline{E}_j = \sum_{k=0}^{w} c_{i,j}(k)\overline{E}_k$$

(recall that these $c_{i,j}(k)$ are the Krein parameters of the association scheme). We will prove by induction on i that these coefficients are independent of C. We have that the set given in (8.10.2) is a basis. Since $\overline{E}_i\overline{I} = \overline{E}_i$ and

$$\overline{E}_i\overline{E}_{w^*+i+1} = \cdots = \overline{E}_i\overline{E}_d = 0,$$

we see that $\overline{E}_i$ is a linear combination of $\overline{E}_0, \ldots, \overline{E}_{i-1}$ and $\overline{I}$. Since the change-of-basis matrix for the bases in (8.10.1) and (8.10.2) is determined by the Krein parameters, it follows by induction that the coefficients $c_{i,j}(k)$ are independent of C for all j and k.   □

An induced subgraph Y of a graph X is *convex* if each vertex that lies on a shortest path (in X) joining two vertices of Y is itself in Y. By way of example, it is not hard to show that the convex subgraphs of the Johnson graph $J(n, k)$ consists of all k-subsets that contain a given subset M and are contained in some subset L.

8.10.3 Lemma. *If C is a narrow subset of the Johnson scheme, it induces a convex subgraph of X_1.*

Proof. Suppose C has width w. Since C is narrow, this implies that $w^* = d - w$. Let U be a subset of $\{1, \ldots, n\}$ of size $k + w$. Then the set C_U of k-subsets of U has width w and is narrow, and therefore the scheme induced by C_U has the same Krein parameters as C. Accordingly C_U is convex.   □

Note that this lemma yields a second characterization of the t-intersecting families of size $\binom{n-t}{k-t}$, for any family of this size must be convex and hence be of the form described earlier. It is also possible to characterize the narrow subsets of the Grassmann scheme and hence determine the t-intersecting families of subspaces of maximum size. We carry this out in the next chapter.

8.11 Exercises

8.1 Assume that $\mathcal{B}$ is a t-$(n, k, 1)$-design. For $i \in \{1, \ldots, t\}$, let λ_i denote the number of blocks that contain an arbitrary i-set (since $\mathcal{B}$ is a design, this number is the same for any i-set, provided that $i \in \{1, \ldots, t\}$).

(a) Show that for $i \in \{1, \ldots, t\}$

$$\lambda_i = \frac{\binom{n-i}{k-i}}{\binom{n-t}{k-t}}.$$

(b) For a fixed block A of the design, let γ_s denote the the number of blocks in the design that intersection size exactly s with A. Show, using a counting argument, that

$$\sum_{s=0}^{k} \gamma_s \binom{s}{i} = \binom{k}{i} \lambda_i$$

where $\lambda_i = 1$ for $i \geq t$. Conclude from this that

$$\sum_{s=0}^{t-1} \gamma_s \binom{s}{i} = \binom{k}{i} (\lambda_i - 1).$$

(c) Using the final equation of the previous part of this question, show that

$$\gamma_s = \sum_{j=s}^{t} (-1)^{j-s} \binom{j}{s} \binom{k}{j} (\lambda_i - 1)$$

(see Lemma 6.2.1 for a hint on how to do this).

(d) If y is the characteristic vector for $\mathcal{B}$, show that the projection of $y^T y$ to the Bose-Mesner algebra of the Johnson scheme $J(n, k)$ is the matrix

$$N = \sum_{s=0}^{t} \frac{\gamma_s}{\binom{n-k}{k-s}\binom{k}{s}} A_{k-s}. \tag{8.11.1}$$

8.2 The matrix N given in (8.11.1) exists even when a t-$(n, k, 1)$ design does not. This is the matrix in our thought experiment in the first section of this chapter. Show that this matrix is equal to Wilson's matrix given in Section 8.4 when $t = 1$ and $t = 2$. Further, show that the coefficients of A_{k-s} of the two matrices agree when $s = t$ and when $s = t - 1$.

8.3 Prove that in the Johnson scheme, the k-subsets that lie in a given subset of size ℓ of V form a set with width $\ell - k$. Similarly, show that the k-subsets that contain a given subset of size $k - \ell$ of V form a set with width $k - \ell$.

8.4 Let $\mathcal{A}$ be a cometric scheme with d classes. If C is a subset of the vertices of the scheme with dual width w^* and degree s, prove that $w^* + s \geq d$.

8.5 Suppose C_0 is a subset of the vertices of a metric association scheme with distance partition $C_0, \ldots, C_r$. Show that C_r is completely regular if C_0 is.

8.6 Let $\mathcal{A}$ be a metric scheme with d classes and let C be a subset of its vertices with width w and covering radius r. Prove that $d \leq w + r$. Deduce that if $d = w + s^*$, then $r = s^*$.

8.7 If λ and β are k-subsets of $\{1, \ldots, n\}$, show that a k-subset lies on a shortest path in $J(n, k)$ from α to β if and only if

$$\alpha \cap \beta \subseteq \gamma \subseteq \alpha \cup \beta.$$

Use this to provide a description of the convex subgraphs of $J(n, k)$.

Notes

We conjecture that the matrix constructed in Exercise 8.2 is actually equal to Wilson's matrix, but we have not yet been able to show that this is true for all values of t, nor have we found a way to calculate all of its eigenvalues.

Our proof of the EKR Theorem for t-intersecting families follows Wilson [172], for the excellent reason that no better way of deriving the bound has been found. However, we do now have a more powerful way of deriving the characterization of the maximal families. The results presented in this chapter are based on important work of Tanaka [160, 161], which extends earlier work by Brouwer et al. [32] on width and related matters.

9

The Grassmann scheme

Let V be a vector space of dimension n over the field of order q (this field is denoted by $GF(q)$). The vertices of the *Grassmann scheme* $G(n, k)$ are the k-dimensional subspaces of V, and two such subspaces are i-related in the scheme if their intersection has dimension $k - i$. The general linear group acts generously transitively on the set of k-dimensional subspaces of V, and therefore we have an association scheme with k classes.

The *Grassmann graph* $J_q(n, k)$ is the graph X_1 in this association scheme; two k-subspaces are adjacent in $J_q(n, k)$ if their intersection has dimension $k - 1$. The graph X_k is called the *q-Kneser graph* (vertices are adjacent in this graph if their intersection has dimension 0). The Grassmann graph is distance regular – to show this it is enough to verify that its diameter is $\min\{k, n - k\}$. However, the q-Kneser graph is not distance regular in general. The Grassmann graph has classical parameters, and we discussed it briefly from this viewpoint in Section 4.7.

If $u \in V$, we define $u^\perp$ by

$$u^\perp = \{w \in V : u^T w = 0\},$$

and if $S \subseteq V$, then

$$S^\perp = \bigcap_{u \in S} u^\perp.$$

For any subspace S of V, the set $S^\perp$ is also a subspace of V. If U is a subspace of V, then

$$\dim(U^\perp) = \dim(V) - \dim(U),$$

and $(U^\perp)^\perp = U$. But we note that $U \cap U^\perp$ need not be the zero subspace. The map $U \mapsto U^\perp$ is an order-reversing bijection on the subspace lattice of V. We can use this bijection to show that $J_q(n, k)$ and $J_q(n, n-k)$ are isomorphic; hence we can assume that $n \geq 2k$.

9.1 q-Binomial coefficients

We need to define some polynomials in a variable q. First, for any integer n,

$$[n]_q = 1 + \cdots + q^{n-1}.$$

We set $[0]!_q = 1$ and if $n > 0$, then

$$[n]!_q = [n]_q[n-1]!_q.$$

We will usually abbreviate our notation and write $[n]$ and $[n]!$ when the value of q is determined by context. For non-negative integers n and k where $k \leq n$, the *q-binomial* or *Gaussian binomial* coefficient $\begin{bmatrix} n \\ k \end{bmatrix}_q$ is defined by

$$\begin{bmatrix} n \\ k \end{bmatrix}_q = \frac{[n]!}{[k]![n-k]!} = \prod_{i=0}^{k-1} \frac{q^{n-i} - 1}{q^{k-i} - 1}.$$

Again, when the value of q is clear, we write $\begin{bmatrix} n \\ k \end{bmatrix}$. Note that if $q = 1$, then $[n] = n$, $[n]! = n!$ and $\begin{bmatrix} n \\ k \end{bmatrix} = \binom{n}{k}$. You may easily verify that

$$\begin{bmatrix} n \\ k \end{bmatrix} = q^k \begin{bmatrix} n-1 \\ k \end{bmatrix} + \begin{bmatrix} n-1 \\ k-1 \end{bmatrix} = \begin{bmatrix} n-1 \\ k \end{bmatrix} + q^{n-k} \begin{bmatrix} n-1 \\ k-1 \end{bmatrix}.$$

Using either of these identities, we see (by induction on n) that $\begin{bmatrix} n \\ k \end{bmatrix}$ is a polynomial in q, not just a rational function.

Just as with binomial coefficients, we can extend the definition of a q-binomial coefficient to negative values of n. For positive integers n and k, we define

$$\begin{bmatrix} -n \\ k \end{bmatrix} = (-1)^k q^{-nk-\binom{k}{2}} \begin{bmatrix} n+k-1 \\ k \end{bmatrix}. \tag{9.1.1}$$

Another similarity with the binomial coefficients is that the q-binomial coefficients also count a type of object.

9.1.1 Lemma. *If q is a prime power, then the number of k-dimensional subspaces of a vector space of dimension n over $GF(q)$ is $\begin{bmatrix} n \\ k \end{bmatrix}$.* □

9.2 q-Commuting variables

One useful and elegant way to work with q-binomial coefficients is to use *q-commuting variables*. Symbols A and B are called q-commuting variables if

$$BA = qAB$$

(you may think of A and B as linear operators on some vector space).

9.2.1 Lemma. *If $BA = qAB$ then*

$$(A + B)^n = \sum_{k=0}^{n} \begin{bmatrix} n \\ k \end{bmatrix} A^k B^{n-k}. \qquad \square$$

We now show how to construct a pair of q-commuting variables, where the variables are operators. Let q and t be independent commuting variables and let V be the vector space of Laurent series in t over $\mathbb{C}(q)$, the field of rational functions in q. Define the operators M and Q on V by

$$Mf(t) = tf(t), \qquad Qf(t) = f(qt).$$

Then

$$(MQ)t^n = q^n t^{n+1}, \qquad (QM)t^n = q^{n+1} t^{n+1},$$

and it follows that $QM = qMQ$.

For a second example, define the *q-derivative* D by

$$Df(t) = \frac{f(qt) - f(t)}{qt - t}.$$

Then

$$Dt^n = [n]t^{n-1},$$

and since

$$(DQ)t^n = q^n[n]t^{n-1}, \qquad (QD)t^n = q^{n-1}[n]t^{n-1},$$

we see that $DQ = qQD$.

Apply Lemma 9.2.1 with $A = MQ$ and $B = Q$ defined earlier. If we denote the function $f(t) = 1$ simply by 1, then we have that

$$A^k B^{n-k} 1 = A^k 1 = q^{\binom{k}{2}} t^k,$$

while

$$(A + B)^n 1 = (1 + t)(1 + qt) \cdots (1 + q^{n-1}t).$$

Hence we have the following identity, which is known as the *q-binomial identity*.

9.2.2 Lemma. *For $n > 0$,*

$$\prod_{i=0}^{n-1}(1 + q^i t) = \sum_{k=0}^{n} \begin{bmatrix} n \\ k \end{bmatrix} q^{\binom{k}{2}} t^k. \qquad \square$$

Setting t to be first -1 and then $-(q^{n-1})^{-1}$ in this lemma yields the following corollary.

9.2.3 Corollary. *Let n be an integer; then*

$$\sum_{j=0}^{n} q^{\binom{j}{2}} \begin{bmatrix} n \\ j \end{bmatrix} (-1)^j = \begin{cases} 1, & n = 0; \\ 0, & n > 0, \end{cases}$$

and

$$\sum_{j=0}^{n} q^{\binom{j+1}{2}-nj} \begin{bmatrix} n \\ j \end{bmatrix} (-1)^j = \begin{cases} 1, & n = 0; \\ 0, & n > 0. \end{cases} \qquad \square$$

The final identity of this section is known as the *q-Vandermonde identity*.

9.2.4 Corollary. *For $k \le m, n$*

$$\begin{bmatrix} m+n \\ k \end{bmatrix} = \sum_{i=0}^{k} q^{(m-i)(k-i)} \begin{bmatrix} m \\ i \end{bmatrix} \begin{bmatrix} n \\ k-i \end{bmatrix}. \qquad \square$$

Proof. From Lemma 9.2.2 we have that

$$\prod_{i=0}^{n+m-1} (1+q^i t) = \sum_{k=0}^{n+m} \begin{bmatrix} n+m \\ k \end{bmatrix} q^{\binom{k}{2}} t^k.$$

But we can also express this left-hand side as

$$\prod_{i=0}^{n+m-1} (1+q^i t) = \prod_{i=0}^{m-1} (1+q^i t) \prod_{j=0}^{n-1} (1+q^j(q^m t))$$

$$= \left(\sum_{i=0}^{m} \begin{bmatrix} m \\ i \end{bmatrix} q^{\binom{i}{2}} t^i \right) \left(\sum_{j=0}^{n} \begin{bmatrix} n \\ j \end{bmatrix} q^{\binom{j}{2}} (q^m t)^j \right).$$

Setting these two expressions equal and considering the coefficient of t^k we get that

$$q^{\binom{k}{2}} \begin{bmatrix} n+m \\ k \end{bmatrix} = \sum_{i=0}^{k} q^{\binom{i}{2}+\binom{k-i}{2}+m(k-i)} \begin{bmatrix} m \\ i \end{bmatrix} \begin{bmatrix} n \\ k-i \end{bmatrix}.$$

Since $\binom{i}{2} + \binom{k-i}{2} + m(k-i) - \binom{k}{2} = (m-i)(k-i)$, the result holds. $\qquad \square$

9.3 Counting subspaces

Just as we use binomial coefficients to express the solutions to counting problems concerning subsets of a set, we use q-binomial coefficients to count subspaces of vector spaces over $GF(q)$. In this section, we determine the parameters of the Grassmann scheme using the q-binomial coefficients. First we need to introduce some terminology.

We say that two subspaces are *skew* if their intersection is the zero subspace. If α and β are skew subspaces of V and $\alpha + \beta = V$, we say that β is a *complement* to α.

9.3.1 Lemma. *If α is a k-dimensional subspace of a vector space V of dimension $k + \ell$ over $GF(q)$, the number of ℓ-dimensional subspaces of V skew to α is $q^{k\ell}$.*

Proof. Suppose α is the row space of a $k \times (k + \ell)$ matrix M and that β is the row space of a $\ell \times (k + \ell)$ matrix N. Then β is skew to α if and only if the stacked matrix

$$\begin{pmatrix} M \\ BN \end{pmatrix}$$

is invertible. Now suppose

$$M = \begin{pmatrix} I_k & 0 \end{pmatrix}, \qquad N = \begin{pmatrix} N_1 & N_2 \end{pmatrix}$$

(where N_2 is $\ell \times \ell$). Then our stacked matrix is invertible if and only if N_2 is. If N_2 is invertible, then the row space of

$$\begin{pmatrix} N_2^{-1} N_1 & I_k \end{pmatrix}$$

is equal to the row space of N and hence is equal to β. It follows now that the number of complements to α is equal to the number of $\ell \times k$ matrices $N_2^{-1} N_1$. $\square$

9.3.2 Lemma. *Let α be a subspace of dimension k in a vector space of dimension n. The number of ℓ-dimensional subspaces whose intersection with α has dimension j is*

$$q^{(k-j)(\ell-j)} \begin{bmatrix} n - k \\ \ell - j \end{bmatrix} \begin{bmatrix} n \\ j \end{bmatrix}.$$

Proof. Fix a k-subspace α and let $\mathcal{M}$ denote the set of ℓ-dimensional subspaces skew to it. Deem two elements γ_1 and γ_2 of $\mathcal{M}$ to be equivalent if

$$\alpha + \gamma_1 = \alpha + \gamma_2.$$

This sum is a subspace of dimension $k + \ell$, and we see that the subspaces of dimension $k + \ell$ that contain α partition the elements of $\mathcal{M}$. By the previous lemma, the number of elements in a class of this partition is $q^{k\ell}$, and the number of classes is $\begin{bmatrix} n-k \\ \ell \end{bmatrix}$. Therefore

$$|\mathcal{M}| = q^{k\ell} \begin{bmatrix} n-k \\ \ell \end{bmatrix}.$$

It follows that the number of ℓ-dimensional subspaces γ such that $\gamma \cap \alpha$ is a fixed subspace of α with dimension j is equal to

$$q^{(k-j)(\ell-j)} \begin{bmatrix} n-k \\ \ell-j \end{bmatrix}.$$

Consequently the number of ℓ-dimensional subspaces whose intersection with α has dimension j is as stated. $\square$

9.3.3 Corollary. *The degree of the graph X_i in the Grassmann scheme is*

$$q^{i^2} \begin{bmatrix} n-k \\ i \end{bmatrix} \begin{bmatrix} k \\ i \end{bmatrix}. \qquad \square$$

Our next goal is to calculate the intersection array for the Grassmann graph. To do this, we calculate the number of vertices at set distances from two fixed vertices in the Grassmann graph.

9.3.4 Lemma. *Let α and β be two vertices in $J_q(n,k)$ at distance r from each other. For any $i \in \{0, \ldots, r\}$ the number of subspaces in the vertex set of $J_q(n,k)$ at distance i from α and distance $r-i$ from β is*

$$\begin{bmatrix} r \\ r-i \end{bmatrix} \begin{bmatrix} r \\ i \end{bmatrix} = \begin{bmatrix} r \\ i \end{bmatrix}^2.$$

Proof. Choose subspaces α and β at distance r in $J_q(n,k)$. Let γ be a subspace at distance i from α and at distance $r-i$ to β. By definition this happens if and only if both

$$\dim(\gamma \cap \alpha) = k-i, \qquad \dim(\gamma \cap \beta) = k-r+i.$$

Set $\dim(\alpha \cap \beta \cap \gamma)$ to be x. Since $\dim(\alpha \cap \beta) = k-r$ this implies that $x \le k-r$. By considering the dimensions of the intersections $\gamma \cap \alpha$ and $\gamma \cap \beta$, it is clear that the subspace γ will have a subspace skew to $\alpha + \beta$ of dimension $r-k+x$. This fact implies that $x \ge k-r$ and thus that $\dim(\alpha \cap \beta \cap \gamma) = k-r$. From this we can conclude that

$$\alpha \cap \beta \subseteq \gamma \subseteq \alpha + \beta. \tag{9.3.1}$$

Conversely, if a k-dimensional subspace γ satisfies the preceding equation, then $\dim(\alpha \cap \gamma) + \dim(\beta \cap \gamma) = 2k-r$, which is equivalent to γ being distance i from α and distance $r-i$ from β for some $i \in \{0, \ldots, r\}$. To prove this result, we only need to count the number of k-dimensional subspaces for which (9.3.1) holds.

Since $\dim(\alpha \cap \beta) = k - r$, the subspace $\gamma \cap \alpha$ contains a subspace of α skew to $\alpha \cap \beta$ with dimension at least $r - i$. Similarly, as $\dim(\gamma \cap \beta) = k - r + i$, we see that $\gamma \cap \beta$ contains a subspace of β skew to $\alpha \cap \beta$ with dimension exactly i. So the number of subspaces γ is equal to the product of the number of subspaces of dimension $r - i$ in the quotient space $\alpha/(\alpha \cap \beta)$, times the number of subspaces of $\beta/(\alpha \cap \beta)$ of dimension i. This number is

$$\begin{bmatrix} r \\ r - i \end{bmatrix}\begin{bmatrix} r \\ i \end{bmatrix} = \begin{bmatrix} r \\ i \end{bmatrix}^2. \qquad \square$$

With this lemma, it is possible to calculate the intersection numbers for the Grassmann graph. Assume that B is at distance r from A. Recall that c_r is the number of neighbors of B that are at distance $r - 1$ from A. There are $[r]^2$ ways to replace a one-dimensional vector space in $B \backslash A$ with a one-dimensional vector space from $A \backslash B$; this implies that $c_r = [r]^2$.

Also recall that if A and B are vertices at distance $r + 1$, then b_r is the number of neighbors of B that are at distance r from A. So if v_r denotes the number of vertices at distance r from a given vertex, then

$$v_r b_r = v_{r+1} c_{r+1}.$$

From Lemma 9.3.2, we have that

$$v_r = q^{r^2} \begin{bmatrix} n - k \\ r \end{bmatrix}\begin{bmatrix} k \\ r \end{bmatrix}$$

and hence

$$b_r = q^{2r+1}[k - r][n - k - r].$$

Finally, a_r can be calculated using the fact that $a_r + b_r + c_r$ equals the degree of the graph (which is $q\begin{bmatrix} n-k \\ 1 \end{bmatrix}\begin{bmatrix} k \\ 1 \end{bmatrix}$).

9.4 Subspace incidence matrices

One advantage of the method used in Sections 6.3 and 6.5 for computing the eigenvalues and eigenvectors of the Johnson scheme is that a very small variation will give us the eigenvalues, eigenvectors and eigenspaces for the Grassmann scheme. In this section we start on this project.

Let V be a fixed vector space of dimension v over $GF(q)$. We first compute the eigenvalues of $J_q(n, k)$. Our strategy is to relate the eigenvalues of $J_q(n, k - 1)$ to the eigenvalues of $J_q(n, k)$, just as we did in Theorem 6.3.2. Since we will be considering different values of k, we use $A_q(n, k, 1)$ to denote the adjacency matrix $J_q(n, k)$.

We use $W_{t,k}$ to denote the incidence matrix of t-subspaces versus k-subspaces of V. So the rows are indexed by the t-dimensional subspace of V, the columns by the k-dimension subspaces, and the (α, β)-entry of this matrix is 1 if $\alpha \subseteq \beta$ and 0 otherwise. If $s \le t$ we have

$$W_{s,t} W_{t,k} = \begin{bmatrix} k - s \\ t - s \end{bmatrix} W_{s,k},$$

from which it follows that the row space of $W_{s,k}$ is a subspace of the row space of $W_{t,k}$.

The key observation is that

$$W_{k-1,k}^T W_{k-1,k} = [k]I + A_q(n, k, 1),$$
$$W_{k-1,k} W_{k-1,k}^T = [n - k + 1]I + A_q(n, k - 1, 1)$$

and that the matrices on the left here have the same nonzero eigenvalues with the same multiplicities (Lemma 6.3.1). Since

$$\mathrm{rk}(W_{k-1,k}^T W_{k-1,k}) \le \begin{bmatrix} n \\ k - 1 \end{bmatrix}$$

it follows that $-[k]$ is an eigenvalue of $A_q(n, k, 1)$ with multiplicity at least

$$\begin{bmatrix} n \\ k \end{bmatrix} - \begin{bmatrix} n \\ k - 1 \end{bmatrix}.$$

For the remaining eigenvalues, if λ is an eigenvalue of $A_q(n, k - 1, 1)$ with multiplicity m, then

$$\lambda + [n - k + 1] - [k]$$

is an eigenvalue of $A_q(n, k, 1)$ with multiplicity m. We can now determine the eigenvalues of $A_q(n, k, 1)$ by induction on k, noting that $A_q(n, 1, 1)$ is the adjacency matrix of the complete graph $K_{[n]}$. We suppress the details and report the result.

9.4.1 Theorem. *The rth eigenspace of $J_q(n, k)$ is the orthogonal complement of $\mathrm{col}(W_{r-1,k}^T)$ in $\mathrm{col}(W_{r,k}^T)$; the eigenvalue of the Grassmann graph on this eigenspace is*

$$-[k] + q^r[k - r][n - k - r + 1]. \qquad \square$$

This implies that if $E_r(k)$ is the rth idempotent for $J_q(n, k)$, then for $i < r$

$$W_{i,k} E_r(k) = 0, \qquad W_{i,k} E_i(k) = E_i(k).$$

It also follows that the dimension of the rth eigenspace is equal to

$$\begin{bmatrix} n \\ r \end{bmatrix} - \begin{bmatrix} n \\ r-1 \end{bmatrix}$$

(where $r = 1, \ldots, k$) and the rows of $W_{r,k}$ are linearly independent.

9.5 Another basis

Analogously to the approach taken in Section 6.5, we now define another basis for the Bose-Mesner algebra, and with this basis we determine the eigenvalues for the matrices in the Grassmann scheme. Let $\overline{W}_{t,k}$ denote the 01-matrix with rows indexed by t-dimensional subspaces, columns by k-dimensional subspaces and with (α, β)-entry equal to 1 if and only if α and β are skew. By Lemma 9.3.2

$$W_{r,s}\overline{W}_{s,t} = q^{t(s-r)}\begin{bmatrix} n-r-t \\ s-r \end{bmatrix}\overline{W}_{r,t}. \tag{9.5.1}$$

It is possible to express $\overline{W}_{t,k}$ as the sum of a product of two of the matrices $W_{i,j}$.

9.5.1 Lemma. *For all integers* t, k *with* $0 \le t \le k \le n$

$$\overline{W}_{t,k} = \sum_{i=0}^{t}(-1)^i q^{\binom{i}{2}} W_{i,t}^T W_{i,k}.$$

Proof. Suppose α is a t-dimensional subspace and β a k-dimensional subspace of our underlying vector space. Then $(\overline{W}_{t,k})_{\alpha,\beta}$ is 1 if α and β are skew, and 0 otherwise. Since

$$(W_{i,t}^T W_{i,k})_{\alpha,\beta} = \begin{bmatrix} \dim(\alpha \cap \beta) \\ i \end{bmatrix},$$

by the first equation in Corollary 9.2.3 we see that the (α, β)-entry of the right-hand side of the equation in the statement of the lemma is

$$\sum_{i=0}^{k}(-1)^i q^{\binom{i}{2}}\begin{bmatrix} \dim(\alpha \cap \beta) \\ i \end{bmatrix} = \begin{cases} 1, & \text{if } \dim(\alpha \cap \beta) = 0; \\ 0, & \text{otherwise.} \end{cases} \qquad \square$$

Using a similar argument with the second equation in Corollary 9.2.3 we have the following lemma.

9.5.2 Lemma. *For all integers* t, k *with* $0 \le t \le k \le n$,

$$W_{t,k} = \sum_{i=0}^{t}(-1)^i q^{\binom{i+1}{2}-ti} W_{i,t}^T \overline{W}_{i,k}. \qquad \square$$

Define $D_i(k) = W_{i,k}^T \overline{W}_{i,k}$. We will prove that these matrices form a second basis for the Bose-Mesner algebra of the Grassmann scheme.

Let $\{A_0, \ldots, A_k\}$ be the matrices in the Grassmann scheme $G(n, k)$. We first show that the matrices A_i can be written as a linear combination of the $D_i(k)$; next we find the eigenvalues of the $D_r(k)$. Then, by finding the eigenvalues of the matrices $D_r(k)$, we have a formula for the eigenvalues of all A_i in $J_q(n, k)$.

A simple counting argument shows that for any r,

$$D_r(k) = \sum_{i=r}^{k} \begin{bmatrix} i \\ r \end{bmatrix} A_i. \tag{9.5.2}$$

(Note that this also implies that $D_r(k)$ is symmetric.) To write the matrix A_i as a linear combination of the matrices $D_r(k)$, we invert this equation.

9.6 Eigenvalues of the Grassmann scheme

We now compute the eigenvalues of the matrices $D_r(k)$, and then use this information to get the eigenvalues of the minimal Schur idempotents in the Grassmann scheme.

9.6.1 Lemma. *If B_j and C_j, where $j = 0, \ldots, k$, are matrices with*

$$B_j = \sum_{i=j}^{k} \begin{bmatrix} i \\ j \end{bmatrix} C_i,$$

then

$$C_i = \sum_{j=i}^{k} (-1)^{j-i} q^{\binom{j-i}{2}} \begin{bmatrix} j \\ i \end{bmatrix} B_j.$$

Proof. Suppose M is the $(k + 1) \times (k + 1)$ matrix, with $M_{i,j} = \begin{bmatrix} j \\ i \end{bmatrix}$. Set $B^T = [B_0, B_1, \ldots, B_k]$ and $C^T = [C_0, C_1, \ldots, C_k]$ so that $B = MC$.

Further suppose that N is the $(k + 1) \times (k + 1)$ matrix, with $N_{i,j} = (-1)^{j-i} q^{\binom{j-i}{2}} \begin{bmatrix} j \\ i \end{bmatrix}$. Then

$$(NM)_{i,j} = \sum_{r=0}^{k} (-1)^{r-i} q^{\binom{r-i}{2}} \begin{bmatrix} r \\ i \end{bmatrix} \begin{bmatrix} j \\ r \end{bmatrix}$$

$$= \sum_{r=0}^{k} (-1)^{r-i} q^{\binom{r-i}{2}} \frac{[j]!}{[j-r]![r-i]![i]!}$$

$$= \begin{bmatrix} j \\ i \end{bmatrix} \sum_{r} (-1)^{r-i} q^{\binom{r-i}{2}} \begin{bmatrix} j-i \\ r-i \end{bmatrix}$$

$$= \begin{cases} 1, & j = i; \\ 0, & j > i. \end{cases}$$

(The final equality follows from Corollary 9.2.3.) Thus $NB = NMC = C$ and the result follows. $\qquad\qquad\qquad\qquad\qquad\qquad\qquad\qquad\qquad\qquad\qquad\qquad\square$

Applying Lemma 9.6.1 to (9.5.2) we have that

$$A_i = \sum_{r=i}^{k} (-1)^{r-i} q^{\binom{r-i}{2}} \begin{bmatrix} r \\ i \end{bmatrix} D_r(k). \qquad (9.6.1)$$

9.6.2 Lemma. *The nonzero eigenvalues of $D_r(k)$ are*

$$(-1)^s q^{r(k-s)} q^{\binom{s}{2}} \begin{bmatrix} n-r-s \\ k-s \end{bmatrix} \begin{bmatrix} k-s \\ r-s \end{bmatrix}$$

for $s = 0, \ldots, r$.

Proof. Consider the following product of matrices:

$$D_r(k)D_s(k) = \overline{W}_{r,k}^T W_{r,k} \overline{W}_{s,k}^T W_{s,k}$$

$$= q^{s(k-r)} \begin{bmatrix} n-r-s \\ k-r \end{bmatrix} \overline{W}_{r,k}^T \overline{W}_{r,s} W_{s,k} \quad \text{by (9.5.1).}$$

If we expand the matrices $\overline{W}_{r,s}$ in this equation using Lemma 9.5.1, we find that $D_r(k)D_s(k)$ is equal to

$$q^{s(k-r)} \begin{bmatrix} n-r-s \\ k-r \end{bmatrix} \overline{W}_{r,k}^T \left(\sum_{i=0}^{s} (-1)^i q^{\binom{i}{2}} W_{i,r}^T W_{i,s} \right) W_{s,k}$$

$$= q^{s(k-r)} \begin{bmatrix} n-r-s \\ k-r \end{bmatrix} \sum_{i=0}^{s} (-1)^i q^{\binom{i}{2}} \begin{bmatrix} k-i \\ s-i \end{bmatrix} \overline{W}_{r,k}^T W_{i,r}^T W_{i,k}.$$

Applying (9.5.1) this becomes

$$q^{s(k-r)} \begin{bmatrix} n-r-s \\ k-r \end{bmatrix} \sum_{i=0}^{s} (-1)^i q^{\binom{i}{2}} q^{k(r-i)} \begin{bmatrix} n-i-k \\ r-i \end{bmatrix} \begin{bmatrix} k-i \\ s-i \end{bmatrix} \overline{W}_{i,k}^T W_{i,k}.$$

Let $E_r(k)$ be the r-th idempotent in $G(n, k)$, then $W_{i,k} E_r(k) = 0$ if $i < r$. This implies that the value of $D_r(k)D_s(k)E_s(k)$ is equal to

$$q^{s(k-r)} \begin{bmatrix} n-r-s \\ k-r \end{bmatrix} (-1)^s q^{\binom{s}{2}} q^{k(r-s)} \begin{bmatrix} n-s-k \\ r-s \end{bmatrix} \overline{W}_{s,k}^T W_{s,k} E_s(k)$$

$$= (-1)^s q^{\binom{s}{2}} q^{k(r-s)} q^{s(k-r)} \begin{bmatrix} n-r-s \\ k-r \end{bmatrix} \begin{bmatrix} n-s-k \\ r-s \end{bmatrix} D_s(k) E_s(k).$$

Thus

$$(-1)^s q^{s(k-r)} q^{\binom{s}{2}} q^{k(r-s)} \begin{bmatrix} n-r-s \\ k-r \end{bmatrix} \begin{bmatrix} n-s-k \\ r-s \end{bmatrix}$$
$$= (-1)^s q^{r(k-s)} q^{\binom{s}{2}} \begin{bmatrix} n-r-s \\ k-s \end{bmatrix} \begin{bmatrix} k-s \\ r-s \end{bmatrix}$$

is an eigenvalue for $D_r(k)$. $\qquad\square$

Putting these eigenvalues into (9.6.1), we get the eigenvalues for every matrix in the Grassmann scheme.

9.6.3 Theorem. *The eigenvalues of* A_i *in* $G(n, k)$ *are*

$$\lambda_j = \sum_{r=i}^{k} (-1)^{r-i+j} q^{\binom{r-i}{2}+\binom{j}{2}+r(k-j)} \begin{bmatrix} r \\ i \end{bmatrix} \begin{bmatrix} k-j \\ r-j \end{bmatrix} \begin{bmatrix} n-r-j \\ k-j \end{bmatrix}$$

where $i, j = 0, \ldots, k$. *The multiplicity of* λ_j *is* $\begin{bmatrix} n \\ k \end{bmatrix} - \begin{bmatrix} n \\ k-1 \end{bmatrix}$. $\qquad\square$

9.7 The EKR bound for q-Kneser graphs

A family $\mathcal{F}$ of subspaces of V is said to be *intersecting* if the intersection of any two elements of $\mathcal{F}$ is a nonzero subspace. If T is a 1-dimensional subspace of V, then the set of all k-dimensional subspaces of V that contain T is an intersecting family of size $\begin{bmatrix} n-1 \\ k-1 \end{bmatrix}$. We call such an intersecting family a *canonical intersecting family* .

The kth graph in the Grassmann scheme is the graph in which vertices are adjacent if the intersection of their corresponding vector spaces has dimension 0. This is the q-Kneser graph, and it is denoted by $K_q(n, k)$. A maximum coclique in this graph is a maximum intersecting family of subspaces. We will prove, using the ratio bound, that the canonical intersecting families are the cocliques of maximum size in $K_q(n, k)$.

From Theorem 9.6.3 we can directly calculate the eigenvalues of the q-Kneser graphs.

9.7.1 Corollary. *The eigenvalues of* $K_q(n, k)$ *are*

$$\lambda_j = (-1)^j q^{\binom{j}{2}+k(k-j)} \begin{bmatrix} n-k-j \\ k-j \end{bmatrix},$$

where $j = 0, \ldots, k$, *with multiplicity*

$$\begin{bmatrix} n \\ k \end{bmatrix} - \begin{bmatrix} n \\ k-1 \end{bmatrix}.$$

$\qquad\square$

In particular, the eigenvalues for $j = 0$ and 1 are

$$\lambda_0 = q^{k^2} \begin{bmatrix} n - k - j \\ k \end{bmatrix}, \qquad \lambda_1 = -q^{k(k-1)} \begin{bmatrix} n - k - j \\ k - 1 \end{bmatrix}.$$

These are the largest and least eigenvalues of $K_q(n, k)$, respectively. When we apply the ratio bound to the q-Kneser graphs, we get that

$$\alpha(K_q(n, k)) \leq \frac{\begin{bmatrix} n \\ k \end{bmatrix}}{1 - \frac{q^2 \begin{bmatrix} n-k \\ k \end{bmatrix}}{-q^{k(k-1)} \begin{bmatrix} n-k-1 \\ k-1 \end{bmatrix}}} = \begin{bmatrix} n - 1 \\ k - 1 \end{bmatrix}.$$

This result is the bound in the EKR Theorem for vector spaces.

9.8 Cocliques in the q-Kneser graphs

As we did with the Kneser graphs, we can use the fact that equality holds in the ratio bound to show that the only maximal cocliques in the q-Kneser graphs are the canonical cocliques. This proof first appeared in [138].

9.8.1 Theorem. *Assume that $n > 2k$ and that V is a vector space of dimension n. If $\mathcal{F}$ is a set of pairwise intersecting k-dimensional subspaces of V, then*

$$|\mathcal{F}| \leq \begin{bmatrix} n - 1 \\ k - 1 \end{bmatrix}$$

and equality holds if and only if $\mathcal{F}$ is a canonical intersecting family.

Proof. We have already shown that the bound holds; we only need to show that the canonical families are the only ones that meet this bound. Since equality holds in the ratio bound (Theorem 2.4.1), the characteristic vector of any coclique of maximum size lies in the sum of the first two eigenspaces. We find the maximum cocliques of the q-Kneser graphs using the same method that was used in Section 6.6 to find the maximum cocliques of the Kneser graphs.

Assume that $\mathcal{F}$ is a family of intersecting k-dimensional vector spaces, and let $v_{\mathcal{F}}$ be the characteristic vector for $\mathcal{F}$. From Theorem 9.4.1, the column space of $W_{1,k}^T$ is the sum of the λ_0-eigenspace and the λ_1-eigenspace, and we conclude that $v_{\mathcal{F}}$ is a linear combination of the columns of $W_{1,k}^T$. So there exists a vector x such that $W_{1,k} x^T = v_{\mathcal{F}}$. We now show that x has exactly one entry equal to 1 and all other entries equal to 0.

Fix a vertex a from $K_q(n, k)$ that is in $\mathcal{F}$. Sort the rows of $W_{1,k}^T$ so that this vertex is first, and all vertices that are adjacent to a are next. Further, sort the columns so that the 1-dimension subspaces of a are first. Then the matrix $W_{1,k}^T$

has the following block structure:

$$W_{1,k}^T = \left(\begin{array}{c|c} \mathbf{1} & 0 \\ \hline 0 & M \\ \hline W_1 & W_2 \end{array} \right).$$

We can write $x = (x_1, x_2)$ (using the same block structure), and since the fixed vertex a is in $\mathcal{F}$ we have that $\mathbf{1}x_1^T = 1$ and $Mx_2^T = 0$.

It suffices to show that M has full column rank, since with this result, we can conclude that $x_2 = 0$. It is not hard to see that W_1 contains a copy of the identity matrix (simply find vector spaces with 1-dimensional intersection with a). Since $v_{\mathcal{F}}$ is a 01-vector this implies that x_1 is also a 01-vector. Finally, the fact that $\mathbf{1}x_1 = 1$ implies that exactly one entry in x_1 is 1 and all other entries are 0. This proves that $\mathcal{F}$ is canonical.

So all that remains is to prove that M has full column rank. Consider the matrix $M^T M$. The rows and columns of this matrix are labeled by the 1-dimensional spaces of V that intersect trivially with a. From Lemma 9.3.2, the diagonal entries of $M^T M$ are all equal to

$$q^{k(k-1)} \begin{bmatrix} n-k-1 \\ k-1 \end{bmatrix}.$$

If b, c are 1-dimensional vector spaces, and the dimension of the span of $\{a, b, c\}$ is equal to $k+2$, then

$$(M^T M)_{b,c} = q^{k(k-2)} \begin{bmatrix} n-k-2 \\ k-2 \end{bmatrix}.$$

If the dimension of the span of $\{a, b, c\}$ is equal to $k+1$, then there are no subspaces that contain both b and c and are disjoint from a. This implies that $(M^T M)_{b,c} = 0$.

Define a graph whose vertices are the 1-dimensional subspaces that are skew to a. Two subspaces b and c are adjacent in this graph if the dimension of the span of $\{a, b, c\}$ is equal to $k+1$. This graph is isomorphic to $[n-k]$ copies of the complete graph K_{q^k}. Denote the adjacency matrix of this graph by L; then

$$M^T M = q^{k(k-1)} \begin{bmatrix} n-k-1 \\ k-1 \end{bmatrix} I + q^{k(k-2)} \begin{bmatrix} n-k-2 \\ k-2 \end{bmatrix} (J - I - L).$$

The eigenvalues of $J - I - L$ are

$$\theta_1 = q^{k+1}[n-k-1], \qquad \theta_2 = q^k, \qquad \theta_3 = 0.$$

Thus we can calculate the eigenvalues of $M^T M$ to be

$$\lambda_1 = q^{k(k-1)}[k]\begin{bmatrix} n-k-1 \\ k-1 \end{bmatrix},$$

$$\lambda_2 = q^{k(k-1)}\left(\begin{bmatrix} n-k-1 \\ k-1 \end{bmatrix} - \begin{bmatrix} n-k-2 \\ k-2 \end{bmatrix}\right),$$

$$\lambda_3 = q^{k(k-1)}\begin{bmatrix} n-k-1 \\ k-1 \end{bmatrix}.$$

Since $n > 2k$, none of these eigenvalues are equal to zero, and we conclude that $M^T M$ is invertible. Thus M has full column rank. $\qquad\square$

9.9 *t*-Intersecting families

A family of k-dimensional subspaces is called *t-intersecting* if the dimension of the intersection of any two subspaces from the family has dimension at least t. Frankl and Wilson [67] proved the following version of the EKR Theorem for t-intersecting families of subspaces.

9.9.1 Theorem. *Assume $n \geq 2k - t$. Then the size of a t-intersecting family of k-subspaces of an n-dimensional vector space over $GF(q)$ is at most*

$$\max\left\{\begin{bmatrix} n-t \\ k-t \end{bmatrix}, \begin{bmatrix} 2k-t \\ k \end{bmatrix}\right\}.$$

We refer to the bound given by Theorem 9.9.1 as the Frankl-Wilson bound. We note the following simplification.

9.9.2 Lemma. *If the Frankl-Wilson bound holds for t-intersecting subspaces when $n \geq 2k$, then it holds when $2k > n \geq 2k - t$.*

Proof. Choose a nondegenerate bilinear form β on our vector space V – the usual dot product will do. If U is a subspace of V, we define $U^\perp$ by

$$U^\perp = \{x \in V : \forall y \in U, \ \beta(x, y) = 0\}.$$

Then $U^\perp$ is a subspace of V and

$$\dim(U) + \dim(U^\perp) = \dim(V).$$

If $\mathcal{F}$ is a family of k-dimensional subspaces of V, then we define a second family

$$\mathcal{F}^\perp = \{U^\perp : U \in \mathcal{F}\}$$

of $(n - k)$-dimensional subspaces.

We claim that if $\mathcal{F}$ is t-intersecting, then $\mathcal{F}^\perp$ is $(n - 2k + t)$-intersecting. For if U_1 and U_2 are k-dimensional subspaces such that $\dim(U_1 \cap U_2) \geq t$, then

$$\dim(U_1 + U_2) \leq 2k - t.$$

Since

$$(U_1 + U_2)^\perp = U_1^\perp \cap U_2^\perp$$

we have

$$\dim(U_1^\perp \cap U_2^\perp) \geq n - 2k + t$$

and our claim follows.

Now suppose $\mathcal{F}$ is a t-intersecting family of k-dimensional subspaces and $n \leq 2k$. Then $\mathcal{F}^\perp$ is an $(n - 2k + t)$-intersecting family of subspaces of dimension $n - k$. Since $n \geq 2(n - k)$ we can apply the Frankl-Wilson bound and deduce that

$$|\mathcal{F}| \leq \max\left\{ \begin{bmatrix} n - (n - 2k + t) \\ n - k - (n - 2k + t) \end{bmatrix}, \begin{bmatrix} 2(n - k) - (n - 2k + t) \\ n - k \end{bmatrix} \right\}.$$

After a little thought we see that this is what we need. $\qquad\square$

In the next section, we prove Theorem 9.9.1 using an approach similar to the one used in Section 8.1.

9.10 The certifying matrix

Following Frankl and Wilson [67], define

$$\gamma_i = (-1)^{t-1-i} q^{(k-t)i + \binom{i}{2}} \begin{bmatrix} k - 1 - i \\ k - t \end{bmatrix} \begin{bmatrix} n - k - t + i \\ k - t \end{bmatrix}^{-1}$$

and

$$\Omega_q(n, k, t) = q^{-k^2 + k + \binom{i}{2}} \sum_{i=0}^{t-1} \gamma_i D_{k-i}(k).$$

Our goal is to use this matrix to define a certificate as described in Section 8.1. This proof is similar to the proof of the exact bound for the EKR Theorem for sets given in Section 8.4. We assume $n \geq 2k$, in which case the bound from Theorem 9.9.1 is $\begin{bmatrix} n-t \\ k-t \end{bmatrix}$, and we aim to show that

$$N = \Omega_q(n, k, t) + I - \begin{bmatrix} n - t \\ k - t \end{bmatrix}^{-1} J$$

is positive semidefinite. In showing that this matrix is positive semidefinite, we show that for the idempotents E_j with $j = 1, \ldots, t$ the corresponding eigenvalue of $\Omega_q(n, k, t)$ is equal to -1. This implies that $N E_j = 0$ for $j = 1, \ldots, t$. These two facts are sufficient for N to be a certificate.

We need a version of Theorem 8.4.1 for the Grassmann scheme. Fortunately, in this case the calculations are easier as the q-binomial coefficients grow more quickly than the binomial coefficients. By Equations (9.6.1) and (9.6.2) the eigenvalues of $\Omega_q(n, k, t)$ are

$$\theta_s = (-1)^s q^{-k^2+k+\binom{t}{2}} \sum_{i=0}^{t-1} \gamma_i q^{(k-i)(k-s)} q^{\binom{s}{2}} \begin{bmatrix} n-k+i-s \\ k-s \end{bmatrix} \begin{bmatrix} k-s \\ i \end{bmatrix} \quad (9.10.1)$$

for $s = 0, \ldots, r$.

9.10.1 Theorem. *If* $1 \le t \le k$, *then*

(a) $\theta_0 = \begin{bmatrix} n \\ k \end{bmatrix} \begin{bmatrix} n-t \\ k-t \end{bmatrix}^{-1} - 1$;

(b) $\theta_i = -1$ *for* $i = 1, \ldots, t$;

(c) $\theta_i > -1$ *for* $i \ge t+1$.

We prove this using four propositions. The first of these implies statements (a) and (b) and uses the matrix $W_{k,t}$ defined in Section 9.4. The next three propositions will together imply that (c) holds.

9.10.2 Proposition. *For* $t \le k$ *we have that* $\Omega_q(n, k, t) W_{t,k}^T = J - W_{t,k}^T$.

Proof. Expanding the matrix $\Omega_q(n, k, t)$, we have that

$$\Omega_q(n, k, t) W_{t,k}^T = q^{-k^2+k+\binom{t}{2}} \sum_{i=0}^{t-1} \gamma_i W_{k-i,k}^T \overline{W}_{k-i,k} W_{k,t}.$$

By (9.5.1) this is equal to

$$q^{-k^2+k+\binom{t}{2}} \sum_{i=0}^{t-1} \gamma_i q^{(k-i)(k-t)} \begin{bmatrix} n-t-k+i \\ k-t \end{bmatrix} W_{k-i,k}^T \overline{W}_{t,k-i}^T.$$

Let U be a k-dimensional subspace and V be a t-dimensional subspace with $\dim(U \cap V) = \ell$. Then, by Lemma 9.3.2 the (U, V)-entry of the above matrix is

$$q^{-k^2+k+\binom{t}{2}} \sum_{i=0}^{t-1} \gamma_i q^{(k-i)(k-t)} \begin{bmatrix} n-t-k+i \\ k-t \end{bmatrix} q^{(k-i)\ell} \begin{bmatrix} k-\ell \\ k-i \end{bmatrix}.$$

If $\ell \geq t$, then $\begin{bmatrix} k-\ell \\ k-i \end{bmatrix} = 0$ for every $i \leq t$, so this equation is clearly equal to 0. If $\ell < t$ then we expand γ_i to get that the value of this entry is

$$\sum_{i=0}^{t-1}(-1)^{t-1-i}q^{-k(t-1)+(k-i)\ell+\binom{i}{2}+\binom{t}{2}}\begin{bmatrix} k-i-1 \\ k-t \end{bmatrix}\begin{bmatrix} k-\ell \\ k-i \end{bmatrix}.$$

This can be written as

$$\sum_{i=0}^{t-1}(-1)^{t-1-i}q^{-k(t-1)+(k-i)\ell+\binom{i}{2}+\binom{t}{2}}\begin{bmatrix} (k-t+1)+(t-i-1)-1 \\ t-i-1 \end{bmatrix}\begin{bmatrix} k-\ell \\ k-i \end{bmatrix},$$

and applying (9.1.1), this becomes

$$\sum_{i=0}^{t-1}(-1)^{t+k-1-i}q^{(k-1)\ell-k(t-1)+\binom{i}{2}+\binom{t}{2}}$$

$$\times \left(q^{(k-t+1)(t-i-1)+\binom{t-i-1}{2}}\begin{bmatrix} t-k-1 \\ t-i-1 \end{bmatrix}\right)\begin{bmatrix} k-\ell \\ i-\ell \end{bmatrix}.$$

Finally, simplifying the exponent on q and applying Corollary 9.2.4, this reduces to $\begin{bmatrix} t-\ell-1 \\ t-\ell-1 \end{bmatrix} = 1$. $\qquad\square$

Part (a) of Theorem 9.10.1 follows from this proposition as

$$\Omega_q(n,k,t)\left(W_{t,k}^T\mathbf{1}\right) = J\mathbf{1} - W_{t,k}^T\mathbf{1} = \left(\begin{bmatrix} n \\ k \end{bmatrix}\begin{bmatrix} n-t \\ k-t \end{bmatrix}^{-1} - 1\right)(W_{t,k}^T\mathbf{1}).$$

Part (b) of Theorem 9.10.1 also holds since for any vector v that is orthogonal to $\mathbf{1}$,

$$\Omega_q(n,k,t)(W_{t,k}^Tv) = Jv - W_{t,k}^Tv = -\mathbf{1}(W_{t,k}^Tv).$$

Since $\mathrm{col}(W_{t,k}^T)$ is the span of all the ith eigenspaces of $J_q(n,k)$ where $i \leq t$, this implies that $\Omega_q(n,k,t)E_i(k) = -E_i(k)$ for $i = 1,\ldots,t$.

To complete the proof of Theorem 9.10.1, we need to prove that $\theta_i > -1$ for all $i > t$. We do this over the next three propositions. For the first of these propositions, we use the following two facts. First, if $b > a$, then

$$\frac{a-1}{b-1} < \frac{a}{b}.$$

Second, if $j \geq 1$, then

$$\frac{q^i-1}{q^j-1} < q^{i-j+1},$$

provided that $q \geq 2$.

9.10.3 Proposition. *The terms in the alternating sum for θ_i from* (9.10.1) *are strictly decreasing in absolute value for any i.*

Proof. The ratio of the j and the $(j + 1)$-term in the alternating sum is

$$q^{i-t+j} \left(\frac{q^{t-j-1} - 1}{q^{k-j-1} - 1} \right) \left(\frac{q^{n-2k+j+1} - 1}{q^{n-k-t+j+1} - 1} \right) \left(\frac{q^{k-i-j} - 1}{q^{j+1} - 1} \right) \left(\frac{q^{n-k-i+j+1} - 1}{q^{n-2k+j+1} - 1} \right)$$

$$= q^{i-t+j} \left(\frac{q^{t-j-1} - 1}{q^{k-j-1} - 1} \right) \left(\frac{q^{k-i-j} - 1}{q^{j+1} - 1} \right) \left(\frac{q^{n-k-i+j+1} - 1}{q^{n-k-t+j+1} - 1} \right)$$

$$< q^{i-t+j} q^{t-k} q^{k-i-2j} q^{t-i} = q^{t-i-j}.$$

Since $i \geq t + 1$ and $j \geq 1$, this is strictly smaller than 1. ☐

This implies that if the first term in the alternating sum for θ_i is positive, then θ_i is positive; and if the first term is negative, then θ_i is larger than the first term. Next we bound the size of this term, by first bounding the size of the first term in θ_{t+1}, and then showing that the first term of θ_{t+1} is larger, in absolute value, than the first term in any θ_i, where $i > t$.

9.10.4 Proposition. *The absolute value of the first term in the alternating sum for θ_{t+1} given in* (9.10.1) *is less than 1.*

Proof. The absolute value of the first term in θ_{t+1} is equal to

$$q^{t(t-k)} \begin{bmatrix} k - 1 \\ k - t \end{bmatrix} \begin{bmatrix} n - k - t \\ k - t \end{bmatrix}^{-1} \begin{bmatrix} n - k - t - 1 \\ k - t - 1 \end{bmatrix} = q^{t(t-k)} \frac{q^{k-1} - 1}{q^{n-k-t} - 1} \begin{bmatrix} k - 1 \\ k - t \end{bmatrix}.$$

Since $\begin{bmatrix} k-1 \\ k-t \end{bmatrix} < q^{t(k-t)}$, this is bounded above by

$$q^{t(t-k)} q^{t(k-t)} \frac{q^{k-1} - 1}{q^{n-k-t} - 1} = \frac{q^{k-1} - 1}{q^{n-k-t} - 1} < 1. \qquad ☐$$

Finally, we show that the first term in the alternating sum of every θ_i is smaller in absolute value than the first term in θ_{t+1}.

9.10.5 Proposition. *The absolute value of the first term in θ_j given in* (9.10.1) *is less than the first term in θ_{t+1}.*

Proof. The first term in θ_s is

$$q^{-k^2+k+\binom{t}{2}} q^{(k)(k-s)} q^{\binom{s}{2}} \begin{bmatrix} k - 1 \\ k - t \end{bmatrix} \begin{bmatrix} n - k - t \\ k - t \end{bmatrix}^{-1} \begin{bmatrix} n - k - s \\ k - s \end{bmatrix}$$

which is equal to

$$q^{k+\binom{t}{2}} q^{\binom{s}{2}-ks} \begin{bmatrix} k - 1 \\ k - t \end{bmatrix} \begin{bmatrix} n - k - t \\ k - t \end{bmatrix}^{-1} \begin{bmatrix} n - k - s \\ k - s \end{bmatrix}.$$

Both $q^{\binom{s}{2}-ks}$ and $\begin{bmatrix} n-k-s \\ k-s \end{bmatrix}$ are decreasing in s, so the entire function is decreasing in s. $\square$

Putting these three facts together gives the final result in Theorem 9.10.1. Combining Propositions 9.10.3, 9.10.4 and 9.10.5 proves Part (c) of Theorem 9.10.1.

9.11 Bilinear forms graphs

So far in this chapter, we have considered only the Grassmann graphs, but there are a number of other interesting graphs that can also be defined over finite vector spaces. In Section 4.7, we introduced dual polar graphs and the bilinear forms graphs. Any of these graphs can be realized as an induced subgraph of a Grassmann graph. In this section and the next, we summarize the known EKR Theorems for these graphs. Our treatment is sketchy and references, rather than proofs, are included.

According to the definition offered in Section 4.7, the vertices of a bilinear forms graph are the $m \times n$ matrices over a finite field, where matrices M and N are adjacent if $\text{rk}(M - N) = 1$. In this graph, two vertices M and N are at distance r exactly when $\text{rk}(M - N) = r$.

There is an alternative definition of this graph which makes it clear that it is an induced subgraph of a Grassmann graph. Let V be a vector space of dimension $m + n$. Choose a subspace W of dimension n, and take the vertices of the bilinear forms graph $B_q(m, n)$ to be the subspaces γ of V with dimension m such that $\dim(\gamma \cap W) = 0$. Two subspaces are adjacent if their intersection has dimension $m - 1$. From this description we see that the bilinear forms graph is an induced subgraph of the Grassmann graph $J_q(m + n, m)$.

If two vertices γ and δ in $B_q(m, n)$ are at distance no more than $m - t$ in $B_q(m, n)$, then the intersection of γ and δ is a vector space with dimension at least t. Such a pair of vector spaces are called t-*intersecting*. We can now state an EKR Theorem for these graphs; this result is due to Tanaka [160].

9.11.1 Theorem. *Let $B_q(m, n)$ be the bilinear forms graph. Let $\mathcal{F}$ be a subset of vertices from $B_q(m, n)$ such that any two are t-intersecting. Then the size of $\mathcal{F}$ is at most $q^{(m-t)n}$. If equality holds, then there are two possible outcomes:*

    (a) *the family $\mathcal{F}$ consists of the m-dimensional subspaces skew to W that contain a given subspace of dimension t;*

    (b) *we have $m = n$ and $\mathcal{F}$ consists of all m-dimensional subspaces skew to W that are contained in a fixed subspace U of dimension $2m - t$, with $\dim(U \cap W) = m - t$.* $\square$

Tanaka's proof of this result makes use of *Singleton systems*, which were defined and constructed by Delsarte [52]. These systems are Delsarte t-designs in the association scheme, and the approach using the clique-coclique bound described in Section 8.1 can be applied in this case.

9.12 Dual polar graphs

We defined the dual polar graphs in Section 4.7, but in this section we provide an alternate definition of these graph. We start by describing the *polar spaces* using the Buekenhout-Shult axioms (we recommend either Shult [152] or Cameron [38] for a more thorough introduction to polar spaces). Recall from Section 5.6 that a *partial linear space* is a collection of points and lines, such that any two points lie on at most one line; a *subspace* of a partial linear space is a subset S of its points such that if a line ℓ contains two points from S, then all points on ℓ belong to one line. A subspace is *singular* if each pair of points in it is collinear. With this terminology established, a polar space is a partial linear space that satisfies the following axioms:

(a) each line contains at least three points;
(b) no point is collinear with all the others;
(c) any chain of singular spaces has finite length;
(d) if a point p is not on a line ℓ, then p is collinear with either exactly one point on ℓ, or all points on ℓ.

One goal of the theory of polar spaces is to axiomatize the concept of orthogonality. One finds that each point in a polar space has an "orthogonal complement" $p^\perp$ that is a hyperplane (which might contain p). Further, condition (d) says for any line ℓ and point p, that $p^\perp$ contains ℓ or meets it in a single point. The maximum length of a chain of singular subspaces is the *rank* of the polar space. Generalized quadrangles are the polar spaces of rank 2. If the rank is at least 3, the singular subspaces are guaranteed to be projective spaces over a field.

We now have the terminology to give an alternate definition of the dual polar graph. This graph has the maximal singular subspaces of a polar space as its vertices; two maximal subspaces are adjacent if their intersection is maximal in each of them. Maximal subspaces are also referred to as *generators*. Dual polar graphs are discussed in more detail in Brouwer, Cohen and Neumaier [31, Chapter 9]; we note that there are six families of dual polar graphs and each one is associated with a group of Lie type.

For polar spaces, we have EKR Theorems for intersecting families of generators. Bounds on the size of intersecting families were first worked out by

Stanton [154]. (These bounds followed from a direct application of the ratio bound.) More recently Pepe, Storme and Vanhove [140] characterized the maximal intersecting families for all but one of the families of dual polar graphs, showing that in most cases these families consisted of the generators containing a given point. Ihringer and Metsch [97] and De Boeck [50] look at problems of EKR type in other projective and polar spaces. Tanaka [161] has characterized the narrow subsets in all known infinite families of distance-regular graphs with classical parameters and with unbounded diameter.

9.13 Exercises

9.1 Prove that the Grassmann graph is distance regular (and thus that the Grassmann scheme is metric).

9.2 Show that the Grassmann scheme is cometric.

9.3 Give an example of a vector space V with a subspace U that has the property that $U \cap U^{\perp}$ is not the zero subspace.

9.4 Verify that

$$\begin{bmatrix} n \\ k \end{bmatrix} = q^k \begin{bmatrix} n-1 \\ k \end{bmatrix} + \begin{bmatrix} n-1 \\ k-1 \end{bmatrix} = \begin{bmatrix} n-1 \\ k \end{bmatrix} + q^{n-k} \begin{bmatrix} n-1 \\ k-1 \end{bmatrix}.$$

9.5 Prove that if A and B are q-commuting variables then

$$(A+B)^n = \sum_{k=0}^{n} \begin{bmatrix} n \\ k \end{bmatrix} A^k B^{n-k}.$$

9.6 Prove the following identity:

$$\sum_{i=0}^{\min\{k,\ell\}} q^{(k-j)(\ell-j)} \begin{bmatrix} n-k \\ \ell-j \end{bmatrix} \begin{bmatrix} k \\ j \end{bmatrix} = \begin{bmatrix} n \\ \ell \end{bmatrix}.$$

9.7 Show that the matrices $D_i(k) = W_{i,k}^T \overline{W}_{i,k}$ form a basis for the Bose-Mesner algebra of the Grassmann scheme.

9.8 Let V be a vector space of dimension n over the finite field of order q. Let k, t be integers with $0 < t \leq k$. Suppose S is a subspace of V with dimension $t + 2i$ where $i = 0, \ldots, (n-t)/2$, and define $\mathcal{F}_i$ to be the set of subspaces α of dimension k such that

$$\dim(\alpha \cap S) \geq t + i.$$

(a) Show that any two subspaces in $\mathcal{F}_i$ intersect in a vector space with dimension at least t.

(b) Show that the size of $\mathcal{F}_i$ is

$$\sum_{j=t+i}^{t+2i} q^{(t+2i-j)(k-j)} \begin{bmatrix} t+2i \\ j \end{bmatrix} \begin{bmatrix} n-(t+2i) \\ k-j \end{bmatrix}.$$

(c) Prove that if $n > 2k$ then $\mathcal{F}_0$ is larger than $\mathcal{F}_{k-t}$. Next show that if $n < 2k$ then $\mathcal{F}_{k-t}$ is larger than $\mathcal{F}_0$. Finally, prove that if $n = 2k$ then these sets are the same size.

9.9 At the start of Section 9.11 we offered two definitions of bilinear forms graphs. Prove that these definition are equivalent.

Notes

Unlike the case for sets, Theorem 9.8.1 holds for $n = 2k$; this is shown and discussed in detail in [138].

Theorem 9.8.1 was first proven by Hsieh; this proof is combinatorial and very technical. Another proof was given by Frankl and Wilson in [67]; their proof is similar to the one presented in this chapter.

There have been several attempts to find a proof of Theorem 9.8.1 using shifting techniques (see [53] and [48]). But in both of these papers the shifting technique is flawed, and there currently is no known appropriate shifting on vector spaces. There is also no known version of Katona's cycle proof for vector spaces (see Exercise 1.7 in Chapter 1). Problem 24E in [166] shows how the obvious generalization of this proof will not give the correct bound.

Several of the extensions of the EKR Theorem also hold for vector spaces. For example, Tanaka [160] proved a version of the EKR Theorem for t-intersecting vector spaces. Further, a result analogous to the Hilton-Milner Theorem holds for vector spaces [21]. See [22] for a survey paper on results from extremal set theory that have extensions to vector spaces.

10

The Hamming scheme

Let Q be an alphabet of q symbols. A *word* of length n over Q is an n-tuple of elements of Q, that is, an element of Q^n. Two words x and y in Q^n are at *Hamming distance* r if they differ at exactly r coordinates. We use $h(x, y)$ to denote the Hamming distance between words x and y. In many cases our alphabet Q will consist of the elements of an abelian group, usually $\mathbb{Z}_q$. In this case Q^n is again a group and

$$h(x, y) = h(0, y - x).$$

The Hamming distance of z from 0 is the *weight* of z.

The *Hamming scheme* $H(n, q)$ is the association scheme with vertex set Q^n, where two words are i-related in the scheme if they are at Hamming distance i. This scheme has n classes and the graphs in the scheme are denoted X_i where $i = 0, \ldots, n$. In general X_1 is known as the *Hamming graph*. For any Q, the graph X_1 is distance regular with diameter n. (This is a nice exercise in the use of Hamming distance.) Equivalently, the Hamming scheme is metric with respect to X_1.

The Hamming scheme depends only on the size q of Q, not on the specific elements of Q, so we denote it by $H(n, q)$. For $H(n, q)$, our basic EKR problem is to determine the maximum possible size of a set of words such that any two agree on at least t coordinates. If two words agree on at least t coordinates, then the distance in the Hamming graph between the words is at most $n - t$. This means that a set of words in which any two agree on at least t-coordinates is a set with width at most $n - t$ in X_1. In this chapter we are interested in determining the maximum size of such sets and characterizing the sets that have this size.

10.1 Eigenvalues of the Hamming scheme

In the binary Hamming scheme $H(n, 2)$, the graph X_1 is the *n-hypercube* and X_n is a matching with 2^{n-1} edges. The graphs in the binary Hamming scheme are the distance matrices on the *n*-cube – two words are *i*-related in $H(n, 2)$ exactly when they are at distance i in the *n*-cube. We will see that this generalizes to the case $q > 2$ and that this generalization provides convenient access to the basic properties of the Hamming scheme.

Consider the *direct product* $X \times Y$ of graphs X and Y. This is the graph whose vertex set is $V(X) \times V(Y)$, and vertices (x_1, y_1) and (x_2, y_2) are adjacent if and only if x_1 and x_2 are adjacent in X and y_1 and y_2 are adjacent in Y. The adjacency matrix of the direct product is

$$A(X \times Y) = A(X) \otimes A(Y).$$

10.1.1 Lemma. *If $A_i(n)$ is the adjacency matrix of the ith graph in $H(n, q)$, then*

$$\sum_{i=0}^{n} t^i A_i(n) = (I_q + t(J_q - I_q))^{\otimes n}.$$

Proof. By induction on n, we have that

$$A_1(n + 1) = (I_q \otimes A_1(n)) + ((J_q - I_q) \otimes A_0(n)).$$

In graph theoretic terms, this says that $X_1(n + 1)$ is the Cartesian product of $X_1(n)$ and the complete graph K_q. More generally, we find that

$$A_i(n + 1) = (I_q \otimes A_i(n)) + ((J_q - I_q) \otimes A_{i-1}(n)).$$

If we assume that $A_m(n) = 0$ when $m < 0$ or $m > n$, then this equation yields

$$\sum_{i=0}^{n} t^i A_i(n + 1) = (I_q + t(J_q - I_q)) \otimes \sum_{i=0}^{n} t^i A_i(n),$$

from which the lemma follows at once. $\qquad \square$

It clearly follows from this lemma that

$$A_n(n) = (J_q - I_q)^{\otimes n},$$

and that the graph X_n in $H(n, q)$ is isomorphic to the direct product of n copies of the complete graph K_q. This isomorphism implies an order on the vertices of X_n, and we assume that the vertices of $H(n, q)$ are ordered this way. Specifically, if we have words $a = (a_1, a_2, \ldots, a_n)$ and $b = (b_1, b_2, \ldots, b_n)$, then a occurs before b if there is some index i for which $a_j = b_j$ when $1 \leq j \leq i - 1$ and $a_i < b_i$. This is the *lexicographic order* on words.

Let $\mathcal{E} = \{x_0, \ldots, x_{q-1}\}$ be an orthogonal set of eigenvectors for K_q with $x_0 = \mathbf{1}$. If

$$y = y_1 \otimes \cdots \otimes y_n, \tag{10.1.1}$$

where each $y_i \in \mathcal{E}$, define the *weight* of y to be the number of indices i such that $y_i \neq \mathbf{1}$. Each y_i is an eigenvector for $I_q + t(J_q - I_q)$. It follows that y is an eigenvector for $(I_q + t(J_q - I_q))^{\otimes n}$, and thus it is also an eigenvector for $\sum_i t^i A_i(n)$. If the weight of y is j, then its eigenvalue is

$$(1 + (q - 1)t)^{n-j}(1 - t)^j.$$

If $p_i(j)$ is the jth eigenvalue for X_i, it follows that

$$\sum_{i=0}^{n} p_i(j)t^i = (1 + (q - 1)t)^{n-j}(1 - t)^j. \tag{10.1.2}$$

10.1.2 Theorem. *The eigenvalues of $H(n, q)$ are*

$$p_i(j) = \sum_{r=0}^{j}(-1)^r(q - 1)^{i-r}\binom{n - j}{i - r}\binom{j}{r}$$

with multiplicity $(q - 1)^j \binom{n}{j}$ for $j = 0, \ldots, n$.

Proof. The values $p_i(j)$ are easily found by expanding the right-hand side of (10.1.2) and comparing the coefficients of t^i.

It remains to determine the multiplicity of $p_i(j)$. This can be found by counting the eigenvectors y of weight i, and noting that they are pairwise orthogonal. (We leave the details as an exercise.) $\square$

The eigenvalues of the Hamming scheme are encoded by a class of orthogonal polynomials. These are the polynomials

$$K_i(x; n) = \sum_{r=0}^{x}(-1)^r(q - 1)^{i-r}\binom{n - x}{i - r}\binom{x}{r},$$

which is a *Krawtchouk polynomial*. By the previous theorem $K_i(j; n) = p_i(j)$.

This result shows that $p_i(j)$ is a polynomial of degree i in j. It follows that the eigenvalues of A_i are given by a polynomial of degree i in the eigenvalues of A_1, and that the matrix A_i is a polynomial in A_1 of degree i. Since the graph X_1 is distance regular, this fact also follows from Corollary 4.1.3.

We note two useful special cases of Theorem 10.1.2. Setting $i = 1$ we find that the eigenvalues of the Hamming graph X_1 are

$$(q - 1)(n - j) - j,$$

where $j = 0, \ldots, n$. Setting $i = n$ shows that the eigenvalues of X_n are

$$(-1)^j (q - 1)^{n-j},$$

where $j = 0, \ldots, n$.

The ratio bound for cliques (Corollary 3.7.2) gives

$$\omega(X_1) \le 1 - \frac{n(q - 1)}{-n} = q.$$

This bound is met by the set of all the words with the same entry in the first $n - 1$ coordinates.

For the graph X_n the ratio bound on the size of a clique is

$$\omega(X_n) \le 1 + \frac{(q - 1)^n}{(q - 1)^{n-1}} = q.$$

This bound is tight, as the set of all the words that contain only one symbol forms a clique of size q. It also follows from the ratio bound for cocliques, Theorem 2.4.1, that

$$\alpha(X_n) \le \frac{q^n}{1 + \frac{(q-1)^n}{(q-1)^{n-1}}} = q^{n-1}.$$

The set of words with ith coordinate equal to some $a \in Q$ is a coclique of this size. Since equality holds for both ratio bounds, equality also holds for the clique-coclique bound for X_n.

A set of words with width at most $n - 1$ in X_1 is a coclique in X_n. From the ratio bound we see that such a set can have size no more than q^{n-1}. Thus, a set of words in which any two agree in at least one coordinate can have size no more than q^{n-1}. This is the bound in the EKR Theorem for words that we will see in Section 10.4.

10.2 Idempotents

In this section we establish relationships between the adjacency matrices of the Hamming scheme and the idempotents. We continue to use E_i to denote the ith principal idempotent of the association scheme.

10.2.1 Theorem. *In $H(n, q)$ we have*

$$\sum_{i=0}^{n} t^i A_i = \sum_{j=0}^{n} (1 + (q - 1)t)^{n-j} (1 - t)^j E_j.$$

Proof. Combining

$$A_i = \sum_{j=0}^{n} p_i(j) E_j,$$

and (10.1.2) gives that

$$\sum_{i=0}^{n} t^i A_i = \sum_{i=0}^{n}\sum_{j=0}^{n} t^i P_i(j) E_j = \sum_{j=0}^{n}(1+(q-1)t)^{n-j}(1-t)^j E_j. \qquad \square$$

10.2.2 Corollary. *In* $H(n,q)$ *we have*

$$q^n \sum_{j=0}^{n} s^j E_j = (s(qI_q - J_q) + J_q)^{\otimes n}.$$

Proof. We know from Lemma 10.1.1 that

$$\sum_{i=0}^{n} t^i A_i = (I + t(J - I))^{\otimes n},$$

and by Theorem 10.2.1 that

$$\sum_{i=0}^{n} t^i A_i = \sum_{j=0}^{n}(1+(q-1)t)^{n-j}(1-t)^j E_j.$$

Since

$$\sum_{j=0}^{n}(1+(q-1)t)^{n-j}(1-t)^j E_j = (1+(q-1)t)^n \sum_{j=0}^{n}\left(\frac{1-t}{1+(q-1)t}\right)^j E_j,$$

we have that

$$(I + t(J - I))^{\otimes n} = (1+(q-1)t)^n \sum_{j=0}^{n}\left(\frac{1-t}{1+(q-1)t}\right)^j E_j.$$

If we set t in this equation equal to $(1-s)(1+(q-1)s)^{-1}$, we find that

$$\left(I + \frac{1-s}{1+(q-1)s}(J-I)\right)^{\otimes n} = \left(\frac{q}{1+s(q-1)}\right)^n \sum_{j=0}^{n} s^j E_j.$$

Simple manipulation of this equation gives the result. $\qquad \square$

This proof also implies that

$$\sum_{i=0}^{n} t^i A_i = (1+(q-1)t)^n \sum_{j=0}^{n}\left(\frac{1-t}{1+(q-1)t}\right)^j E_j. \qquad (10.2.1)$$

Setting $t = (1-s)(1+(q-1)s)^{-1}$ in this equation gives the following corollary.

10.2.3 Corollary. *In* $H(n,q)$,

$$q^n \sum_{j=0}^{n} s^j E_j = \sum_{i=0}^{n}(1+(q-1)s)^{n-i}(1-s)^i A_i. \qquad \square$$

Again, we can expand the polynomials on the right-hand side and compare the coefficient of s^j. This gives an expression for the idempotent E_i in terms of the matrices A_i, namely

$$E_j = \frac{1}{q^n} \sum_{i=0}^{n} \sum_{r=0}^{j} (-1)^r \binom{n-i}{j-r} \binom{i}{r} (q-1)^{j-r} A_i.$$

10.2.4 Theorem. *The dual eigenvalues $q_j(i)$ of $H(n,q)$ are given by*

$$q_j(i) = \sum_{r=0}^{j} (-1)^r \binom{n-i}{j-r} \binom{i}{r} (q-1)^{j-r}. \qquad \square$$

Just as $p_i(j)$ is a polynomial of degree i in j, we see from this theorem that $q_j(i)$ is a polynomial of degree j in i. Similar to the case for the matrices A_i, we can conclude that the matrix E_j is a polynomial of degree j in E_1. This shows that the association scheme $H(n,q)$ is cometric relative to E_1.

10.3 Partial words

In this section, we establish the existence of a third basis for the Bose-Mesner algebra of $H(n,q)$.

If we set

$$N = \{1, \ldots, n\}$$

we can view a word of length n over an alphabet Q as a function from N to Q. If w is a word and $S \subseteq N$, then we call the restriction of w to S a *partial word*. For a partial word f, the support of f (denoted supp(f)) is the subset of N where f is defined. The *size* of a partial word is the size of the support.

If f and g are partial words, we write $f \leq g$ if supp(f) $\subseteq$ supp(g) and f and g agree on supp(f). This is a partial order on partial words, and as an exercise you might show that any two partial words f and g have a greatest lower bound (which we denote by $f \wedge g$). A partial word f is *incident* with the word w if $f \leq w$. We use $W_{r,n}$ to denote the incidence matrix of partial words of size r versus words of length n.

With all this notation at hand, we see that if α and β are vertices of $H(n,q)$ whose greatest lower bound is a partial word with size s, then

$$(W_{r,n}^T W_{r,n})_{\alpha,\beta} = \binom{s}{r}.$$

We define matrices $C_0, \ldots, C_n$ by

$$C_r = W_{r,n}^T W_{r,n}$$

and note that

$$C_r = \sum_{i=0}^{n-r} \binom{n-i}{r} A_i. \tag{10.3.1}$$

This shows that the matrices C_r lie in the Bose-Mesner algebra of the Hamming scheme. From (10.3.1) we find that

$$\sum_{r=0}^{n} s^r C_r = \sum_{r=0}^{n}\sum_{i=0}^{n-r} \binom{n-i}{r} s^r A_i = \sum_{i=0}^{n}\sum_{r=0}^{n-i} \binom{n-i}{r} s^r A_i = \sum_{i=0}^{n}(1+s)^{n-i} A_i.$$

If we set $s + 1 = t$, this becomes

$$\sum_{i=0}^{n} t^{n-i} A_i = \sum_{r=0}^{n}(t-1)^r C_r.$$

This shows that each A_i can be expressed as a linear combination of the matrices $C_0, \ldots, C_n$, and we have the following result.

10.3.1 Lemma. *The matrices $C_0, \ldots, C_n$ form a basis for the Bose-Mesner algebra of the Hamming scheme.* □

Further, the next result shows that the column space of C_r is the direct sum of the first r eigenspaces of $H(n, q)$.

10.3.2 Lemma. *In $H(n, q)$ for any $r = 1, \ldots, n$,*

$$C_r = q^{n-r} \sum_{j=0}^{r}(-1)^j(q-1)^{r-j} \binom{n-j}{r-j} E_j.$$

Proof. From (10.2.1), we have that

$$\sum_{j}(1+(q-1)t)^{n-j}(1-t)^j E_j = \sum_{i} t^i A_i = \sum_{r}(t-1)^r C_r.$$

Setting t equal to $1 + s$ and rearranging yields that

$$\sum_{r} s^r C_r = \sum_{j}(q + (q-1)s)^{n-j}(-s)^j E_j.$$

The eigenvalue of C_r on the jth eigenspace equals the coefficient of s^r in

$$(q + (q-1)s)^{n-j}(-s)^j,$$

which is

$$(-1)^j q^{n-r}(q-1)^{r-j} \binom{n-j}{n-r}.$$ □

10.4 The EKR Theorem for words

In Section 10.1 we saw that the clique-coclique bound holds with equality in X_n. The maximum cliques have size q. So any set of words such that any two agree on at least one coordinate has size at most q^{n-1}. The set of all words with the ith coordinate equal to a is an example of a set of this size. We call such a set of words a *canonical coclique*, and we aim to prove that any coclique in X_n of size q^{n-1} is canonical. If we consider two words that agree on at least one coordinate to be intersecting, then this is the natural analog of the EKR Theorem for words.

First we need to set up some machinery. We view a function on $V(H(n, q))$ as a vector of length $|V(H(n, q))|$ and say that such a function is *affine* if it lies in the sum of the first two eigenspaces of $H(n, q)$. If $u \in \mathbb{R}^q$, let $T_r(u)$ denote the Kronecker product of u with $n - 1$ copies of $\mathbf{1}$, with the u in the rth position. So

$$T_1(u) = u \otimes \mathbf{1} \otimes \cdots \otimes \mathbf{1}.$$

Our next lemma is an immediate consequence of our description of a basis of eigenvectors for $H(n, q)$, which we presented in Section 10.1.

10.4.1 Lemma. *A function f on the vertices of $H(n, q)$ is affine if and only if there are vectors $u_1, \ldots, u_n$ in $\mathbb{R}^q$ such that $\mathbf{1}^T u_j = 0$ and*

$$f = f_0 \mathbf{1} + \sum_{i=1}^{n} T_i(u_i). \qquad \square$$

Suppose ρ is a partial word for $H(n, q)$. Define $W(\rho)$ to be the set of words w such that $w \geq \rho$. If $s = |\operatorname{supp}(\rho)|$, the words in $W(\rho)$ form an association scheme isomorphic to $H(n - s, q)$. More precisely, the restriction to the complement of $\operatorname{supp}(\rho)$ of the words in $W(\rho)$ can be identified with the vertices of $H(n - s, q)$, and the Hamming distance between two such restrictions is equal to the Hamming distance between the original words in $H(n, q)$.

10.4.2 Lemma. *If ρ is a partial word for $H(n, q)$ and φ is an affine function on the vertices of $H(n, q)$, then the restriction of φ to the words in $W(\rho)$ is affine.*

Proof. Suppose

$$w = (a_1, \ldots, a_n)$$

is a word and f is as in Lemma 10.4.1. Then (recalling the words are ordered in the lexicographic ordering) we see that the wth entry of f (viewed as a length-q^n vector) is

$$f(w) = f_0 + \sum_{i=1}^{n} u_i(a_i),$$

where $u_i(a_i)$ is the a_i-th entry of the vector u_i.

The entry in the final coordinate of the word w is a_n, so we can rewrite this as

$$f(w) = f_0 + u_n(a_n) + \sum_{i=1}^{n-1} u_i(a_i),$$

and we see that f is affine on the set of all words w that have the same entry in the final coordinate. The lemma follows directly from this fact. □

10.4.3 Lemma. *Any coclique of size q^{n-1} in the nth distance graph of $H(n, q)$ is canonical.*

Proof. Let x be the characteristic vector of a coclique S of size q^{n-1}. Since the ratio bound is tight, x lies in the sum of the first two eigenspaces and accordingly it is affine.

Let ρ be a partial word with support of size $n - 2$ and assume $W = W(\rho)$. Then x is affine on the complement of supp(ρ) as well. So there are vectors $u_1, u_2 \in \mathbb{R}^q$ with $\mathbf{1}^T u_1 = \mathbf{1}^T u_2 = 0$ and

$$x = x_0 \mathbf{1} + T_1(u_1) + T_2(u_2).$$

Thus if $w \in W$ and the restriction of w to the complement of supp(ρ) is (a_i, a_j), then

$$x(w) = x_0 + u_1(a_i) + u_2(a_j),$$

where x_0 is constant and $u_1(a_i)$ and $u_2(a_j)$ are the a_ith and a_jth entries in the vectors u_1 and u_2. Since this holds for all $w \in W$, there is an entry of x equal to $x_0 + u_1(a_i) + u_2(a_j)$ for all a_i and a_j in Q.

Since x is a characteristic vector, $x(w)$ is either 0 or 1. This implies that the entries of u_1 and u_2 take at most two different values. If the entries of u_1 and u_2 each take two distinct values, then $u_1(a_i) + u_2(a_j)$ takes at least three values as a_i and a_j vary. We conclude that one of u_1 and u_2 is constant.

By considering partial words with smaller support, we see that this applies to x outside the complement of supp(ρ), so we can conclude that

$$x = x_0 \mathbf{1} + \sum_i T_i(u_i)$$

where all but one of the vectors u_i is constant and all the vectors u_i are orthogonal to $\mathbf{1}$. Since the only constant vector that sums to zero is the zero vector, we can assume that

$$x = x_0(\mathbf{1} \otimes \mathbf{1} \otimes \cdots \otimes \mathbf{1}) + (u_1 \otimes \mathbf{1} \otimes \cdots \otimes \mathbf{1})$$
$$= (x_0\mathbf{1} + u_1) \otimes \mathbf{1} \otimes \cdots \otimes \mathbf{1}.$$

As $\mathbf{1}^T x = |S| = q^{n-1}$ we have $x_0 = 1/q$, and since x is a 01-vector, this implies that the entries of u_1 must be $(q-1)/q$ and $-1/q$. Since $\mathbf{1}^T u_1 = 0$, exactly one entry is equal to $(q-1)/q$. Accordingly we can assume that

$$u_i = e_i - \frac{1}{q}\mathbf{1},$$

and therefore

$$x = e_i \otimes \mathbf{1} \cdots \otimes \mathbf{1}. \qquad \square$$

10.5 The complete EKR for words

Any two words that agree in at least t positions are called *t-intersecting*. In keeping with our terminology, any set of words in which any two words are t-intersecting is called a *t-intersecting sets of words*. These sets are exactly the sets of vertices from $H(n, q)$ that have width at most $n - t$.

Ahlswede and Khachatrian [4] determined the largest sets of t-intersecting words from Q^n for all values of n, t and q. The method used by Ahlswede and Khachatrian to prove this result for words is similar to the method they used to prove the complete EKR Theorem in [3]. They showed that the largest t-intersecting sets in $H(n, q)$ are the sets of all words that have the same entry in at least $t + i$ coordinates from a fixed set of $t + 2i$ coordinates, where i is an integer that depends on n, q and t. We now define this more precisely.

For a word w in $H(n, q)$, define the *zeros* of w to be the set

$$Z(w) = \{j : w_j = 0\}.$$

For $i \in \{0, \ldots, (n - t)/2\}$ define the set of words

$$\mathcal{K}_i = \{w : |Z(w) \cap \{1, 2, \ldots, t + 2i\}| \geq t + i\}.$$

A set of words K is *isomorphic* to $\mathcal{K}_i$ if it can be obtained by simultaneously permuting the indices of the words and permuting the elements in Q in all the words in $\mathcal{K}_i$. Any set isomorphic to $\mathcal{K}_i$ is t-intersecting. Further, any set isomorphic to $\mathcal{K}_0$ consists of all the words that have the same entry in t fixed coordinates. Any set of words that is isomorphic to $\mathcal{K}_0$ will be called a *canonical t-intersecting set of words*.

10.5.1 Theorem. *For $q \geq 2$, let r be the largest non-negative integer such that*

$$t + 2r < \min \left\{ n + 1, t + 2 \left(\frac{t-1}{q-2} \right) \right\}$$

(with $t + 2r < n + 1$, when $q = 2$). The size of the largest set of t-intersecting words of length n with entries from Q is $|\mathcal{K}_r|$. If $s = (t-1)/(q-2) \in \mathbb{Z}$ and

$$t + 2 \left(\frac{t-1}{q-2} \right) \leq n,$$

then $|\mathcal{K}_s| = |\mathcal{K}_{s-1}|$. Otherwise, the only sets that have size $|\mathcal{K}_r|$ are isomorphic to $\mathcal{K}_r$. □

This theorem is also known as the diametric theorem, and a proof of this theorem can be found in [2, Lecture 2].

In the case where $t = 1$, this theorem reduces to Lemma 10.4.3. If $q = 2$, then the largest set of t-intersecting words is isomorphic to $\mathcal{K}_r$ where $r = \lfloor (n-t)/2 \rfloor$. If $q \geq t + 2$, then by this theorem, the largest t-intersecting set has size q^{n-t} and the only sets that meet this bound are the canonical t-intersecting sets. This result can be considered the natural generalization of the strength-t EKR Theorem for words and is stated next.

10.5.2 Corollary. *Let $\mathcal{A}$ be a set of t-intersecting words from $H(n, q)$. If $q \geq t + 2$, then*

$$|\mathcal{A}| \leq q^{n-t}.$$

Moreover, the only sets that meet this bound are canonical t-intersecting sets.

In the following sections we give a proof of a stronger result, due to Moon [135], that implies this corollary.

10.6 Cross-intersecting sets of words

If $\mathcal{A}$ and $\mathcal{B}$ are two sets of words from $H(n, q)$, then $(\mathcal{A}, \mathcal{B})$ are *cross t-intersecting* if every word from $\mathcal{A}$ is t-intersecting with every word from $\mathcal{B}$ (this is the same idea of cross intersection that is discussed in Section 1.7). A pair of two such sets of words is called a *cross t-intersecting pair*. A cross 1-intersecting pair is a cross coclique in the graph X_n from the Hamming scheme. Further, a cross t-intersecting pair is a cross coclique in

$$\bigcup_{i=n-t}^{n} X_i.$$

We will prove that the maximum value of $|\mathcal{A}|\,|\mathcal{B}|$ for a cross t-intersecting pair $(\mathcal{A}, \mathcal{B})$ is achieved if and only if $\mathcal{A}$ and $\mathcal{B}$ are the same canonical t-intersecting set of words (we will see that Corollary 10.5.2 follows directly from this). This result was proven by Moon in [135] and the proof we give in the following sections is the one presented in Moon's paper.

10.6.1 Theorem. *Let $(\mathcal{A}, \mathcal{B})$ be a cross t-intersecting pair from Q^n. If $q \geq t + 2$, then*

$$|\mathcal{A}|\,|\mathcal{B}| \leq q^{2(n-t)}.$$

Moreover, if this bound is met, then $\mathcal{A}$ and $\mathcal{B}$ are equal to the same canonical t-intersecting set of words.

This theorem gives an alternate proof of Corollary 10.5.2. Indeed, if $\mathcal{C}$ is any set of t-intersecting words, then $(\mathcal{C}, \mathcal{C})$ is a cross t-intersecting pair. Theorem 10.6.1 implies that $|\mathcal{C}| \leq q^{n-t}$ and equality holds if and only if $\mathcal{C}$ a canonical set of t-intersecting words.

10.7 An operation on words

Our proof of Theorem 10.6.1 uses induction on the length of the words. We introduce a simple operation to move from words of length n to words of length $n - 1$. For a word x, let $\partial_i(x)$ be the word of length $(n - 1)$ obtained by removing the ith entry of x. Similarly, for a set $\mathcal{A}$ of words, define

$$\partial_i(\mathcal{A}) = \{\partial_i(x) : x \in \mathcal{A}\}.$$

For a cross t-intersecting pair $(\mathcal{A}, \mathcal{B})$ and $i \in \{1, \ldots, n\}$, we define $\mathcal{A}_i$ to be the subset of words $x \in \mathcal{A}$ for which $\partial_i(x)$ and $\partial_i(y)$ are still t-intersecting for all $y \in \mathcal{B}$. We call the set $\mathcal{A}_i$ a *derived set* of $\mathcal{A}$. The set $\mathcal{A}_i$ is the set of words from $\mathcal{A}$ for which the ith entry of the word is not essential to the intersection property of the sets. Consider the example where $\mathcal{A}$ and $\mathcal{B}$ are both $\mathcal{K}_0$. In this case, $\mathcal{A}_i = \emptyset$ and $\mathcal{B}_i = \emptyset$ for $i = 1, \ldots, t$, while $\mathcal{A}_i = \mathcal{A}$ and $\mathcal{B}_i = \mathcal{B}$ for all $i > t$. In this example, the first t-entries of any word are all essential for the intersection property, whereas the final $n - t$ entries are never essential.

In the next section, we prove that if $|\mathcal{A}|\,|\mathcal{B}| \geq q^{2(n-t)}$, then there are $n - t$ indices i such that $\mathcal{A}_i = \mathcal{A}$ and $\mathcal{B}_i = \mathcal{B}$; this implies that $\mathcal{A}$ and $\mathcal{B}$ are the same canonical t-intersecting set.

We start with two simple lemmas that will be used to bound the size of $\partial_i(\mathcal{A})$ where $(\mathcal{A}, \mathcal{B})$ is a cross t-intersecting pair with $|\mathcal{A}|\,|\mathcal{B}|$ of maximum size.

10.7.1 Lemma. *Let $(\mathcal{A}, \mathcal{B})$ be a cross t-intersecting pair. Assume $x, y \in \mathcal{A}$. If $x \notin \mathcal{A}_i$ and $\partial_i(x) = \partial_i(y)$, then $x = y$.*

Proof. Assume that $x \notin A_i$. Then there is a word $z \in B$ such that $\partial_i(x)$ and $\partial_i(z)$ are not t-intersecting. Since x and z are t-intersecting, $\partial_i(x)$ and $\partial_i(z)$ are $(t - 1)$-intersecting and the ith entry of x must equal the ith entry of z. By assumption, $\partial_i(x) = \partial_i(y)$ so it must be that $\partial_i(y)$ and $\partial_i(z)$ are also $(t - 1)$-intersecting. Similarly, the ith entry of y must equal the ith entry of z. Thus $x = y$. $\qquad\square$

10.7.2 Lemma. *Let* (A, B) *be a cross t-intersecting pair with* $|A| \, |B|$ *of maximum size. If* $x \in A_i$, *and* y *is any word such that* $\partial_i(x) = \partial_i(y)$, *then* $y \in A$.

Proof. Clearly y is t-intersecting with every word in B, and since A is maximal, $y \in A$. $\qquad\square$

This lemma implies that each word in $\partial_i(A_i)$ is a subword of exactly q words in A_i, and thus that

$$|\partial_i(A_i)| = \frac{1}{q}|A_i|.$$

10.8 Bounds on the size of a derived set

Throughout this section we assume that (A, B) is a maximal cross t-intersecting pair with

$$|A| \, |B| \geq q^{2(n-t)}.$$

To prove Theorem 10.6.1, we need to show that A and B are the same canonical t-intersecting sets. To do this, we show that $A_i = A$ and $B_i = B$ for $n - t$ indices.

Our approach is to use induction on the length of the words. The base case for our induction is the words of length t. In this case it is clear that this result holds. For the induction hypothesis, we assume that Theorem 10.6.1 is true for words of length $n - 1 \geq t$. Assuming this induction hypothesis, we will show for length-n words that $A_i = A$ and $B_i = B$ for $n - t$ indices. To do this, we first prove three lemmas. The first of these shows that for all i, either $A_i = A$, or A_i is small.

10.8.1 Proposition. *For each index* $i \in \{1, \ldots, n\}$, *either:*

(a) $A_i = A$ *and* $B_i = B$; *or*
(b) $|A_i| < \frac{1}{q-1}|A|$ *and* $|B_i| < \frac{1}{q-1}|B|$.

Proof. By the definition of A_i, the set $(\partial_i(A_i), \partial_i(B))$ is a cross t-intersecting pair. Since all the words in these two sets have length $(n - 1)$, we have by the induction hypothesis that

$$|\partial_i(\mathcal{A}_i)|\,|\partial_i(\mathcal{B})| \le q^{2(n-1-t)} \le \frac{1}{q^2}|\mathcal{A}|\,|\mathcal{B}|.$$

Applying Lemma 10.7.1 to the elements in $\mathcal{B}$ that are not in $\mathcal{B}_i$, and Lemma 10.7.2 to the elements that are in $\mathcal{B}_i$, we have that

$$|\partial_i(\mathcal{B})| = (|\mathcal{B}| - |\mathcal{B}_i|) + \frac{1}{q}|\mathcal{B}_i| = |\mathcal{B}| + \left(\frac{1-q}{q}\right)|\mathcal{B}_i|.$$

Similarly, we can apply Lemma 10.7.2 to the elements in $\mathcal{A}_i$ to get that

$$|\partial_i(\mathcal{A}_i)| = \frac{1}{q}|\mathcal{A}_i|.$$

Putting these three equations together (and multiplying by q^2) gives

$$|\mathcal{A}_i|\,(q|\mathcal{B}| + (1-q)|\mathcal{B}_i|) \le |\mathcal{A}|\,|\mathcal{B}|, \qquad (10.8.1)$$

which can be rearranged to

$$|\mathcal{B}|\,(q|\mathcal{A}_i| - |\mathcal{A}|) \le (q-1)|\mathcal{A}_i|\,|\mathcal{B}_i|. \qquad (10.8.2)$$

By considering the sets $\partial_i(\mathcal{A})$ and $\partial_i(\mathcal{B}_i)$ (rather than $\partial_i(\mathcal{B})$ and $\partial_i(\mathcal{A}_i)$) we get the inequality

$$|\mathcal{B}_i|\,(q|\mathcal{A}| + (1-q)|\mathcal{A}_i|) \le |\mathcal{A}|\,|\mathcal{B}|. \qquad (10.8.3)$$

If we multiply the previous equation by $q|\mathcal{A}_i| - |\mathcal{A}|$ and multiply (10.8.2) by $|\mathcal{A}|$, we see that

$$|\mathcal{B}_i|\,(q|\mathcal{A}_i| - |\mathcal{A}|)(q|\mathcal{A}| + (1-q)|\mathcal{A}_i|) \le (q-1)|\mathcal{A}|\,|\mathcal{A}_i|\,|\mathcal{B}_i|.$$

Note that if $|\mathcal{B}_i| = 0$ then the lemma holds (by (10.8.1)), so we assume that $\mathcal{B}_i \ne \emptyset$. With this assumption we can rearrange the previous equation to get

$$(q|\mathcal{A}| - q|\mathcal{A}_i|)((q-1)|\mathcal{A}_i| - |\mathcal{A}|) \le 0.$$

Since $|\mathcal{A}_i| \le |\mathcal{A}|$, this means that either $|\mathcal{A}_i| = |\mathcal{A}|$ or $|\mathcal{A}_i| \le \frac{1}{q-1}|\mathcal{A}|$. The same conclusion can be made for the sets $\mathcal{B}$ and $\mathcal{B}_i$.

If $|\mathcal{A}_i| = \frac{1}{q-1}|\mathcal{A}|$, then

$$|\partial_i(\mathcal{A}_i)|\,|\partial_i(\mathcal{B})| = q^{2(n-1-t)}.$$

By the induction hypothesis, this implies that $|\partial_i(\mathcal{A}_i)| = q^{n-1-t}$. But since $|\partial_i(\mathcal{A}_i)| = \frac{1}{q}|\mathcal{A}_i|$, it must be the case that $\mathcal{A}_i = \mathcal{A}$. Thus the strict inequality $|\mathcal{A}_i| < \frac{1}{q-1}|\mathcal{A}|$ holds. Finally, if $|\mathcal{A}_i| = |\mathcal{A}|$, then by (10.8.2) it is the case that $|\mathcal{B}| = |\mathcal{B}_i|$. Similarly, by (10.8.3), if $|\mathcal{B}| = |\mathcal{B}_i|$ then $|\mathcal{A}| = |\mathcal{A}_i|$. $\square$

Next we show, still using our induction hypothesis, that for each index i, either $\mathcal{A}_i = \mathcal{A}$ or a large number of the words in $\mathcal{A}$ have the same entry in the ith coordinate. Denote the subset of words from $\mathcal{A}$ that have entry a in position i by $\mathcal{A}(i, a)$.

10.8.2 Proposition. *For any index $i \in \{1, \ldots, n\}$, the following are equivalent:*

(a) $\mathcal{A} = \mathcal{A}_i$;
(b) $|\mathcal{A}(i, a)| = \frac{1}{q}|\mathcal{A}|$ and $|\mathcal{B}(i, a)| = \frac{1}{q}|\mathcal{B}|$ for all $a \in Q$.

Proof. Assume $\mathcal{A} = \mathcal{A}_i$. Then by Lemma 10.7.2, $|\mathcal{A}(i, a)| = \frac{1}{q}|\mathcal{A}|$ for all values of a. Further, by Proposition 10.8.1, $\mathcal{B} = \mathcal{B}_i$ so it also follows that $|\mathcal{B}(i, a)| = \frac{1}{q}|\mathcal{B}|$ for all values of $a \in Q$.

Conversely, assume that $|\mathcal{A}(i, a)| = \frac{1}{q}|\mathcal{A}|$ and $|\mathcal{B}(i, a)| = \frac{1}{q}|\mathcal{B}|$ for all $a \in Q$. To prove that $\mathcal{A} = \mathcal{A}_i$, it is enough to show that $|\mathcal{A}| = |\mathcal{A}_i|$. To do this we show that $\partial_i(\mathcal{A}(i, a)) = \partial_i(\mathcal{A}(i, b))$ for any $a, b \in Q$.

Consider the pair $(\partial_i(\mathcal{A}(i, a)), \partial_i(\mathcal{B}(i, c)))$ with $a \neq c$. This pair is cross t-intersecting and

$$|\partial_i(\mathcal{A}(i, a))| \, |\partial_i(\mathcal{B}(i, c))| = \frac{1}{q^2}|\mathcal{A}| \, |\mathcal{B}| \geq q^{2(n-1-t)}.$$

By the induction hypothesis $\partial_i(\mathcal{A}(i, a)) = \partial_i(\mathcal{B}(i, c))$. Similarly, we have that $\partial_i(\mathcal{B}(i, c)) = \partial_i(\mathcal{A}(i, b))$, so we have that $\partial_i(\mathcal{A}(i, a)) = \partial_i(\mathcal{A}(i, b))$. □

In the case that $\mathcal{A} \neq \mathcal{A}_i$, there must be an $a \in Q$ such that $\mathcal{A}(i, a)$ is larger than $\frac{1}{q}|\mathcal{A}|$. We can assume that this $a = 0$, and we next show that in this case a significant portion of the words in $\mathcal{A}$ have the entry 0 in coordinate i.

10.8.3 Proposition. *If $\mathcal{A}(i, 0) > \frac{1}{q}|\mathcal{A}|$, then $\mathcal{A}(i, 0) > \frac{q-1}{q}|\mathcal{A}|$ and $\mathcal{B}(i, 0) > \frac{q-1}{q}|\mathcal{B}|$.*

Proof. Assume $\mathcal{A}(i, 0) > \frac{1}{q}|\mathcal{A}|$. Then, by Proposition 10.8.2, $\mathcal{A} \neq \mathcal{A}_i$ and consequently, by Proposition 10.8.1 $|\mathcal{A}_i| < \frac{1}{q-1}|\mathcal{A}|$.

The pair

$$(\partial_i(\mathcal{B}(i, 0)), \, \partial_i(\mathcal{A} \setminus \mathcal{A}(i, 0)))$$

is cross t-intersecting; thus by the induction hypothesis,

$$|\partial_i(\mathcal{B}(i, 0))| \, |\partial_i(\mathcal{A} \setminus \mathcal{A}(i, 0))| \leq q^{2(n-1-t)} \leq \frac{1}{q^2}|\mathcal{A}| \, |\mathcal{B}|. \tag{10.8.4}$$

Since

$$\mathcal{A} \setminus \mathcal{A}(i, 0) = ((\mathcal{A} \setminus \mathcal{A}(i, 0)) \cap \mathcal{A}_i) \bigcup ((\mathcal{A} \setminus \mathcal{A}(i, 0)) \setminus \mathcal{A}_i),$$

the following lower bound holds:

$$|\partial_i(\mathcal{A} \setminus \mathcal{A}(i,0))| = \frac{1}{q}|\mathcal{A}_i| + \left(|\mathcal{A}| - |\mathcal{A}(i,0)| - \frac{q-1}{q}|\mathcal{A}_i|\right)$$
$$> \left(1 - \frac{q-2}{q(q-1)}\right)|\mathcal{A}| - |\mathcal{A}(i,0)|.$$

Since $|\partial_i(\mathcal{B}(i,0))| = |\mathcal{B}(i,0)|$, this inequality, together with (10.8.4), yields (with some rearranging) that

$$q - 1 + \frac{1}{q-1} - \frac{|\mathcal{B}|}{q|\mathcal{B}(i,0)|} < q\frac{|\mathcal{A}(i,0)|}{|\mathcal{A}|}. \qquad (10.8.5)$$

Alternatively, we could have started with the pair

$$(\partial_i(\mathcal{A}(i,0)), \ \partial_i(\mathcal{B} \setminus \mathcal{B}(i,0))),$$

and derived the inequality

$$q - 1 + \frac{1}{q-1} - \frac{|\mathcal{A}|}{q|\mathcal{A}(i,0)|} < q\frac{|\mathcal{B}(i,0)|}{|\mathcal{B}|}.$$

We need to prove that $|\mathcal{A}(i,0)| > \frac{q-1}{q}|\mathcal{A}|$. To do this we set

$$x_0 = \frac{q|\mathcal{A}(i,0)|}{|\mathcal{A}|}$$

and show that $x_0 > q - 1$. By assumption we already have that $x_0 > 1$. Similarly, we set

$$y_0 = \frac{q|\mathcal{B}(i,0)|}{|\mathcal{B}|}$$

and show that $y_0 > q - 1$.

Assume that $x_0 \le y_0$ (alternatively we could assume that $y_0 \le x_0$). Then by (10.8.5),

$$x_0 > q - 1 + \frac{1}{q-1} - \frac{1}{x_0}.$$

Consider the function

$$f(x) = x - \left(q - 1 + \frac{1}{q-1} - \frac{1}{x}\right).$$

We know that $f(q-1) = 0$ and $f(x_0) > 0$. Since the derivative is strictly positive, f is a strictly increasing function and $x_0 > q - 1$. Since $y_0 \ge x_0$, we also have that $y_0 > q - 1$. $\qquad \square$

We now have all the tools to complete the induction step in the proof of Theorem 10.6.1. Assume that there are fewer than $n - t$ coordinates i with

$\mathcal{A}_i = \mathcal{A}$. By Proposition 10.8.1, this means that there are at least $t + 1$ coordinates j with $|\mathcal{A}_j| < \frac{1}{q-1}|\mathcal{A}|$. By Proposition 10.8.2, for each of these indices j we may assume that $|\mathcal{A}(j, 0)| > \frac{1}{q}|\mathcal{A}|$, and then by Proposition 10.8.3 $|\mathcal{A}(j, 0)| > \frac{q-1}{q}|\mathcal{A}|$.

Consider the intersection of the sets $\mathcal{A}(j, 0)$. Since every word in this set has $t + 1$ indices fixed,

$$\left| \bigcap_j \mathcal{A}(j, 0) \right| \le q^{n-t-1}.$$

Finally, $|\mathcal{A}(j, 0)| \ge \frac{q-1}{q}|\mathcal{A}|$, so we also know that

$$\left(1 - (t + 1)\frac{1}{q} \right) |\mathcal{A}| \le \left| \bigcap_j \mathcal{A}(j, 0) \right|.$$

Putting this together gives that

$$\left(1 - (t + 1)\frac{1}{q} \right) q^{n-t} \le q^{n-t-1},$$

but this is a contradiction with $t \le q - 2$.

Thus we conclude that if $(\mathcal{A}, \mathcal{B})$ is a cross t-intersecting pair of words from $H(n, q)$ with $|\mathcal{A}||\mathcal{B}|$ maximum, then there are $n - t$ coordinates i such that $\mathcal{A}_i = \mathcal{A}$ and $\mathcal{B}_i = \mathcal{B}$. This implies that all the words in $\mathcal{A}$ and $\mathcal{B}$ must take the same value for the remaining t coordinates, and Theorem 10.6.1 holds for length-n words.

10.9 Cocliques in power graphs

In the first section of this chapter we saw that the graph X_n in the Hamming scheme is the direct product of n copies of K_q; and in Section 10.4 we proved that all the cocliques of maximum size in X_n are the sets of all words that have the same entry in a fixed coordinate. These cocliques can be formed by replacing a single copy of K_q in the direct product of n copies of K_q by a coclique in K_q. In this section we show that this can be generalized to any coclique in the direct product of a connected regular graph.

Throughout this section we denote the direct product of n copies of a graph X by X^n; such a graph is a *power graph of X*.

10.9.1 Lemma. *The fractional chromatic number of X and X^n are equal.*

Proof. We show that X and X^n are *homomorphically equivalent*, that is, there are graph homomorphisms $X \to X^n$ and $X^n \to X$. Once this is established, then the result follows directly from Lemma 2.13.2.

The map that sends a vertex x in X to the n-tuple $(x, \ldots, x)$ is known as the *diagonal map*. The diagonal map is an isomorphism to an induced subgraph of X^n, and thus is a homomorphism of X to X^n. Conversely, the map that sends an n-tuple w to its ith coordinate is a homomorphism of X^n to X (known as the *coordinate homomorphism*). $\square$

10.9.2 Corollary. *If X is a vertex-transitive graph, then*

$$\alpha(X^n) = \alpha(X)|V(X)|^{n-1}.$$

Proof. If X is a vertex-transitive graph, then X^n is also vertex transitive and its fractional chromatic number is $|V(X)|/\alpha(X)$. With Lemma 10.9.1 we have the equation

$$\frac{\alpha(X)}{|V(X)|} = \chi_f(X) = \chi_f(X^n) = \frac{\alpha(X^n)}{|V(X)|^n}$$

which completes the proof. $\square$

It is easy to find cocliques in X^n of size $\alpha(X)|V(X)|^{n-1}$. Simply take π_i to be a coordinate homomorphism on X^n and let S be a coclique in X. Then the preimage $\pi_i^{-1}(S)$ of S in X^n is a coclique of the right size. (This holds because π_i is a homomorphism, not because it is a coordinate map.) The problem is to decide whether there are cocliques of maximum size that do not arise as preimages of coordinate maps.

10.9.3 Lemma. *Let X be a connected regular graph. If S is a coclique in X that meets the ratio bound and π is a coordinate map from X^n to X, then $\pi^{-1}(S)$ is a coclique that meets the ratio bound in X^n.*

Proof. The eigenvalues of X^n are all the products of n eigenvalues of X. In particular, if d is the largest eigenvalue of X, then the largest eigenvalue of X^n is d^n (this is also the degree of X^n). Similarly, if τ is the least eigenvalue of X, then τd^{n-1} is the least eigenvalue of X^n. So we have

$$|\pi^{-1}(S)| = |V(X)|^{n-1}|S| = |V(X)|^{n-1}\frac{|V(X)|}{1 - \frac{d}{\tau}} = \frac{|V(X)|^n}{1 - \frac{d^n}{\tau d^{n-1}}}. \qquad \square$$

It follows that $\alpha(K_q^n) = q^{n-1}$, and once again we see the EKR bound! We can also prove this quickly and directly as follows. The image of the diagonal map $K_q \to K_q^n$ is a clique in K_q^n with size q. If we identify $V(K_q)$ with the cyclic group $\mathbb{Z}_q$, this clique is a subgroup of $\mathbb{Z}_q^n$ and its cosets form a partition of $V(K_q^n)$ into cliques of size q. Accordingly, by the clique-coclique bound,

$$\alpha(K_q^n) \le q^{n-1}.$$

The following theorem, due to Alon, Dinur, Friedgut and Sudakov [10], characterizes the cocliques of maximum size in the direct powers of a class of vertex-transitive graphs.

10.9.4 Theorem. *Assume X is a connected vertex-transitive graph with least eigenvalue τ. If X is not bipartite and the ratio bound is tight for X, then any coclique in X^n of size $\alpha(X)(V(X))^{n-1}$ is the preimage, with respect to a coordinate map, of a coclique in X of size $\alpha(X)$.*

Proof. Assume X is a d-regular graph on v vertices with least eigenvalue τ. Define

$$U = \ker(\tau I - A(X)).$$

Let S be a coclique in X^n with size $\alpha(X)v^{n-1}$ and with characteristic vector x. Since the ratio bound is tight, the vector

$$x - \frac{|S|}{v^n}\mathbf{1}$$

lies in the $d^{n-1}\tau$-eigenspace of X^n.

Using the notation from Section 10.4, this eigenspace is spanned by vectors of the form $T_i(u)$ where $u \in U$ and $i = 1, \dots, n$. Hence in the terminology of Section 10.4, the vectors in this eigenspace are affine functions on $V(X^n)$. So, arguing as in the proof of Lemma 10.4.3, we deduce that we can write x as

$$x = (x_0\mathbf{1} + u) \otimes \mathbf{1} \otimes \cdots \otimes \mathbf{1}.$$

Here $x_0 = \alpha(X)/v$ and the entries of u take exactly two values and sum to zero. Since x is a 01-vector, it follows that the entries of u must be

$$-\frac{\alpha(X)}{v}, \qquad 1 - \frac{\alpha(X)}{v}.$$

Further, the entries of u sum to zero, so these entries occur in $v - \alpha(X)$ and $\alpha(X)$ coordinates, respectively. Consequently $x_0\mathbf{1} + u$ has exactly $\alpha(X)$ entries equal to 1 (and the rest are 0). Accordingly, the set S is the preimage under a coordinate map of a subset of X of size $\alpha(X)$. Since S is a coclique, this subset is also a coclique of X. $\square$

There are many classes of graphs for which the conditions of this theorem hold. Some examples are line graphs of graphs that contain a perfect matching;

Kneser graphs; Paley graphs of square order (Section 5.9); the complement of orthogonal array graphs (Section 5.5); block graphs of sufficient size (Section 5.3); matching graph (Section 15.2); and the derangement graph (Section 14.1). For more examples see [10]. We conjecture that the foregoing theorem also holds for the skew partition graph (Section 15.7).

Note that we only use vertex transitivity to deduce that

$$\frac{\alpha(X)}{|V(X)|} = \frac{\alpha(X^n)}{|V(X)|^n}.$$

Presumably there are many cases where X is not vertex transitive, but this condition still holds.

10.10 Cocliques in product graphs

If X and Y are arbitrary graphs, then the bound

$$\alpha(X \times Y) \geq \max\{\alpha(X)|Y|, \alpha(X)|Y|\} \qquad (10.10.1)$$

clearly holds, since for any coclique $S \subseteq V(X)$ the set $\{(x, y) : x \in S, y \in Y\}$ is in a coclique in $X \times Y$ (similarly for any coclique in Y). There are graphs for which the inequality (10.10.1) is strict (see [99]). Tardif, in [162], proposed the following problem: does equality hold in (10.10.1) if both X and Y are vertex-transitive graphs? If equality does hold, then the next question to ask is: what is the structure of the maximum cocliques in $X \times Y$?

The first of these questions was answered by Zhang [175], who proved that (10.10.1) does indeed hold with equality for vertex-transitive graphs. He also addressed the second question by giving conditions when there exist maximum cocliques in $X \times Y$ that are not the preimage of a maximum coclique under projection to either X or Y.

We use $N(x)$ to denote the *neighborhood* of a vertex x. For a set of vertices S in X the *neighborhood* of S, denoted $N(S)$, is the set of all vertices in X that are adjacent to some vertex in S. Define the *closed neighborhood* of S, denoted $N[S]$, to be the union of $N(S)$ and S. If S is a maximal coclique in X, then $N[S] = V(X)$, and the following inequality holds:

$$\frac{|S|}{|N[S]|} \leq \frac{\alpha(X)}{|V(X)|}.$$

It was shown by Zhang [175] that this inequality actually holds for all cocliques in a vertex-transitive graph. We use Lemma 2.13.3, the no-homomorphism lemma, to prove this fact.

10.10.1 Lemma. *Let S be a coclique in a vertex-transitive graph X. Then*

$$\frac{|S|}{|N[S]|} \leq \frac{\alpha(X)}{|V(X)|}.$$

If equality holds, then $|N[S] \cap C| = |S|$ for any maximum coclique C, and S is contained in a maximum coclique of X.

Proof. Let $\overline{N[S]}$ denote the complement of $N[S]$ in $V(X)$. For a set $A \subseteq V(X)$, we use X_A to denote the subgraph of X induced by A.

Since S is a coclique,

$$\frac{|S| + \alpha(X_{\overline{N[S]}})}{|N[S]| + |\overline{N[S]}|} \leq \frac{\alpha(X)}{|V(X)|}. \tag{10.10.2}$$

The embedding map from $X_{\overline{N[S]}}$ to X is a homomorphism, so by Lemma 2.13.3 the inequality

$$\frac{\alpha(X)}{|V(X)|} \leq \frac{\alpha(X_{\overline{N[S]}})}{|\overline{N[S]}|} \tag{10.10.3}$$

also holds. Putting (10.10.2) and (10.10.3), together yields that

$$|V(X)|\,|S| \leq \alpha(X)\,|N[S]|,$$

and the inequality in the lemma holds.

If equality holds in the final inequality, then equality must hold in both of the first two inequalities. Since (10.10.2) holds with equality, the size of maximum coclique in $X_{\overline{N[S]}}$ is $\alpha(X) - |S|$. By the no-homomorphism lemma the preimage of any maximum coclique in X (under the embedding map) is a maximum coclique in $X_{\overline{N[S]}}$. In particular, if C is any maximum coclique in X, then the preimage of C in $X_{\overline{N[S]}}$ is $\overline{N[S]} \cap C$. Thus

$$|\overline{N[S]} \cap C| = \alpha(X) - |S|,$$

which implies that $|N[S] \cap C| = |S|$. If S is any coclique in X, then the union of S with a maximum coclique in $X_{\overline{N[S]}}$ is a coclique of size $\alpha(X)$. □

Before proving that equality holds in (10.10.1) for all vertex-transitive graphs, we introduce some notation. Assume that X and Y are vertex-transitive graphs and that S is a coclique in $X \times Y$. For $x \in V(X)$ define

$$S_x = \{y \in V(Y) : (x, y) \in S\}.$$

The set S_x induces a subgraph of Y and $|S| = \sum_{x \in V(X)} |S_x|$. Further, assume that S is a maximal coclique in $X \times Y$ and S', the image of S under projection to X, is also a coclique. Then for every $x \in S'$ we have $S_x = V(Y)$, and for every $x \notin S'$ it is the case that $S_x = \emptyset$. Conversely, if x and x' are adjacent in X then, since S is a coclique, no vertex of S_x is adjacent to an element in $S_{x'}$.

The set S_x can be decomposed into the following two sets:

$$S_x^* = \{y \in S_x : N(y) \cap S_x = \emptyset\}, \qquad S_x' = V(Y) \setminus S_x^*.$$

The set S_x^* is the collection of isolated vertices in the subgraph of Y induced by S_x. It is possible that different vertices x and x' will have $S_x^* = S_{x'}^*$.

The following proof is due to Zhang [175].

10.10.2 Theorem. *Let X and Y be vertex-transitive graphs. Then*

$$\alpha(X \times Y) = \max\{\alpha(X)|Y|, \alpha(Y)|X|\}.$$

Proof. Assume that $\alpha(X)|Y| \geq \alpha(Y)|X|$ and that S is a coclique in $X \times Y$ with $|S| \geq \alpha(X)|Y|$. For any $x \in V(X)$, let S_x, S_x' and S_x^* be as defined earlier.

The size of the coclique is

$$|S| = \sum_{x \in V(X)} |S_x| = \sum_{x \in V(X)} |S_x^*| + \sum_{x \in V(X)} |S_x'|. \qquad (10.10.4)$$

We will find bounds on the final two summands to bound the size of S.

We can express the first summand on the right-hand side of (10.10.4) as follows. Define $Y_1, \ldots, Y_k$ to be the distinct sets S_x^* as x ranges over $V(X)$. Define B_i to be the set of all $x \in V(X)$ with $S_x^* = Y_i$. Then

$$\sum_{x \in V(X)} |S_x^*| = \sum_{i=0}^{k} |B_i||Y_i|. \qquad (10.10.5)$$

The next three claims provide a bound on the size of the second summand in the right-hand side of (10.10.4). In these claims we consider the following two sets:

$$A_y = \{x \in V(X) : y \in S_x'\}, \qquad S' = \bigcup_{x \in V(X)} S_x'.$$

Claim (a)

$$\sum_{x \in V(X)} |S_x'| \leq \sum_{y \in S'} \frac{\alpha(X)}{|V(X)|} |N[A_y]|. \qquad (10.10.6)$$

The left-hand side of (10.10.6) counts the number of pairs $(x, y) \in S$ where $y \in V(Y)$ and $N(y) \cap S_x \neq \emptyset$. Alternatively, we can count these pairs by summing the sizes of the sets A_y for $y \in S'$.

For each $y \in V(Y)$ the set A_y is a coclique in X. To see this, assume that $x, x' \in A_y$. Then $(x, y), (x', y) \in S$. Further, since $y \in S_x'$ there exists a $y' \in S_x$

that is adjacent to y. If x and x' are adjacent, then (x', y) and (x, y') are adjacent and both in S. Since S is a coclique, this is impossible.

Apply Lemma 10.10.1 to the set A_y to get the bound

$$|A_y| \le \frac{\alpha(X)}{|V(X)|} |N[A_y]|. \tag{10.10.7}$$

Claim (b)

$$|N[A_y]| \le \sum_{\substack{i \\ y \in \overline{N[Y_i]}}} |B_i|. \tag{10.10.8}$$

We will show that if $y \in N[Y_i]$, then for any $i \in \{1, \dots, k\}$ the set $B_i \subseteq \overline{N[A_y]}$. This implies that

$$\bigcup B_i \subseteq \overline{N[A_y]}$$

where the union is taken over all i for which $y \in N[Y_i]$. This gives the bound

$$\sum |B_i| \le |\overline{N[A_y]}| = |V(X)| - |N[A_y]|$$

(again, the sum is over all i with $y \in N[Y_i]$). This implies that

$$|N[A_y]| \le |V(X)| - \sum |B_i|,$$

which is equal to the sum of all $|B_i|$ where $y \in \overline{N[Y_i]}$.

We only need to prove that if $y \in N[Y_i]$, then for any $i \in \{1, \dots, k\}$ the set $B_i \subseteq \overline{N[A_y]}$. To show this, we consider two cases: $y \in Y_i$, and $y \in N(Y_i)$. Note that if A_y is empty, then the claim is true.

Assume that $y \in Y_i$ and let $x \in B_i$ (so $(x, y) \in S$). By definition, $S_x^* = Y_i$. So $y \notin S_x'$, which implies that $x \notin A_y$. Next, we need to show that $x \notin N(A_y)$. Assume there is an $x' \in A_y$ such that x and x' are adjacent. Since $x' \in A_y$, we know that $(x', y) \in S$ and there exists a $y' \in S_{x'}$ adjacent to y. As $y' \in S_{x'}$, we have both $(x', y') \in S$ and $(x, y) \in S$, which is a contradiction with S being a coclique. Thus $x \notin N[A_y]$.

Assume that $y \in N(Y_i)$ and that y is adjacent to $y' \in Y_i$. Let $x \in B_i$. Then $S_x^* = Y_i$, so $(x, y') \in S$. Since y and y' are adjacent and $y' \in S_x^*$, we have that $(x, y) \notin S$ so $x \notin A_y$. Further, if x is adjacent to some $x' \in A_y$, then (x, y') and (x', y) are adjacent vertices in S. So again we have that $x \notin N[A_y]$.

Claim (c)

$$\sum_{x \in V(X)} |S_x'| \le \frac{\alpha(X)}{|V(X)|} \sum_{i=1}^{k} (|V(Y)| - |N[Y_i]|) |B_i|.$$

Let $Y_i = S_x^*$. Then $y \notin N[Y_i]$, so we can conclude that $y \in \overline{N[Y_i]}$. Applying this argument to each $x \in V(X)$ gives

$$S' \subset \bigcup_{i=1}^{k} \overline{N[Y_i]}. \tag{10.10.9}$$

Then using (10.10.6) and (10.10.8) we have that

$$\sum_{x \in V(X)} |S'_x| \leq \sum_{y \in S'} \frac{\alpha(X)}{|V(X)|} |N[A_y]|$$

$$\leq \frac{\alpha(X)}{|V(X)|} \sum_{i=1}^{k} \sum_{y \in \overline{N[Y_i]}} |B_i|$$

$$= \frac{\alpha(X)}{|V(X)|} \sum_{i=1}^{k} |\overline{N[Y_i]}| |B_i|$$

$$= \frac{\alpha(X)}{|V(X)|} \sum_{i=1}^{k} (|V(Y)| - |N[Y_i]|) |B_i|.$$

Using the previous equation with (10.10.4) and (10.10.5), we get that

$$|S| \leq \sum_{i=1}^{k} |Y_i||B_i| + \frac{\alpha(X)}{|V(X)|} \sum_{i=1}^{k} (|V(Y)| - |N[Y_i]|) |B_i|$$

$$\leq \sum_{i=1}^{k} |B_i| \left(|Y_i| + \frac{\alpha(X)}{|V(X)|} |V(Y)| - \frac{\alpha(X)}{|V(X)|} |N(Y_i)| \right).$$

Since $\alpha(X)|V(Y)| \geq \alpha(Y)|V(X)|$ and Y_i is a coclique in Y, we have from Lemma 10.10.1 that

$$|Y_i| - \frac{\alpha(X)}{|V(X)|} |N(Y_i)| \leq |Y_i| - \frac{\alpha(Y)}{|V(Y)|} |N(Y_i)| \leq 0.$$

Thus we conclude that

$$|S| \leq \sum_{i=1}^{k} |B_i| \frac{\alpha(X)}{|V(X)|} |V(Y)| = \alpha(X)|V(Y)|. \qquad \square$$

Further, if $|S| = \alpha(X)|V(Y)|$, then equality holds in all the inequalities just shown. This implies that either Y_i is empty for all i, or $\alpha(X)|V(Y)| = \alpha(Y)|V(X)|$. In this second case, either Y_i is a maximum coclique in Y, or Y_i satisfies the two conditions

$$|Y_i| < \alpha(Y), \qquad \frac{|Y_i|}{|N[Y_i]|} = \frac{\alpha(Y)}{|V(Y)|}.$$

Zhang calls a coclique that satisfies these last two conditions *imprimitive*.

10.10.3 Corollary. *If* $X_1, \ldots, X_n$ *are vertex-transitive graphs and* $X = X_1 \times \cdots \times X_n$, *then*

$$\alpha(X) = \alpha(X_j) \prod_{\substack{i=1,\ldots,n \\ i \neq j}} |V(X_i)|$$

for some $j \in \{1, \ldots, n\}$

Zhang [175] also gives a characterization of the maximum cocliques in the product graph $X \times Y$.

10.10.4 Theorem. *Let X and Y be vertex-transitive graphs, and suppose* $\alpha(X \times Y) = \alpha(X)|V(Y)|$. *Then one of the following holds:*

(a) *every maximum coclique in $X \times Y$ is the preimage of a maximum coclique in X or Y under projection;*

(b) $\frac{\alpha(X)}{|V(X)|} = \frac{\alpha(Y)}{|V(Y)|}$ *and either X or Y contains a coclique S that is not maximum but satisfies the equation* $\frac{|S|}{N[S]} = \frac{\alpha(X)}{|V(X)|}$;

(c) $\frac{\alpha(X)}{|V(X)|} > \frac{\alpha(Y)}{|V(Y)|}$ *and Y is disconnected.* $\qquad\square$

10.11 Exercises

10.1 Prove that the Hamming graph in $H(n, q)$ is distance regular with diameter n (and thus $H(n, q)$ is metric).

10.2 Find the intersection numbers a_r, b_r, and c_r for $H(n, q)$.

10.3 Show that the Hamming scheme is cometric.

10.4 Let C be a canonical coclique in the nth distance graph of $H(n, q)$. Show that $w + s^* = d$ and $w^* + s = d$. (In other words, show that equality holds for C in both Theorem 8.6.1 and Theorem 8.6.2.)

10.5 Prove that the $(q - 1)\binom{n}{i}$ eigenvectors y of weight i defined in (10.1.1) are pairwise orthogonal. From this deduce the multiplicities given in Theorem 10.1.2.

10.6 Show that any two partial words have a greatest lower bound.

10.7 Assume $n > 2k$ and prove that a version of the EKR Theorem holds for partial words with length n and domain of size k.

10.8 A *strength-t orthogonal array*, denoted $OA(t, n, q)$, is a set of words of length n over an alphabet of size q, with the property that for each partial word f of size t there is exactly one word w in the array such that $f \leq w$. Show that the words from an $OA(t, n, q)$ form a clique of size q^t in the union of the graphs X_i in $H(n, q)$ where $i \in \{n - t + 1, \ldots, n\}$.

10.9 Show that if there exists an $OA(t, n, q)$ then the largest set of words with width at most $n - t$ has size q^{n-t}. (See the previous question for the definition of an $OA(t, n, q)$.) In particular, show that a subset of $H(n, q)$ with width at most 1 has size at most q.

10.10 Prove that Theorem 10.6.1 holds for $t = 1$ using Theorem 2.7.1.

10.11 Let X and Y be two vertex-transitive graphs in which the ratio bound holds with equality. Show that the ratio bound holds with equality in the graph $X \times Y$.

10.12 Find the size of the maximum coclique in the direct product

$$K(n_1, k_1) \times \cdots \times K(n_\ell, k_\ell)$$

where $2k_j \le n_j$ for each $j = 1, \ldots, \ell$.

Notes

Our approach to determining the eigenvalues and idempotents of the Hamming scheme in Sections 10.1 and 10.2 follows [82].

In this chapter we considered *cross-intersecting* pairs of sets of words. For any object that can be considered to be intersecting, we can consider a cross-intersecting pair of these objects. There has been extensive work on cross-intersecting pairs; see for example [11, 74, 126].

Larose and Tardif [114] proved that if X is a vertex-transitive graph, then $\alpha(X^n) = \alpha(X)|V(G)|^{n-1}$ for all $n > 1$.

The proof of Theorem 10.9.4 given by Alon, Dinur, Friedgut and Sudakov in [10] is particularly interesting as it uses discrete Fourier analysis on the group $\mathbb{Z}_q^n$. Their paper also includes a "stability" result, for which there is currently no purely combinatorial proof. Specifically, they show that if a coclique in K_q^n (or, more generally, the graph X^n where X is a graph that satisfies the condition of Theorem 10.9.4) is close to the maximum size of a coclique, then it is "close" to one of the maximum cocliques in the sense that the symmetric difference between the two cocliques is small.

The result in Exercise 10.12 is due to Frankl [64]. He proved the bound using an eigenvalue argument but also gave a combinatorial proof for the result for the direct product of two Kneser graphs. Ahlswede, Aydinian and Khachatrian [1] gave the value of $\alpha(\Pi_{1 \le j \le \ell} K(n_j, k_j, t_j))$.

11

Representation theory

Many of the graphs we consider in this book admit a large group of automorphisms. For example, the symmetric group $\text{Sym}(n)$ acts as a group of automorphisms of the Johnson graph $J(n, k)$. In some situations it is possible to use information about the group to derive information about the eigenvalues and eigenvectors of the graph, and this can allow us to derive theorems of EKR type. The group theoretic information needed comes from representation theory, and this chapter provides a reasonably self-contained introduction to this subject.

11.1 Representations

A *representation* Φ of a group G over the field $\mathbb{F}$ is a homomorphism from G into the group of invertible linear maps of some vector space V over $\mathbb{F}$. For our purposes G is finite and V is a finite-dimensional vector space over $\mathbb{C}$. The dimension of V is called the *dimension* of the representation. If Φ is a representation of G, we will denote the image of an element v of V under the action of an element g of G by $\Phi(g)v$, or simply by gv when the choice of representation is irrelevant. As a representation corresponds to a homomorphism of the group algebra $\mathbb{F}[G]$ into $\text{End}(V)$, it follows that V is a G-module. Hence we can express our thoughts on representations using the language of modules; this is often more convenient for developing the theory, but can be less useful when we have to do calculations.

We consider some examples. The first and simplest is the *trivial representation*. This representation maps each element of G to the identity map on a 1-dimensional vector space. We will often denote the trivial representation of G by 1_G.

For our next example, let G be a permutation group on a set Ω and let $V = \mathbb{F}^{\Omega}$. We can identify the points of Ω with the standard basis of V. Each element g of G then determines a permutation of this basis, and therefore it determines an endomorphism $\Phi(g)$ of V. We say Φ is the *permutation representation* of G. (The term permutation representation is also used to denote a homomorphism from a group in the symmetric group; we will provide a warning if there is a risk of confusion.)

If $S \subseteq \text{End}(V)$, the *commutant* of S is the set of endomorphisms of V that commute with each element of S. The commutant of S is an algebra. The commutant of a representation Φ of G is the commutant of the image of G under Φ. One reason we are interested in commutants is that if G is the automorphism group of a graph X and Λ is the corresponding representation on $\mathbb{F}^{V(X)}$, then A, the adjacency matrix of X, lies in the commutant of Λ. This can be used to get information about the eigenspaces of A. In particular, if z is an eigenvector for A with eigenvalue θ, then

$$A\Lambda(g)z = \Lambda(g)Az = \Lambda(g)\theta z = \theta \Lambda(g)z$$

and it follows that the θ-eigenspace of A is invariant under $\Lambda(g)$. From this we conclude that each eigenspace of A is a G-module, and this raises the hope that we can use representation theory to get information about the eigenspaces of A.

11.2 The group algebra

We denote the space of complex functions on a group G by $L(G)$. We define the *convolution* $\varphi * \psi$ of two elements φ, ψ of $L(G)$ by

$$(\varphi * \psi)(g) = \sum_{x \in G} \varphi(gx^{-1})\psi(x).$$

(Some authors prefix the sum in this definition by $|G|^{-1}$; some do not.) The algebra formed by $L(G)$ with this multiplication is the *group algebra*. This space has an inner product defined by

$$\langle \varphi, \psi \rangle_G = \sum_{x \in G} \overline{\varphi(x)}\psi(x)$$

(if the group G is clear from context, the subscript is omitted).

We define $\Lambda(g)$ for any $\psi \in L(G)$ to be

$$(\Lambda(g)\psi)(x) = \psi(g^{-1}x).$$

(Informally $\Lambda(g)$ maps the function $\psi(x)$ to the function $\psi(g^{-1}x)$.) Note that

$$(\Lambda(g)\Lambda(h)\psi)(x) = (\Lambda(h)\psi)(g^{-1}x) = \psi(h^{-1}g^{-1}x) = (\Lambda(gh)\psi)(x)$$

and therefore the map $g \mapsto \Lambda(g)$ is a representation of G on the vector space $L(G)$. It is the *left-regular representation* of G.

The characteristic function of $g \in G$ is the function defined by

$$\delta_g(h) = \begin{cases} 1, & \text{if } h = g; \\ 0, & \text{otherwise.} \end{cases}$$

Clearly, the set $\{\delta_g : g \in G\}$ is a basis of $L(G)$ and

$$(\delta_g * \delta_h)(a) = \sum_{x \in G} \delta_g(ax^{-1})\delta_h(x) = \delta_g(ah^{-1}) = \delta_{gh}(a).$$

Also, for any $\psi \in L(G)$

$$(\psi * \delta_g)(a) = \sum_{x \in G} \psi(ax^{-1})\delta_g(x) = \psi(ag^{-1}),$$

while

$$(\delta_g * \psi)(a) = \sum_{x \in G} \delta_g(x^{-1})\psi(xa) = \psi(g^{-1}a).$$

Note that the final equation implies that $\delta_g * \psi = \Lambda(g)\psi$. In Exercise 11.10 we ask you to prove that, if for $\psi \in L(G)$ we define the matrix M_ψ by

$$(M_\psi)_{g,h} = \psi(hg^{-1}),$$

then $M_{\psi*\varphi} = M_\psi M_\varphi$ for any $\varphi \in L(G)$. Thus M_{δ_g} is the matrix representing $\Lambda(g)$ in its action on $L(G)$.

From the equations just given, we can also conclude that a function $\psi \in L(G)$ lies in the center of the group algebra if and only if $\psi(ag^{-1}) = \psi(g^{-1}a)$ for all a and g. This is equivalent to requiring that

$$\psi(g^{-1}ag) = \psi(a)$$

for all a and g – that is, requiring ψ to be constant on each conjugacy class of G. An element of $L(G)$ that is constant on the conjugacy classes of G is called a *class function*.

11.3 Operations on representations

Suppose Φ and Ψ are representations of G on vector spaces V and W respectively. We say that Φ and Ψ are *equivalent* if there is an invertible linear map

$M : V \to W$ such that, for all g in G and all v in V,

$$M(\Phi(g)v) = \Psi(g)Mv.$$

In other words, Φ and Ψ are equivalent if there exists M such that

$$\Phi = M^{-1}\Psi M.$$

When working with vector spaces we may take subspaces. We may also form direct sums, tensor products and dual spaces. These operations all extend naturally to G-modules. Thus if V is a G-module then a submodule is simply a subspace of V that is mapped to itself by each element of G. Similarly, the direct sum of two G-modules is a G-module.

Tensor products and duals are trickier. If U and V are modules for any algebra $\mathcal{A}$, then the tensor product $U \otimes V$ is a module for $\mathcal{A} \otimes \mathcal{A}$ but it is not, in general, a module for $\mathcal{A}$. The dual space $V^* = \hom(V, \mathbb{F})$ is a module for the opposite algebra of $\mathcal{A}$ (the algebra with the product $a \circ b = ba$), but this usually fails to be a module for $\mathcal{A}$. For our purposes we need to consider only G-modules, and in this case, these difficulties vanish.

For tensor products, if U and V are G-modules then $U \otimes V$ is a module for $G \times G$, where

$$(g, h) : u \otimes v \mapsto gu \otimes hv.$$

Since the *diagonal* $\{(g, g) : g \in G\}$ of $G \times G$ is a subgroup isomorphic to G, we can view $U \otimes V$ as a G-module.

For an element φ of the dual space V^* and $g \in G$, consider the function

$$(\Lambda(g)\varphi)(v) = \varphi(g^{-1}v)$$

for all $v \in V$. It is easy to verify that Λ is a representation of G on V^*. (If we had used g in place of g^{-1} when we defined the action, we would have $\Lambda(gh) = \Lambda(h)\Lambda(g)$ and so Λ would not be a representation in general.)

Now that we have introduced submodules, we need to make a remark about the appearance of our representations. Suppose U is a G-submodule of V affording a representation Φ, and let β be an ordered basis for V obtained by extending an ordered basis of U. If M is the invertible linear map with the elements of β as its columns, then $M^{-1}\Phi(g)M$ has the form

$$\begin{pmatrix} \Phi_{1,1}(g) & \Phi_{1,2}(g) \\ 0 & \Phi_{2,2}(g) \end{pmatrix},$$

and the map $g \mapsto \Phi_{1,1}(g)$ is the representation belonging to the submodule U.

If U has a G-invariant complement U', then we choose M so its columns consist of a basis for U followed by a basis for U'. Now $M^{-1}\Phi(g)M$ has the form

$$\begin{pmatrix} \Phi_{1,1}(g) & 0 \\ 0 & \Phi_{2,2}(g) \end{pmatrix}$$

and in this case $\Phi_{2,2}$ is the representation of G on U'. This observation is the key to the decomposition of representations.

11.4 Sums and idempotents

In this section, we develop a correspondence between idempotents and direct sums. It will follow that G-invariant direct sums correspond to idempotents in the commutant of the representation. In the next section we use this to show that any G-invariant subspace has a G-invariant complement (Theorem 11.5.1).

Suppose V_1 and V_2 are a complementary pair of subspaces of V. Then any $v \in V$ can be expressed uniquely as a sum $v = v_1 + v_2$ where $v_i \in V_i$. This gives a map $P : V \to V$ defined by $Pv = v_1$, and it is straightforward to verify that P is linear and $P^2 = P$. We also note that $(I - P)v = v_2$, so $I - P$ is also idempotent, and the subspaces V_1 and V_2 are the images of P and $I - P$, respectively. Note that if $P \in \mathrm{End}(V)$ is any idempotent, then $I - P$ is also an idempotent and

$$\mathrm{im}(P) = \ker(I - P), \quad \ker(P) = \mathrm{im}(I - P).$$

We conclude that a direct sum decomposition of V gives rise to two idempotent elements of $\mathrm{End}(V)$, namely P and $I - P$. The next lemma shows that the converse also holds.

11.4.1 Lemma. *Each idempotent P in* $\mathrm{End}(V)$ *determines a direct sum decomposition of V.*

Proof. Since P is idempotent, its minimal polynomial divides $t^2 - t$ and thus it is diagonalisable. Since $\ker(P)$ is one eigenspace and $\mathrm{im}(P)$ is the other, V can be expressed as the following direct sum:

$$V = \ker(P) \oplus \mathrm{im}(P). \qquad \square$$

Our correspondence between idempotents and direct sums is not a bijection. However, there is a natural bijection if V is an inner product space. For in this case, if $U \leq V$, we can choose P to represent orthogonal projection onto U, and then $I - P$ represents orthogonal projection onto $U^\perp$.

We may view the results just presented as statements about modules for the identity group, and then these results generalize to G-modules, for any group G. The key is the following.

11.4.2 Lemma. *If A and P are operators on V and P is an idempotent, then* $AP = PA$ *if and only if* $\text{im}(P)$ *and* $\ker(P)$ *are both A-invariant.*

Proof. Suppose A and P commute. If $x \in \text{im}(P)$, then

$$PAx = APx = Ax,$$

which shows that $Ax \in \text{im}(P)$ and $\text{im}(P)$ is A-invariant. Since A also commutes with $I - P$, it follows that $\ker(P) = \text{im}(I - P)$ is A-invariant.

Conversely, if $\text{im}(P)$ is A-invariant, then

$$P(APx) = APx$$

for all x and therefore $PAP = AP$. If $\ker(P)$ is A-invariant as well, then

$$(I - P)A(I - P) = A(I - P).$$

Putting these equations together yields

$$A - AP = (I - P)(A - AP) = A - PA - AP + PAP = A - PA,$$

and we conclude that A and P commute. $\qquad\qquad\qquad\square$

If a G-module V can be decomposed into the direct sum of two G-modules, then the idempotent that this decomposition gives rise to is in the commutant of the representation of G. Conversely, every idempotent in the commutant of the representation gives rise to a direct sum decomposition of V, and both subspaces in this decomposition are G-modules. Thus direct sum decompositions of a G-module correspond to idempotent elements in the commutant of the representation.

11.5 Irreducible representations

A representation of a group G is *irreducible* if the corresponding G-module is simple. Equivalently, if V is a representation of G, then it is irreducible if no nonzero proper subspace of V is fixed by the action of G. By way of example, no permutation representation with dimension greater than 1 is irreducible, because the constant functions always form a submodule with dimension 1.

A module is *indecomposable* if it cannot be written as the direct sum of two nonzero submodules. An irreducible module is necessarily indecomposable, but the converse does not hold in general. For example, if $V = \mathbb{R}^2$ and G is the group of invertible upper-triangular matrices, then V is indecomposable but not irreducible. One of our main tasks in this section is to show that, in the cases of interest to us, indecomposable modules are irreducible.

11.5.1 Theorem. *If the group G acts on a vector space V and U is a G-submodule of V, then U has a G-invariant complement.*

Proof. Let P be an idempotent such that $\operatorname{im}(P) = U$. Define

$$Q = \frac{1}{|G|} \sum_{g \in G} g^{-1} P g.$$

If $u \in U$, then for any $g \in G$ we have $gu \in U$, whence $gu = Pgu$ and $u = g^{-1}Pgu$. Thus $Qu = u$ for all $u \in U$ and $\operatorname{im}(Q) = \operatorname{im}(P) = U$.

If $v \in V$, then for any $g \in G$ the vector Pgv is in U. Since U is a G-module the vector $hg^{-1}Pgv$ is also in U for any $h \in G$. Thus $Phg^{-1}Pgv = hg^{-1}Pgv$ and for any v in V,

$$Q^2 v = \frac{1}{|G|^2} \sum_{g \in G} \sum_{h \in G} h^{-1} P h g^{-1} P g v$$

$$= \frac{1}{|G|^2} \sum_{g \in G} \sum_{h \in G} h^{-1} h g^{-1} P g v$$

$$= \frac{1}{|G|} \sum_{g \in G} g^{-1} P g v.$$

We conclude that Q is also an idempotent.

By Lemma 11.4.1

$$V = \ker(Q) \oplus \operatorname{im}(Q)$$

so we only need to prove that $\ker(Q)$ is G-invariant. To do this we use Lemma 11.4.2. If $h \in G$ then

$$hQ = \frac{1}{|G|} \sum_{g \in G} h g^{-1} P g$$

and since

$$h g^{-1} P g = (g h^{-1})^{-1} P (g h^{-1}) h$$

it follows that $hQ = Qh$. As Q commutes with each element in G, its kernel is G-invariant.   □

We now sketch an alternative approach to this result. An inner product on a G-module V is *G-invariant* if for all u, v in V and g in G,

$$\langle ug, vg \rangle = \langle u, v \rangle.$$

If $\langle u, v \rangle$ is any inner product on V, then the average

$$\frac{1}{|G|} \sum_{g \in G} \langle ug, vg \rangle$$

is a G-invariant inner product. If U is a G-submodule of G, then $U^\perp$ (relative to this G-invariant inner product) is a submodule and is a complement to U in V.

We have the same conclusion from either approach: if V is a representation of a group G, then for any submodule U of V there is another module U' such that U' is G-invariant and $V = U \oplus U'$. Since this can be done for any representation, V can be expressed as the direct sum of irreducible G-modules. Thus we have the following.

11.5.2 Theorem. *Any representation of a finite group over the complex field is the direct sum of irreducible representations.* □

In fact, this result holds for any finite group over a field of characteristic zero, and it follows from this that a representation over a field of characteristic zero is irreducible if and only if it is indecomposable.

We say that an irreducible representation is a *constituent* of a representation if it is one of the irreducible representations in the direct sum.

11.6 Schur's Lemma

The next result is known as *Schur's Lemma*. This lemma plays a central role in representation theory. It holds for any representation over an algebraically closed field, but we continue to assume that our representations are over $\mathbb{C}$.

11.6.1 Lemma. *If V and W are irreducible representations of a group G and $\Phi : V \to W$ is a nonzero module homomorphism from V to W, then Φ is an isomorphism.*

Proof. Both $\ker(\Phi)$ and $\mathrm{im}(\Phi)$ are G-invariant submodules of V and W, respectively. Since these modules are simple and $\Phi \neq 0$, we must have $\ker(\Phi) = 0$ and $\mathrm{im}(\Phi) = W$. Therefore Φ is an isomorphism. □

11.6.2 Corollary. *If a representation Φ of G is irreducible, its commutant consists of scalar multiples of the identity.*

Proof. Let V be the G-module corresponding to Φ and suppose A is a nonzero element in the commutant of Φ. Then A is a module homomorphism from V to itself. Since the field is algebraically closed, A must have an eigenvalue λ.

Then $A - \lambda I$ is an endomorphism of V with nonzero kernel, and therefore by Schur's Lemma it must be the zero map.   □

If U and V are two vector spaces, then $\hom(U, V)$ is the set of all homomorphisms from U to V; if both U and V are G-modules, then we can consider $\hom_G(U, V)$ the subset of G-invariant homomorphisms from $\hom(U, V)$. Assume that Φ and Ψ are the representations of G corresponding to U and V. Then a homomorphism R is in $\hom_G(U, V)$ if and only if the following diagram commutes.

$$
\begin{array}{ccc}
U & \xrightarrow{\ \Phi\ } & U \\
{\scriptstyle R}\downarrow & & \downarrow{\scriptstyle R} \\
V & \xrightarrow[\ \Psi\ ]{} & V
\end{array}
$$

Our next lemma gives a method for constructing elements of $\hom_G(U, V)$ from representations on U and V.

11.6.3 Lemma. *Let U and V be G-modules corresponding to representations Φ and Ψ, respectively. If $R \in \hom(U, V)$ we define the map $\widehat{R}$ in $\hom(U, V)$ by*

$$
\widehat{R} = \sum_{g \in G} \Psi(g^{-1}) R \Phi(g).
$$

Then $\widehat{R} \in \hom_G(U, V)$.

Proof. If $a \in G$, then

$$
\begin{aligned}
\widehat{R}\Phi(a) &= \sum_{g \in G} \Psi(g^{-1}) R \Phi(g) \Phi(a) \\
&= \sum_{g \in G} \Psi(g^{-1}) R \Phi(ga) \\
&= \sum_{g \in G} \Psi(a) \Psi((ga)^{-1}) R \Phi(ga) \\
&= \Psi(a) \widehat{R}.
\end{aligned}
$$
                                                              □

11.7 Coordinate functions

If Φ is a representation of G on the vector space V and $a, b \in V$, the map

$$
g \mapsto \langle a, \Phi(g)b \rangle
$$

is a function on G that we call a *coordinate function* of the representation Φ. For an example, consider the left-regular representation Λ. This is a representation on $L(G)$, and for any $\varphi, \psi \in L(G)$ we have

$$g \mapsto \langle \varphi, \Lambda(g)\psi \rangle = \sum_{x \in G} \overline{\varphi(x)}(\Lambda(g)\psi)(x) = \sum_{x \in G} \overline{\varphi(x)}\psi(g^{-1}x).$$

Our next lemma shows how to evaluate a coordinate function involving $\widehat{R}$ for a specific function R.

11.7.1 Lemma. *Let U and V be G-modules with corresponding representations Φ and Ψ. Choose elements $a_1 \in U$ and $b_1 \in V$ and define $R : U \to V$ to be the map*

$$R(u) = \langle a_1, u \rangle b_1.$$

For any $a_2 \in U$ and $b_2 \in V$, we have that

$$\langle b_2, \widehat{R}a_2 \rangle = \sum_{g \in G} \langle a_1, \Phi(g)a_2 \rangle \, \overline{\langle b_1, \Psi(g)b_2 \rangle}.$$

Proof. To prove this we simply use the definitions of $\widehat{R}$ and R and properties of inner products:

$$\begin{aligned}
\langle b_2, \widehat{R}a_2 \rangle &= \sum_{g \in G} \langle b_2, \Psi(g^{-1})R\Phi(g)a_2 \rangle \\
&= \sum_{g \in G} \langle \Psi(g)b_2, R\Phi(g)a_2 \rangle \\
&= \sum_{g \in G} \langle \Psi(g)b_2, \langle a_1, \Phi(g)a_2 \rangle b_1 \rangle \\
&= \sum_{g \in G} \langle a_1, \Phi(g)a_2 \rangle \, \langle \Psi(g)b_2, b_1 \rangle \\
&= \sum_{g \in G} \langle a_1, \Phi(g)a_2 \rangle \, \overline{\langle b_1, \Psi(g)b_2 \rangle}. \qquad \square
\end{aligned}$$

11.7.2 Theorem. *Coordinate functions of distinct irreducible representations of G are orthogonal.*

Proof. Let U and V be simple G-modules with corresponding distinct irreducible representations Φ and Ψ. Suppose $a_1 \in U$ and $b_1 \in V$ and that R maps u in U to $\langle a_1, u \rangle b_1$ in V. By Schur's Lemma, $\widehat{R}$ is the zero map and so for any $a_2 \in U$ and any $b_2 \in V$, it follows that $\langle b_2, \widehat{R}a_2 \rangle = 0$. Orthogonality follows at once from the previous lemma. $\qquad \square$

If Φ is a representation of G on V and $\{v_1, \ldots, v_n\}$ is an orthonormal basis for V, we define coordinate functions $\Phi_{i,j}$ by

$$\Phi_{i,j}(g) = \langle v_i, \Phi(g)v_j \rangle.$$

11.7.3 Theorem. *Suppose Φ is an irreducible representation of G on V. Then the coordinate functions $\Phi_{i,j}$ are orthogonal, and*

$$\langle \Phi_{i,j}, \Phi_{i,j} \rangle = \frac{|G|}{\dim(V)}.$$

Proof. Let $\{v_1, \ldots, v_d\}$ be an orthonormal basis for V and take $R_{k,i}$ to be the map that sends $u \in V$ to $\langle v_i, u \rangle v_k$. The matrix representation of $R_{k,i}$ will have a 1 in the (k, j) position if $\langle v_i, v_j \rangle = 1$ and zero otherwise, so the (k, i)-entry will be one and all other entries zero. Then Corollary 11.6.2 yields that $\widehat{R_{k,i}} = \lambda I$ for some λ. To determine λ note that from the definition of $\widehat{R_{k,i}}$,

$$\mathrm{tr}(\widehat{R_{k,i}}) = \sum_{g \in G} \mathrm{tr}(\Phi(g^{-1})R_{k,i}\Phi(g)) = \sum_{g \in G} \mathrm{tr}(R_{k,i}) = |G|\,\mathrm{tr}(R_{k,i}) = \delta_{k,i}|G|,$$

and consequently

$$\widehat{R_{k,i}} = \frac{|G|}{\dim(V)}\delta_{k,i}.$$

By Lemma 11.7.1,

$$\langle v_\ell, \widehat{R_{k,i}}v_j \rangle = \sum_{g \in G} \langle v_i, \Phi(g)v_j \rangle \overline{\langle v_k, \Phi(g)v_\ell \rangle} = \langle \Phi_{i,j}, \Phi_{k,\ell} \rangle.$$

If $i \neq k$ then this inner product is zero. Similarly, if $j \neq \ell$ the inner product is zero and the theorem follows immediately. $\qquad\square$

For the proof of the next result we refer the reader to [94, Chapter I, Section 5].

11.7.4 Corollary. *If Φ is an irreducible representation of G on a vector space V with dimension d, then the algebra spanned by $\Phi(G)$ is isomorphic to $\mathrm{Mat}_{d \times d}(\mathbb{C})$.* $\qquad\square$

11.8 Characters

If V is a G-module with homomorphism Φ, the map

$$g \mapsto \mathrm{tr}(\Phi(g))$$

is a function on G, called the *character* associated with Φ. It is a sum of coordinate functions belonging to Φ. A character is *irreducible* if the corresponding representation is irreducible. The *degree* of a character φ is the dimension of the representation associated with it; the dimension is equal to $\varphi(1)$. Since

$$\text{tr}(M^{-1}\Phi(g)M) = \text{tr}(\Phi(g)),$$

we see that equivalent representations have the same character. We will soon find that representations with the same character are equivalent (this is Corollary 11.9.3).

The *trivial character* is the character of the trivial representation; it takes the value 1 on each group element. If G is a permutation group and $\text{fix}(g)$ denotes the number of elements fixed by g, then $g \mapsto \text{fix}(g)$ is a character of G; this character is called the *permutation character*. It is the character for the permutation representation.

Our next lemma is a summary of the basic properties of characters. The proofs are left as useful exercises.

11.8.1 Theorem. *Let* Φ *and* Ψ *be representations of G with characters φ and ψ, respectively. Then:*

(a) *$\varphi + \psi$ is the character of the representation $\Phi + \Psi$.*
(b) *$\varphi\psi$ is the character of $\Phi \otimes \Psi$.*
(c) *$\varphi(g^{-1}) = \overline{\varphi(g)}$.*
(d) *If $g^n = 1$, then $\varphi(g)$ lies in the extension of $\mathbb{Q}$ generated by the complex nth roots of unity.* $\square$

If Φ is a representation of G and $a, g \in G$, then

$$\text{tr}(\Phi(g^{-1}ag)) = \text{tr}(\Phi(g)^{-1}\Phi(a)\Phi(g)) = \text{tr}(\Phi(a)).$$

This shows that the character of a representation is constant on the conjugacy classes of G, and thus it is a class function. If φ and ψ are characters, their difference is a class function but is not, in general, a character. Similarly, if φ is a character, then the function $g \mapsto \varphi(g^2)$ is a class function but it is not a character if the degree of φ is greater than 1.

11.9 Orthogonality of characters

In dealing with characters, it is convenient to normalize our inner product on $L(G)$, so we define

$$\langle \varphi, \psi \rangle_G = \frac{1}{|G|} \sum_{x \in G} \overline{\varphi(x)}\psi(x) = \frac{1}{|G|} \sum_{x \in G} \varphi(x^{-1})\psi(x).$$

In other terms, $\langle \varphi, \psi \rangle_G = |G|^{-1}(\varphi * \psi)(1)$.

11.9.1 Theorem. *Let φ and ψ be the characters of the irreducible representations Φ and Ψ of G. Then*

$$\varphi * \varphi = \frac{|G|}{\varphi(1)}\varphi$$

and if $\Phi \neq \Psi$, then $\varphi * \psi = 0$.

Proof. First

$$(\varphi * \psi)(a) = \sum_{x \in G} \operatorname{tr}(\overline{\Phi(x)}) \operatorname{tr}(\Psi(xa)) = \sum_{x \in G} \operatorname{tr}\left(\overline{\Phi}(x) \otimes \Psi(x)\Psi(a)\right),$$

where

$$\operatorname{tr}\left(\overline{\Phi}(x) \otimes \Psi(x)\Psi(a)\right) = \sum_{i,j,k} \overline{\Phi}_{i,i}(x)\Psi_{j,k}(x)\Psi_{k,j}(a).$$

If we sum this over x in G and appeal to the orthogonality of coordinate functions, we get that $(\varphi * \psi)(a) = 0$ if $\Phi \neq \Psi$. Again using Theorem 11.7.3, if $\Phi = \Psi$ we have

$$(\varphi * \varphi)(a) = \sum_{x \in G} \sum_{i} \overline{\Phi}_{i,i}(x)\Phi_{i,i}(x)\Phi_{i,i}(a)$$

$$= \sum_{i} \frac{|G|}{\varphi(1)}\Phi_{i,i}(a)$$

$$= \frac{|G|}{\varphi(1)}\varphi(a). \qquad \square$$

This theorem shows that if φ is irreducible, then $|G|^{-1}\varphi(1)\varphi$ is an idempotent element of $L(G)$, and that the product of idempotents associated to distinct irreducible representations is zero. Such idempotent elements are said to be *orthogonal*. We will make much use of this theorem, starting with the following.

11.9.2 Corollary. *If φ and ψ are irreducible characters of G, then*

$$\langle \varphi, \psi \rangle_G = \begin{cases} 1, & \textit{if } \varphi = \psi; \\ 0, & \textit{otherwise.} \end{cases} \qquad \square$$

Proof. Simply recall that

$$\frac{1}{|G|}(\varphi * \psi)(1) = \langle \varphi, \psi \rangle_G. \qquad \square$$

It follows that the irreducible characters of G form an orthonormal subset in the space of class functions on G. Since the dimension of the space of class functions is the number of conjugacy classes of G, by this corollary the number of conjugacy classes of G is an upper bound on the number of distinct irreducible representations of G. (We will see that equality holds.)

One consequence of the previous corollary is that the characters of distinct irreducible representations are distinct. A general and useful way of stating this is the following.

11.9.3 Corollary. *Any representation is determined by its character.* $\square$

The next corollary follows directly from the fact that any representation is a direct sum of irreducible representations (Theorem 11.5.2).

11.9.4 Corollary. *Any character of G is a non-negative integer combination of irreducible characters.* $\square$

The irreducible characters with nonzero coefficients in this linear combination are called the *constituents* of the character.

If $\chi_1, \ldots, \chi_d$ are all the irreducible characters of a group and ψ is a character for the group with decomposition

$$\psi = \sum_i m_i \chi_i,$$

then the integer m_i is the *multiplicity* of χ_i in ψ. Since distinct irreducible characters are orthogonal, for any representation of a group G the multiplicities can be calculated as follows:

$$m_i = \langle \psi, \chi_i \rangle_G.$$

11.9.5 Corollary. *If $\psi = \sum_i m_i \psi_i$ then $\langle \psi, \psi \rangle_G = \sum_i m_i^2$. In particular, a character ψ is irreducible if and only if $\langle \psi, \psi \rangle_G = 1$.* $\square$

11.10 Decomposing the regular representation

We determine the decomposition of the regular representation of a group into irreducibles. We see that each irreducible representation is a constituent of the regular representation, and we use this to deduce that the number of distinct irreducible representations of a group is equal to the number of conjugacy classes.

11.10.1 Theorem. *Let G be a group and let $\psi_1, \ldots, \psi_d$ be the distinct irreducible characters of G. If λ is the character of the left-regular representation Λ of G, then*

$$\lambda = \sum_{i=1}^{d} \psi_i(1)\psi_i.$$

Proof. Since the set $\{\delta_g : g \in G\}$ is a basis of $L(G)$ and since $\Lambda(a)\delta_g = \delta_{ag}$, we have

$$\lambda(a) = \text{tr}(\Lambda(a)) = \sum_g \langle \delta_g, \Lambda(a)\delta_g \rangle = \sum_g \langle \delta_g, \delta_{ag} \rangle.$$

Hence $\lambda(1) = |G|$ and, if $a \neq 1$, then $\lambda(a) = 0$. This gives the exact value of the character λ; using this we can calculate that the multiplicity of φ_i in λ is equal to

$$
\begin{aligned}
m_i = \langle \psi_i, \lambda \rangle_G &= \frac{1}{|G|}(\psi_i * \lambda)(1) \\
&= \frac{1}{|G|}\sum_x \psi_i(x^{-1})\lambda(x) \\
&= \frac{\psi_i(1)}{|G|}\lambda(1) \\
&= \psi_i(1).
\end{aligned}
$$
$\qquad\square$

One consequence of the previous result is that a finite group has only a finite number of distinct irreducible representations. Our next result implies that the number of irreducible representations is equal to the number of conjugacy classes.

11.10.2 Corollary. *The irreducible characters of G form an orthogonal basis for the space of class functions on G.*

Proof. The issue here is to show that the irreducible characters span the space of class functions. If this was not the case, then there would be another class function that is orthogonal to all the irreducible characters. Assume that f is such a class function; so for any irreducible representation ψ we have $\langle \psi, f \rangle = 0$. If we can show that this implies that f is the zero function, then we are done.

For each irreducible representation Ψ on V, define

$$\Psi_f = \sum_{x \in G} f(x^{-1})\Psi(x).$$

Since f is a class function, Ψ_f commutes with $\Psi(g)$ for every $g \in G$, and by Schur's Lemma Ψ_f is a scalar multiple of the identity on V. We can determine the value of this scalar by taking the trace of Ψ_f:

$$\text{tr}(\Psi_f) = \sum_{x \in G} f(x^{-1})\psi(x) = |G|\langle f, \psi \rangle = 0.$$

We conclude that $\Psi_f = 0$ when Ψ is irreducible, and hence that $\Psi_f = 0$ for any representation Ψ of G.

In particular, if Λ is the left-regular representation of G, then $\Lambda_f = 0$ and

$$0 = \Lambda_f \delta_e = \sum_{x \in G} f(x^{-1})\Lambda(x)\delta_e = \sum_{x \in G} f(x^{-1})\delta_x.$$

Since the characteristic functions δ_x are linearly independent, we conclude that f is the zero function. $\square$

11.11 The conjugacy class scheme: idempotents

For any group G, there is an association scheme based on the conjugacy classes of the group – the conjugacy class scheme defined in Section 3.3. The conjugacy class scheme has the elements of G as its vertices, and two vertices x and y are i-related if and only if yx^{-1} is in the ith conjugacy class. We denote the conjugacy class scheme on G by $\mathcal{A}_G$. In this section we show how the eigenvalues and the dual eigenvalues of the conjugacy class scheme of G can be calculated using the irreducible characters of G.

If $x \in G$, then right multiplication by x permutes the elements of G, and we use $P(x)$ to denote the corresponding permutation matrix. The adjacency matrix A_i belonging to the conjugacy class C_i is given by

$$A_i = \sum_{x \in C_i} P(x).$$

If $x \in C_i$, define $C_{i'}$ to be the conjugacy class that contains x^{-1}. With this definition we have that $A_{i'} = A_i^T$.

If we set X_i to be the graph whose adjacency matrix is A_i, then, by construction, each graph X_i is a Cayley graph for G, and hence G is a subgroup of the automorphism group of $\mathcal{A}_G$. As we saw in Section 3.3, the product $G \times G$ also acts as a group of automorphisms of the scheme, and the stabilizer in $G \times G$ of a vertex in this association scheme is the diagonal subgroup of elements (g, g) (for g in G). One consequence of this is that each eigenspace of the scheme is a module for both $G \times G$ and G.

If ψ is an irreducible representation for G, then, as in Section 11.2, define a matrix M_ψ by

$$(M_\psi)_{g,h} = \psi(hg^{-1})$$

for g, h in G. From Theorem 11.9.1 we then get the following.

11.11.1 Lemma. *If ψ is an irreducible character for a group G, then*

$$M_\psi^2 = \frac{|G|}{\psi(1)} M_\psi.$$

If φ is an irreducible character of G and $\varphi \neq \psi$, then $M_\psi M_\varphi = 0$. $\square$

The matrices M_ψ lie in the Bose-Mesner algebra $\mathbb{C}[\mathcal{A}_G]$ of $\mathcal{A}_G$, and there is one for each irreducible representation of G. The matrix E_ψ, given by

$$E_\psi = \frac{\psi(1)}{|G|} M_\psi, \qquad (11.11.1)$$

is an idempotent matrix in $\mathbb{C}[\mathcal{A}_G]$. By appeal to Exercise 3.3, we conclude that the matrices E_i are the matrix idempotents for $\mathcal{A}_G$. Finally we note that

$$\mathrm{tr}(E_\psi) = \frac{\psi(1)}{|G|} \mathrm{tr}(M_\psi) = \frac{\psi(1)}{|G|} \sum_{g \in G} \psi(g^{-1}g) = \psi(1)^2.$$

11.12 The conjugacy class scheme: eigenvalues

In this section we express the eigenvalues of the scheme $\mathcal{A}_G$ in terms of the values of the irreducible characters of G. Throughout this section, we use the matrices E_ψ defined in the previous section.

11.12.1 Lemma. *Let* $\mathcal{A}_G = \{A_0, \ldots, A_d\}$ *be the conjugacy class scheme for a group* G. *If* Ψ *is an irreducible representation of* G *with* ψ *as its character and* c *is any element of the* i*th conjugacy class, then*

$$A_i E_\psi = \frac{|C_i| \overline{\psi(c)}}{\psi(1)} E_\psi.$$

Proof. Assume that the matrix A_i is the adjacency matrix for the ith class in the conjugacy class scheme. Thus, for $g, h \in G$ the (g, h)-entry of A_i is 1 if $hg^{-1} \in C_i$ and 0 otherwise. Therefore

$$(A_i E_\psi)_{g,h} = \frac{\psi(1)}{|G|} \sum_{xg^{-1} \in C_i} \psi(hx^{-1}) = \frac{\psi(1)}{|G|} \sum_{c \in C_i} \psi(hg^{-1}c^{-1}).$$

Since ψ is the character of the representation Ψ, the summation in this last term is equal to

$$\sum_{c \in C_i} \mathrm{tr}(\Psi(hg^{-1}c^{-1})) = \mathrm{tr}\left(\sum_{c \in C_i} \Psi(hg^{-1})\Psi(c^{-1}) \right)$$

$$= \mathrm{tr}\left(\Psi(hg^{-1}) \sum_{c \in C_i} \Psi(c^{-1}) \right).$$

Since C_i is a conjugacy class, the sum

$$\sum_{c \in C_i} \Psi(c^{-1})$$

lies in the commutant of Ψ and therefore, by Schur's Lemma (see Corollary 11.6.2) we can conclude that

$$\sum_{c \in C_i} \Psi(c^{-1}) = \lambda_\psi I$$

for some constant λ.

Putting all this together, we have that

$$(A_i E_\psi)_{g,h} = \frac{\psi(1)}{|G|} \operatorname{tr}\left(\Psi(hg^{-1})(\lambda_\psi I)\right) = \lambda_\psi \frac{\psi(1)}{|G|} \psi(hg^{-1}) = \lambda_\psi (E_\psi)_{g,h}.$$

To complete the proof of the theorem, we need to find the exact value of λ_ψ. Taking the trace of both sides of $\sum_{c \in C_i} \Psi(c^{-1}) = \lambda_\psi I$ we see that

$$|C_i|\psi(c^{-1}) = \lambda_\psi \psi(1), \tag{11.12.1}$$

and thus $\lambda_\psi = |C_i|\overline{\psi(c)}\psi(1)^{-1}$. $\qquad\square$

This lemma gives the eigenvalues of $\mathcal{A}_G$, and we can use these to find the dual eigenvalues. Since the dual eigenvalues $q_\psi(i)$ satisfy the relation

$$E_\psi = \frac{1}{|G|} \sum_{i=0}^{d} q_\psi(i) A_i,$$

we have that

$$A_i \circ E_\psi = \frac{q_\psi(i)}{|G|} A_i.$$

Then

$$\langle A_i, E_\psi \rangle = \langle E_\psi^*, A_i \rangle = \operatorname{sum}(E_\psi \circ A_i) = \frac{q_\psi(i)}{|G|} \operatorname{sum}(A_i) = q_\psi(i)|C_i|.$$

Since

$$\langle A_i, E_\psi \rangle = \operatorname{tr}(A_i^T E_\psi) = \operatorname{tr}(\overline{\lambda_\psi} E_\psi) = \overline{\lambda_\psi} \psi(1)^2,$$

we can use Lemma 11.12.1 to conclude that if $c_i \in C_i$, then

$$q_\psi(i) = \psi(c_i)\psi(1).$$

Note that the dimension of the module associated to an irreducible character ψ is given by $m_\psi = q_\psi(1) = \psi(1)^2$.

11.12.2 Corollary. *Let G be a group and and let $A_0, \ldots, A_d$ be the Schur idempotents of the conjugacy class scheme of G. For each irreducible representation ψ of G, the corresponding eigenvalue of A_i is*

$$p_i(\psi) = \frac{|C_i|\overline{\psi(c)}}{\psi(1)}.$$

If $c \in C_i$, the dual eigenvalue is

$$q_\psi(i) = \psi(c)\psi(1). \qquad \square$$

Finally, we conclude by confirming that $PQ = vI$ (as expected from (3.5.3)). The (ϕ, ψ)-entry of PQ is

$$\sum_i p_i(\phi)q_\psi(i) = \sum_i \frac{|C_i|\overline{\phi(c_i)}}{\phi(1)}\psi(c_i)\psi(1)$$

$$= \sum_{x \in G} \frac{\overline{\phi(x)}}{\phi(1)}\psi(x)\psi(1)$$

$$= \frac{|G|\psi(1)}{\phi(1)}\langle \phi, \psi \rangle.$$

(Here the first two summations are over all conjugacy classes C_i of G.)

A Cayley graph $\Gamma(G, C)$ is a *normal Cayley graph* if C is closed under taking conjugates. The eigenvalues of any normal Cayley graph can be calculated from the irreducible characters of the group.

11.12.3 Theorem. *If $\Gamma(G, C)$ is a normal Cayley graph, then the eigenvalues are*

$$\eta_\psi = \frac{1}{\psi(1)} \sum_{c \in C} \overline{\psi(c)},$$

where ψ runs over all irreducible characters of G. Further, the multiplicity of η_ψ is the sum of $\varphi(1)^2$ taken over all irreducible characters φ with $\eta_\varphi = \eta_\psi$. $\square$

This follows immediately from the observation that the adjacency matrix of a normal Cayley graph is the sum of distinct Schur idempotents of the scheme $\mathcal{A}_G$.

11.13 Restriction and induction

If Ψ is a representation of a group G, then for any subgroup H of G, define the *restriction* of Ψ on H to be the representation $\mathrm{res}_H(\Psi)$ of H with $\mathrm{res}_H(\Psi)(h) = \Psi(h)$. If ψ is the character of Ψ, then $\mathrm{res}_H(\psi)$ is the character belonging to $\mathrm{res}_H(\Psi)$.

Let H be a subgroup of G. Any representation Φ of H on V can be used to build a representation $\mathrm{ind}_G(\Phi(g))$ of G. First we define the space on which G will act – it is the subspace of functions F from G to V that have the property that

$$F(xh) = \Phi(h^{-1})F(x).$$

We call this subspace Z. For a function F and an element g of G, we define

$$(\text{ind}_G (\Phi)(g)F)(x) = F(g^{-1}x).$$

If $h \in H$ and $F \in Z$, then

$$F(g^{-1}xh) = \Phi(h^{-1})F(g^{-1}x)$$

and this implies that $\text{ind}_G (\Phi)(g)F \in Z$.

The simplest case of an induced representation is when the representation of H is trivial. Then Z consists of the functions F from G to V such that

$$F(xh) = F(x).$$

Equivalently, Z consists of the functions in $L(G)$ that are constant on left cosets of H.

There is an explicit formula for the character of an induced representation.

11.13.1 Theorem. *Let φ be a character of a subgroup H of G and let $\text{ind}_G (\varphi)$ denote the induced character of G. Then*

$$(\text{ind}_G (\varphi)) (g) = \frac{1}{|H|} \sum_{x \in G} \varphi(x^{-1}gx),$$

with the understanding that $\varphi(x^{-1}gx) = 0$ if $x^{-1}gx \notin H$. $\square$

With this formulation of the induced character it is easy to determine the degree.

11.13.2 Corollary. *The degree of $\text{ind}_G (\varphi)$ is $[G : H]\varphi(1)$.* $\square$

The next theorem is known as *Frobenius reciprocity*. For the proof we refer the reader to [147, Theorem 1.12.6].

11.13.3 Theorem. *Suppose $H \leq G$ and assume that ψ and φ are characters of H and G, respectively. Then*

$$\langle \varphi, \text{ind}_G(\psi)\rangle_G = \langle \text{res}_H(\varphi), \psi \rangle_H.$$ $\square$

For a simple example of Frobenius reciprocity, consider the representation of G induced from the trivial representation on H. Since restriction of the trivial representation of G to a subgroup H is the trivial representation on H, we have that

$$\langle \text{ind}_G (1_H), 1_G\rangle_G = \langle 1_H, 1_H\rangle_H = 1.$$

This means that the trivial representation always occurs with multiplicity one in $\text{ind}_G (1_H)$.

11.14 Exercises

11.1 Let G be a permutation group acting on a set V and for an element ψ
$L(\Omega)$, define $(\Lambda(g)\psi)(v) = \psi(g^{-1}v)$. Prove that $\Lambda(g)$ is a linear map
and that $\Lambda(gh) = \Lambda(g)\Lambda(h)$.

11.2 Show that the group G in Exercise 1 has a second representation R
given by

$$(R(g)\psi)(x) = \psi(xg).$$

This is the *right-regular representation* of G. Prove that this is a permutation representation.

11.3 Prove that the Euclidean inner product on V is G-invariant for a given
representation if and only if the representation maps G to a subgroup
of the group of orthogonal matrices.

11.4 Assume $G = \mathbb{Z}_n$. If θ is a complex nth root of unity, show that the map

$$m \mapsto \theta^m$$

is a character (and a representation) of G. Prove that all characters of
G can be obtained in this way.

11.5 Assume $G = \mathbb{Z}_2^d$, and view the elements of G as binary vectors. If
$a \in G$, show that the map $\tau_a : V \to V$, where V is a d-dimensional
vector space, given by

$$\tau_a(u) = (-1)^{a^T u}$$

is a character of G, and that all characters of G can be obtained in this
way.

11.6 Prove that the set of characters of G with degree 1 form an abelian
group. If G is finite and abelian, show that this group is isomorphic to
G.

11.7 If ψ is a character with degree greater than 1, prove that the class
function $g \mapsto \psi(g^2)$ is not a character.

11.8 Let G and H be subgroups with $H \leq G$ and φ be a character of a
subgroup H. Show that

$$(\text{ind}_G(\varphi))(g) = \frac{1}{|H|} \sum_{x \in G} \varphi(x^{-1}gx)$$

where $\varphi(x^{-1}gx) = 0$ if $x^{-1}gx \notin H$.

11.9 Suppose G is a group that is 2-transitive on the set V, and let ρ be
the associated permutation character. Prove that $\psi = \rho - 1$ is an irreducible character. Find the degree of ψ and describe the value ψ on the
conjugacy classes of G.

11.10 If $\psi \in L(G)$, let $M = M_\psi$ denote the matrix with rows and columns indexed by G and with entries

$$M_{g,h} = \psi(hg^{-1}).$$

Prove for any $\varphi \in L(G)$ that $M_\varphi M_\psi = M_{\varphi * \psi}$.

11.11 Suppose the group G has permutation representations on the sets X and Y, with characters φ and ψ, respectively. Then G acts a group of permutations on $X \times Y$. Prove that the character of this representation is $\varphi\psi$.

11.12 Using the expression for an induced character given in Theorem 11.13.1, verify that Frobenius reciprocity holds.

Notes

For a more complete treatment of the representation theory of groups, there are many possibilities. We have made use of Fulton and Harris [73], Ceccherini-Silberstein, Scarabotti and Tolli [43] and Thomas [163]. We highly recommend the last for further details and additional exercises.

12

Representation theory of the symmetric group

The representation theory of the symmetric group has been more extensively studied than that of other groups, and the symmetric group plays a special role in many of the problems we consider. In this chapter we give more detail about the irreducible representations of the symmetric group. We frequently omit proofs, or just provide outlines.

12.1 Permutation representations

If G is a permutation group on a set Ω, our standard approach yields a representation Φ of G on the space $L(\Omega)$ of functions on Ω given by

$$(\Phi(g)\psi)(x) = \psi(g^{-1}x),$$

where $\psi \in L(\Omega)$. Let δ_i denote the characteristic function of the element i of Ω. With respect to the inner product

$$\langle \varphi, \psi \rangle = \sum_{i \in \Omega} \overline{\varphi(i)}\psi(i),$$

the functions δ_i form an orthonormal basis of $L(\Omega)$. Further,

$$(\Phi(g)\delta_i)(x) = \delta_i(g^{-1}x) = \delta_{gi}(x),$$

and hence $\Phi(g)$ simply permutes the functions δ_i. Using this basis, we can calculate

$$\mathrm{tr}(\Phi(g)) = \sum_{i \in \Omega} \langle \delta_i, \Phi(g)\delta_i \rangle = \sum_{i \in \Omega} \langle \delta_i, \delta_{gi} \rangle,$$

which equals the number of points fixed by g in its action on Ω.

232

12.1.1 Theorem. *If $H \leq G$, then the representation of G induced by the trivial representation of H is equivalent to the permutation representation of G on the left cosets of H.*

Proof. Let Φ be the representation of G induced by the trivial representation of H. As we noted in Section 11.13, Φ provides an action of G on the subspace of $L(G)$ formed by the functions that are constant on the left cosets of H. The dimension of this subspace is $[G : H]$. If we let $g_i H$, with $i = 1, \ldots, [G : H]$, be the left cosets of H in G, and ψ_i the characteristic function on $g_i H$, then the functions

$$\frac{1}{\sqrt{|H|}} \psi_i$$

form an orthonormal basis for the space of functions constant on left cosets of H.

Therefore

$$\mathrm{tr}(\Phi(g)) = \sum_{i=1}^{[G:H]} \langle \psi_i, \Phi(g)\psi_i \rangle.$$

Here

$$\langle \psi_i, \Phi(g)\psi_i \rangle = \frac{1}{|H|} \sum_{x \in G} \overline{\psi_i(x)} \psi_i(g^{-1}x),$$

which is equal to 0, unless $g g_i H = g_i H$, in which case it is equal to 1. Therefore $\mathrm{tr}(\Phi(g))$ is equal to the number of fixed points of g in its action as a permutation of left cosets of H. Since representations with the same character are equivalent, we are done. ☐

The left-regular representation is a example of a permutation representation, and the character of an element is simply the number of fixed points (where the action is left multiplication on the group). In particular, the value of the character on the identity of the group is the size of the group, while the value of the character for any other element is equal to zero.

12.2 Examples: orbit counting

If G is a permutation group on a set Ω, with permutation character χ, then for any $g \in G$, the value of $\chi(g)$ is equal to $\mathrm{fix}(g)$, the number of points in Ω left fixed by g. Burnside's Lemma informs us that

$$\langle 1_G, \chi \rangle = \frac{1}{|G|} \sum_{g \in G} \mathrm{fix}(g)$$

is equal to the number of orbits of G on Ω. We also have that

$$\langle \chi, \chi \rangle = \frac{1}{|G|} \sum_{g \in G} \text{fix}(g)^2.$$

Since $\text{fix}(g)^2$ is the number of elements of $\Omega \times \Omega$ fixed by g, by Burnside's Lemma, $\langle \chi, \chi \rangle$ is the number of orbits of G on $\Omega \times \Omega$. Such orbits are known as *orbitals*, and are studied at length in the next chapter.

For an example, let χ_1 denote the character of $\text{Sym}(n)$ acting as usual on $\Omega = \{1, \ldots, n\}$. The action of $\text{Sym}(n)$ on Ω has just one orbit, so if $1_{\text{Sym}(n)}$ denotes the trivial character for $\text{Sym}(n)$, then $\langle \chi_1, 1_{\text{Sym}(n)} \rangle = 1$. Further, $\text{Sym}(n)$ has just two orbits on $\Omega \times \Omega$: one is the diagonal $\{(i, i) : i \in \Omega\}$ and the other is the complement of the diagonal in $\Omega \times \Omega$. It follows that $\langle \chi_1, \chi_1 \rangle = 2$. We then have

$$\langle \chi_1 - 1_{\text{Sym}(n)}, \chi_1 - 1_{\text{Sym}(n)} \rangle = \langle \chi_1, \chi_1 \rangle - 2 \langle \chi_1, 1_{\text{Sym}(n)} \rangle + \langle 1_{\text{Sym}(n)}, 1_{\text{Sym}(n)} \rangle = 1.$$

This shows that $\chi_1 - 1_{\text{Sym}(n)}$ is an irreducible character for $\text{Sym}(n)$. (Here we are making use of the result of Exercise 12.9.) Since the degree of χ_1 is n, the degree of $\chi_1 - 1_{\text{Sym}(n)}$ is $n - 1$.

For a second example, let Ω_2 be the set of unordered pairs of distinct elements of $\{1, \ldots, n\}$, and let χ_2 denote the character for the permutation representation of $\text{Sym}(n)$ acting on Ω_2. Here $\text{Sym}(n)$ has three orbits on $\Omega_2 \times \Omega_2$ – two pairs may have 0, 1 or 2 elements in common – therefore $\langle \chi_2, \chi_2 \rangle = 3$. We also see that $\text{Sym}(n)$ has two orbits on $\Omega_2 \times \Omega_1$ (the element from Ω_1 can either be in the pair from Ω_2, or not) whence $\langle \chi_2, \chi_1 \rangle = 2$. It follows that

$$\langle \chi_2 - \chi_1, \chi_2 - \chi_1 \rangle = 3 - 2 - 2 + 2 = 1,$$

and therefore $\chi_2 - \chi_1$ is also an irreducible character. Since Ω_2 has size $\binom{n}{2}$, the degree of this character is $\binom{n}{2} - n$.

We can extend this procedure. If Ω_k is the set of all k-subsets of $\{1, \ldots, n\}$ and χ_k is the permutation character for $\text{Sym}(n)$ on Ω_k and $n \geq 2k$, you may prove that $\chi_k - \chi_{k-1}$ is irreducible with degree $\binom{n}{k} - \binom{n}{k-1}$. If we set ϕ_k equal to $\chi_k - \chi_{k-1}$ when $k \geq 1$, and equal to the trivial character if $k = 0$, we also see that

$$\chi_k = \phi_k + \phi_{k-1} + \cdots + \phi_0.$$

Effectively, we have decomposed the permutation characters for the representations of $\text{Sym}(n)$ on the cosets of the subgroup $\text{Sym}(n - k) \times \text{Sym}(k)$ (when $n \geq 2k$), and in doing so we identified a number of irreducible representations of $\text{Sym}(n)$. To make real progress, however, we need to consider a larger class of subgroups, and we need substantially more machinery.

12.3 Young subgroups

Two elements α and β of Sym(n) are conjugate if they have the same cycle structure. More precisely, for each non-negative integer k, the number of cycles of length k in α is equal to the number of cycles of length k in β.

An *integer partition* λ of n is a non-increasing sequence of positive integers $\lambda_1, \ldots, \lambda_k$ that sums to n. We write $\lambda \vdash n$ to denote that λ is a partition of n. If there are j parts of the partition that are equal to λ_i, we may write λ_i^j rather than writing λ_i repeated j times. Writing the cycle lengths of a permutation as a non-decreasing sequence gives a partition of n; so we see that the conjugacy classes of Sym(n) are parameterized by the partitions of n. This already gives us an expression for the number of irreducible representations of Sym(n).

Suppose $\lambda \vdash n$, with parts $\lambda_1, \ldots, \lambda_k$. A *tabloid of shape* λ is a partition π of the set $\{1, \ldots, n\}$ into cells $C_1, \ldots, C_k$, where $|C_i| = \lambda_i$. Any permutation γ in Sym(n) maps a tabloid to a tabloid of the same shape. Further, if τ_1 and τ_2 are two tabloids of shape λ, then the permutations γ that send τ_1 to τ_2 form a coset of the subgroup of Sym(n) that fixes each cell of τ_1 (and thus fixes τ_1). The group that fixes a tabloid of shape λ is a *Young subgroup*, Sym(λ) and is defined to be the direct product

$$\text{Sym}(\lambda) = \text{Sym}(\lambda_1) \times \text{Sym}(\lambda_2) \times \cdots \times \text{Sym}(\lambda_k).$$

Thus there is a bijection between the tabloids of shape λ and the cosets of the Young subgroup Sym(λ) in Sym(n).

For each partition λ of n, there is a permutation representation of Sym(n) given by its action on the cosets Sym(n)/ Sym(λ). By Theorem 12.1.1, this is the representation of Sym(n) induced by the trivial representation on the Young subgroup for λ. We denote its character by χ_λ, and the corresponding module is denoted by M_λ. We will see that each irreducible module for Sym(n) is a submodule of some M_λ.

Throughout this chapter, the trivial representation on a Young subgroup Sym(λ) is denoted by 1_λ (including the trivial representation on Sym(n), which is denoted by $1_{[n]}$). If Φ is any representation of Sym(λ), then the representation of Sym(n) induced by Φ is denoted by $\Phi \uparrow^n$ rather than $\text{ind}_{\text{Sym}(n)}(\Phi)$. Similarly if Φ is a representation of Sym(n), the restriction of Φ to the group Sym(λ) is denoted by $\Phi \downarrow_\lambda$ rather than $\text{res}_{\text{Sym}(\lambda)}(\Phi)$.

12.4 Irreducible representations

We use the modules M_λ defined in the previous section to find the irreducible representations of Sym(n). To start we define the *dominance ordering* on

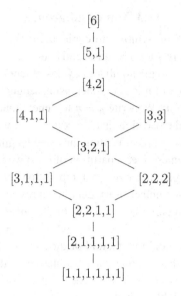

Figure 12.1 Dominance ordering on partitions of 6.

partitions $\lambda = [\lambda_1, \lambda_2, \ldots, \lambda_k]$ and $\mu = [\mu_1, \mu_2, \ldots, \mu_\ell]$ by $\lambda \leq \mu$ if

$$\sum_{i=0}^{j} \lambda_i \leq \sum_{i=0}^{j} \mu_i$$

for all j. (We assume $\lambda_i = 0$ for $i > k$ and $\mu_i = 0$ for $i > \ell$.) This is a partial order on the set of integer partitions of n; the largest partition under this ordering is $[n]$ and the least is $[1^n]$. Figure 12.1 shows the partial ordering on the partitions of 6. This is indeed a partial ordering, not a total ordering, because the partitions $[4, 1, 1, 1]$ and $[3, 3]$ are not comparable.

The key property of the dominance order is expressed in the following.

12.4.1 Lemma. *For partitions λ and μ of n, we have $\lambda \leq \mu$ if and only if $M_\lambda = M_\mu \oplus N$, for some submodule N of M_λ.* □

One corollary of this is that if ϕ_λ and ϕ_μ are the characters associated to λ and μ, respectively, and $\lambda \leq \mu$, then $\phi_\lambda - \phi_\mu$ is also a character. We also see that if $\lambda \vdash n$, then $\lambda \leq [n]$ and so $M_{[n]}$ is a submodule of M_λ, confirming the fact that the trivial module is a summand of any permutation module M_λ.

12.4.2 Lemma. *Suppose $\lambda \vdash n$ and let $\mathcal{N}$ denote the sum of the submodules M_μ where $\lambda < \mu$. Then $\mathcal{N}$ is a proper submodule of M_λ, and its complement S_λ in M_λ is irreducible.* □

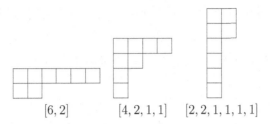

Figure 12.2 Young diagrams for three partitions of 8.

One consequence of this result is that M_λ can be expressed as a sum of irreducible submodules S_μ, where $\mu \geq \lambda$. The multiplicity of S_μ as a summand of M_λ is the *Kostka number* $K_{\mu\lambda}$. These are non-negative integers, and the previous lemma tells us that $K_{\lambda\lambda} = 1$ and $K_{\mu\lambda} = 0$ if $\mu < \lambda$. If we use ϕ_μ to denote the irreducible character belonging to S_μ, we have the following.

12.4.3 Theorem. *If $\lambda \vdash n$, then*

$$\chi_\lambda = \phi_\lambda + \sum_{\mu > \lambda} K_{\mu\lambda}\phi_\mu. \qquad \square$$

In the next section we present an algorithm for computing the Kostka numbers.

12.5 Kostka numbers

Any partition $\lambda = [\lambda_1, \lambda_2, \dots, \lambda_k]$ of n can be represented by a diagram called the *Young diagram*. In this diagram there are n boxes arranged in k left-justified rows, and the ith row contains exactly λ_i boxes. Figure 12.2 has the Young diagrams for three partitions of 8; the *shape of a Young diagram* refers to the partition it represents.

If the Young diagram of a partition λ has ℓ columns of each of length μ_i then the partition $\lambda' = [\mu_1, \mu_2, \dots, \mu_\ell]$ is called the *conjugate partition* of λ. The first and third Young diagrams in Figure 12.2 represent conjugate partitions.

Suppose λ and μ are two partitions of n and that λ has exactly ℓ parts. A *semi-standard tableau* with shape μ and content λ is a Young diagram whose boxes have been filled with integers $1, \dots, \ell$ such that:

(a) i occupies λ_i boxes, for $i = 1, \dots, \ell$;
(b) the numbers in each column are strictly increasing, and the numbers in each row are non-decreasing.

Figure 12.3 shows all the semi-standard Young tableaux with shape $[6, 2]$ and content $[4, 2, 2]$.

1	1	1	1	3	3
2	2				

1	1	1	1	2	3
2	3				

1	1	1	1	2	2
3	3				

Figure 12.3 All semi-standard tableaux with shape [6, 2] and content [4, 2, 2].

7	6	4	3	2	1
2	1				

Figure 12.4 The hook lengths of each of the boxes in the partition [6, 2].

12.5.1 Theorem. *The Kostka number $K_{\mu\lambda}$ is the number of semi-standard Young tableaux of shape μ and content λ.* □

The next result is a straightforward application of Theorem 12.4.3, and the proof is left as an exercise.

12.5.2 Lemma. *For integers n and k with $n \geq 2k$,*

$$1_{[n-k,k]} = \sum_{i=0}^{k} \phi_{[n-i,i]}.$$ □

12.6 Hook length formula

There is a reasonably straightforward expression for the degree of the irreducible character ϕ_λ, which we we present here.

The *hook length* for a box in a Young diagram is the number of boxes in the diagram that are either directly below the box or directly to the right of the box, counting the box itself only once.

The hook length of the box in the ith row (from the top) and the jth column (from the left) of a Young diagram is denoted by $h_{i,j}$. For any partition λ of n, the dimension of the λ-module can be calculated using the hook lengths of the boxes in the Young diagram.

12.6.1 Theorem. *If $\lambda \vdash n$, then the dimension of the irreducible module S_λ is*

$$\frac{n!}{\prod h_{i,j}},$$

where the product is taken over all boxes in the Young diagram for λ. □

Applying the hook length formula we can calculate the dimension of some of the irreducible Sym(n)-modules. For example, the hook lengths of the partition

$[n-1, 1]$ are given here:

n	n -2	$\cdots$	n -3	2	1
1

and the dimension of the irreducible module $S_{[n-1,1]}$ is

$$\frac{n!}{n(n-2)!} = n - 1.$$

In general the dimension of $S_{[n-k,k]}$ is

$$\frac{n!}{(n-k+1)(n-k)\cdots(n-2k+2)(n-2k)!k!} = \binom{n}{k} - \binom{n}{k-1}. \quad (12.6.1)$$

Finally, we note that if λ and λ' are conjugate partitions, then the dimensions of S_λ and $S_{\lambda'}$ are equal.

12.7 Branching rule

In this section we introduce the *branching rule*. For a proof of this rule see [44, Corollary 3.3.11]. We use this rule to find bounds on the degree of some irreducible representations of the symmetric group.

12.7.1 Lemma. *If $\lambda \vdash n$, then*

$$\phi_\lambda \downarrow_{n-1} = \sum \phi_{\lambda^-},$$

where the sum is taken over all partitions λ^- of $n-1$ that have a Young diagram which can be obtained by the deletion of a single box from the Young diagram of λ. Further,

$$\phi_\lambda \uparrow^{n+1} = \sum \phi_{\lambda^+},$$

where the sum is taken over partitions λ^+ of $n+1$ that have a Young diagram which can be obtained by the addition of a single box to the Young diagram of λ. $\square$

The next two results use the branching rule, combined with the hook length formula, to determine which irreducible representations of Sym(n) have degree smaller than a given value. The first result is used in Exercise 14.2 and the second in Section 15.3. We prove only the second result and leave the first as an exercise. The first result is due to Ellis [54], and the second result is proved using the same method as Ellis.

12.7.2 Lemma. *For $n \geq 9$, let χ be a representation of Sym(n) with degree less than $(n^2 - 3n)/2$. If ϕ_λ is a constituent of χ, then λ is one of the following*

Table 12.1 *Constituents of χ, if $\chi \downarrow_{[n]}$ has a constituent with degree less than $(n^2 - n)/2$.*

Constituent of $\chi \downarrow_{[n]}$	Constituents of χ
$[n]$	$[n+1], [n, 1]$
$[n-1, 1]$	$[n, 1], [n-1, 2], [n-1, 1, 1]$
$[n-2, 2]$	$[n-1, 2], [n-2, 3], [n-2, 2, 1]$
$[n-2, 1, 1]$	$[n-1, 1, 1], [n-2, 2, 1], [n-2, 1, 1, 1]$
$[1^n]$	$[2, 1^{n-1}], [1^{n+1}]$
$[2, 1^{n-2}]$	$[3, 1^{n-2}], [2, 2, 1^{n-3}], [2, 1^{n-1}]$
$[2, 2, 1^{n-4}]$	$[3, 2, 1^{n-4}], [2, 2, 2, 1^{n-5}], [2, 2, 1^{n-3}]$
$[3, 1^{n-3}]$	$[4, 1^{n-3}], [3, 2, 1^{n-4}], [3, 1^{n-2}]$

partitions of n:

$$[n], [1^n], [n-1, 1], [2, 1^{n-2}]. \qquad \square$$

12.7.3 Lemma. *For $n \geq 9$, let χ be a representation of $\mathrm{Sym}(n)$ with degree less than $(n^2 - n)/2$. If ϕ_λ is a constituent of χ, then λ is one of the following partitions of n:*

$$[n], [1^n], [n-1, 1], [2, 1^{n-2}], [n-2, 2], [2, 2, 1^{n-4}], [n-2, 1, 1], [3, 1^{n-3}].$$

Proof. We prove this by induction. The result for $n = 9, 10$ can be read directly from the character table for $\mathrm{Sym}(n)$ (these character tables are available in [98], or from the GAP character table library [76]). We assume that the lemma holds for n and $n - 1$.

Assume that χ is a representation of $\mathrm{Sym}(n + 1)$ that has dimension less than $(n^2 - n)/2$, but does not have one of the eight irreducibles representations listed in the statement of the theorem as a constituent.

Consider the restriction of χ to $\mathrm{Sym}(n)$, this representation will be denoted by $\chi \downarrow_{[n]}$. If one of the eight irreducible representations of $\mathrm{Sym}(n)$ with dimension less than $(n^2 - n)/2$ is a constituent of $\chi \downarrow_{[n]}$, then, using the branching rule, we can determine the constituents of χ. The first column of Table 12.1 gives the irreducible representations of $\mathrm{Sym}(n)$ that have dimension less than $(n^2 - n)/2$. The second column lists the irreducible representations that must be constituents of χ, if the representation in the first column is a constituent of $\chi \downarrow_{[n]}$. These are determined by the branching rule.

Using the table above, we see that either one of the eight representations of $\mathrm{Sym}(n + 1)$ with dimension less than $((n + 1)^2 - (n + 1))/2$ is a constituent of χ; or, using the hook length formula, the dimension of χ is larger than

Table 12.2 *Degrees of the representations from Table 12.1 that are larger than $((n + 1)^2 - (n + 1))/2$.*

Representation	Degree
$[n - 2, 3]$	$n(n + 1)(n - 4)/6$
$[n - 2, 2, 1]$	$(n + 1)(n - 1)(n - 3)/3$
$[n - 2, 1, 1, 1]$	$n(n - 1)(n - 2)/6$
$[3, 2, 1^{n-4}]$	$(n + 1)(n - 1)(n - 3)/3$
$[2, 2, 2, 1^{n-5}]$	$n(n + 1)(n - 4)/6$
$[4, 1^{n-3}]$	$n(n - 1)(n - 2)/6$

$((n + 1)^2 - (n + 1))/2$ (provided that $n > 8$, see Table 12.2). Either case is a contradiction. So we can assume that $\chi \downarrow_n$ does not contain any of the representations from Sym(n) with dimension less than $(n^2 - n)/2$.

If the decomposition of $\chi \downarrow_n$ contains two irreducible representations of Sym(n), neither of which is one of the eight irreducible representations with dimension less than $(n^2 - n)/2$, then the dimension of χ must be at least

$$2((n^2 - n)/2) = n^2 - n.$$

But since $n > 3$, this is strictly larger than $((n + 1)^2 - (n + 1))/2$.

Thus $\chi \downarrow_n$ must be an irreducible representation of Sym(n). The branching rule then implies that if ϕ_λ is a constituent of χ, then there is only one block whose removal from the Young diagram of λ will produce a Young diagram for a parition on n. (The irreducible representation of Sym(n) corresponding to this Young diagram is equal to $\chi \downarrow_n$.) The only such Young diagrams are rectangular, so the constituents of χ must be of the form $\phi_{[s^t]}$, for some s and t. Finally, since $n > 2$, there can only be one constituent of χ (or else $\chi \downarrow_n$ will not be irreducible), thus we can assume that $\chi = \phi_{[s^t]}$.

Next consider the restriction of $\chi = \phi_{[s^t]}$ to Sym($n - 1$); this representation will be denoted by $\chi \downarrow_{n-1}$. By the branching rule, this can contain only the irreducible representations of $n - 1$ that correspond to the partitions $\lambda' = [s^{t-1}, s - 2]$ and $\lambda'' = [s^{t-2}, s - 1, s - 1]$.

If $\lambda' = [n - 2, 1]$ or $[n - 3, 2]$, then $t = 2$ and $s = 3$ or 4. Neither case can happen, since $n = s^{t-1} + 2$ or $n = s^{t-1} + 3$, and n is greater than 10. If λ' is any of the other eight partitions of $n - 1$ that correspond to irreducible representations of Sym($n - 1$) dimension less than $((n - 1)^2 - (n - 1))/2$, we again have a contradiction with $n = s^t$ and $n > 10$. Similarly, λ'' cannot be any of the partitions corresponding to the eight representations of Sym($n - 1$) that have dimension less than $((n - 1)^2 - (n - 1))/2$.

Thus the dimension of χ is at least

$$2((n-1)^2 - (n-1))/2 = (n-1)^2 - (n-1),$$

which is strictly greater than $((n+1)^2 - (n+1))/2$ for any $n \geq 7$.   □

12.8 Wreath products

We say that a partition π of the set $\{1, \ldots, n\}$ is *uniform* if its cells are all of the same size. If k divides n, then $\mathrm{Sym}(n)$ acts transitively on the set of uniform partitions with k cells, each of size n/k. In this section we study this permutation representation arising from this action.

Assume that $n = k\ell$ and π is a partition of $\{1, \ldots, \ell k\}$ into k cells, each of size ℓ. The stabilizer of π in $\mathrm{Sym}(n)$ has cardinality $(\ell!)^k k!$ and is isomorphic to the automorphism group of the graph kK_ℓ. This group could be described as the *wreath product* of the symmetric group $\mathrm{Sym}(\ell)$ by the symmetric group $\mathrm{Sym}(k)$, which is denoted by $\mathrm{Sym}(\ell) \wr \mathrm{Sym}(k)$. (Although we use the term "wreath product," we have no desire to discuss it: the identification with $\mathrm{Aut}(kK_\ell)$ will suffice.)

In this section we consider only two families of uniform partitions: those with cells of size 2, and partitions with cells of size $n/2$. We start with the former.

The number of partitions of $\{1, \ldots, 2k\}$, in which all of the cells have size 2, is equal to the index of $\mathrm{Sym}(2) \wr \mathrm{Sym}(k)$ in $\mathrm{Sym}(2k)$, that is,

$$\frac{2k!}{2^k(k!)} = (2k-1)(2k-3)\cdots 1.$$

The cosets of $\mathrm{Sym}(2) \wr \mathrm{Sym}(k)$ in $\mathrm{Sym}(2k)$ correspond to the perfect matchings of K_{2k}. By Theorem 12.1.1, the representation of $\mathrm{Sym}(2k)$ induced from the trivial representation on $\mathrm{Sym}(2) \wr \mathrm{Sym}(k)$ is equivalent to the permutation representation from the action of $\mathrm{Sym}(2k)$ on the perfect matchings of K_{2k}. The decomposition of this representation is given in the next lemma. For a proof of this result, see Saxl [148, Example 2.2]. But first we need the following notation: if λ is a partition of k with $\lambda = [\lambda_1, \lambda_2, \ldots, \lambda_a]$, then $2\lambda = [2\lambda_1, 2\lambda_2, \ldots, 2\lambda_a]$ is a partition of $2k$.

12.8.1 Lemma. *The character of the permutation representation of* $\mathrm{Sym}(2k)$ *on perfect matchings is equal to*

$$\sum_{\lambda \vdash k} \Phi_{2\lambda}.$$ □

We turn now to Sym(2k) acting on partitions with two cells each of size k. Here the stabilizer of a partition is the wreath product Sym$(k) \wr$ Sym(2), and the number of such partitions is

$$\frac{(2k)!}{2(k!)^2} = \frac{1}{2}\binom{2k}{k} = \binom{2k-1}{k-1}.$$

The permutation representation of Sym(n) from its action on these partitions is equivalent to the representation induced by the trivial representation on Sym$(k) \wr$ Sym(2). To find the decomposition of this permutation character, we first need a simple lemma.

12.8.2 Lemma. *Let $K \leq H \leq G$ be groups. Any irreducible representation that is a constituent of* $\text{ind}_G(1_H)$ *is also a constituent of* $\text{ind}_G(1_K)$.

Proof. To see this consider the representation $\text{ind}_H(1_K)$ (details are left as an exercise). $\qquad\square$

12.8.3 Lemma. *The character of the permutation representation of* Sym$(2k)$ *acting on partitions with two cells each of size k is equal to*

$$\sum_{i=0}^{\lfloor k/2 \rfloor} \phi_{[2k-2i,2i]}.$$

Proof. From Lemma 12.8.2 and the observation that the Young subgroup Sym$([k, k])$ is a subgroup of Sym$(k) \wr$ Sym(2), we know that the only possible constituents of this character are of the form $\phi_{[n-k,k]}$. To determine exactly which of these are constituents, consider

$$\langle 1_{\text{Sym}(k)\wr\text{Sym}(2)} \uparrow^{2k}, 1_{\text{Sym}([n-i,i])} \uparrow^{2k} \rangle.$$

Since Sym$([n - i, i])$ has exactly $\lceil (i + 1)/2 \rceil$ orbits on the partitions of $\{1, \ldots, 2k\}$ into two cells of size k, we see that $\phi_{[n-i,i]}$ is a constituent if and only if i is even. $\qquad\square$

We will meet these groups and their representations again in Chapter 15.

12.9 Exercises

12.1 Suppose that ρ is a linear combination of characters with integer coefficients. If $\langle \rho, \rho \rangle = 1$, prove that ρ is a character (and thus it is an irreducible character).

12.2 Prove that $\lambda \leq \mu$ if and only if $\mu' \leq \lambda'$.

12.3 Prove that $K_{\mu\lambda} = 0$ if $\mu < \lambda$ in the dominance order.

12.4 Prove that $K_{[n]\lambda} = 1$ and $K_{\lambda\lambda} = 1$ for all $\lambda \vdash n$.

12.5 List all semi-standard Young tableaux of shape λ and content $[n - 2, 1, 1]$ with $[n - 2, 1, 1] \leq \lambda$ in the dominance ordering.

12.6 Using Theorem 12.4.3, prove that for integers n and k with $n \geq 2k$

$$\chi_{[n-k,k]} = \sum_{i=0}^{k} \phi_{[n-i,i]}.$$

(This is Lemma 12.5.2.)

12.7 Find all partitions $\lambda \vdash n$ for which the dimension of V_λ is equal to 1. Next, find all partitions $\lambda \vdash n$ for which the dimension of V_λ is equal to $n - 1$.

12.8 Using the hook length formula, find the dimension of V_λ when $\lambda = [\lambda_1, 1, 1 \ldots, 1]$ (such a partition is called a *hook*).

12.9 Assume that $n \geq 9$ and let ϕ_λ be an irreducible representation of $\mathrm{Sym}(n)$. Show that if the dimension of ϕ_λ is less than $(n^2 - 3n)/2$, then λ is one of the following partitions of n:

$$[n], \quad [1^n], \quad [n - 1, 1], \quad [2, 1^{n-2}].$$

(This is very similar to the proof of Lemma 12.7.3.)

12.10 Let $G \leq \mathrm{Sym}(n)$. Assume the Young subgroup $\mathrm{Sym}([n - 1, 1])$ acts transitively on the cosets $\mathrm{Sym}(n)/G$. Prove that $[n - 1, 1]$ is a constituent of $1_G \uparrow^n$ if and only if the group $\mathrm{Sym}([n - 1, 1])$ has at least two orbits on the cosets $\mathrm{Sym}(n)/G$.

12.11 Consider $G = \mathrm{Sym}(k) \wr \mathrm{Sym}(\ell)$ as a subgroup of $\mathrm{Sym}(k\ell)$ with both k and ℓ strictly greater than 1. Show that the decomposition of $\mathrm{ind}_{\mathrm{Sym}(k\ell)}(1_G)$ includes the representation $[n]$ with multiplicity 1 and the representation $[n - 2, 2]$ with multiplicity 1, but neither of the representations $[n - 1, 1]$ or $[n - 2, 1, 1]$ is a constituent.

Notes

As we noted in the introduction, our treatment of the representation theory of the symmetric group follows the usual path. The missing proofs appear in the standard treatments, in particular in any one of Ceccherin-Silberstein et al. [43, 44], Fulton and Harris [73, Chapters 1–4], Sagan [147] and Sternberg [156, Chapter 2].

We offer a warning. The representation theory of $\mathrm{Sym}(n)$ employs a lot of fairly baroque terminology, and even if two writers are using the same term, they may not be using exactly the same definition for it.

13

Orbitals

Our goal in this chapter is to use group representation theory to get information about the eigenvalues and eigenspaces of vertex-transitive graphs.

Suppose X is a vertex-transitive graph with adjacency matrix A and automorphism group G. As defined, G is a permutation group, but we may also view it as the set of permutation matrices that commute with A. Given this, we see that A must lie in the commutant of the set of permutation matrices that form G. The key observation is that information about the representations of G can be used to provide information about its commutant, and this will lead us to the eigenvalues and eigenspaces of X.

13.1 Arc-transitive directed graphs

Suppose G is a permutation group acting on a set V. Then G acts on the product $V \times V$; for $g \in G$ and $(i, j) \in V \times V$, the action is

$$(i, j)^g := (i^g, j^g).$$

The orbits of this action are called the *orbitals* of G. The group G is transitive on V if and only if the pairs

$$\{(i, i) : i \in V\}$$

form a single orbit; this orbital is called the *diagonal*. The number of orbitals of a transitive permutation group is called its *permutation rank*. If $|V| \geq 2$, then there must be at least one orbital besides the diagonal and therefore the permutation rank of G is at least 2. The permutation rank is equal to 2 exactly when G is 2-transitive.

If G is transitive, each non-diagonal orbital can be viewed as the arc set of a non-empty directed graph on which G acts arc-transitively. (It is also convenient to identify the diagonal with the directed graph with no arcs.) The union of any set of distinct non-diagonal orbitals is the arc set of a directed graph, and again G acts on this directed graph, but in this case the action is vertex transitive and not, in general, arc transitive. Conversely, if X is a vertex-transitive directed graph, then it is a union of orbitals of $\text{Aut}(X)$.

If the rank of G is k, we see that there are 2^{k-1} vertex-transitive directed graphs on which G acts. Note, though, that some of these graphs could be isomorphic, and also G need not be the full automorphism group of each orbital.

Our next lemma provides useful information about the rank of a transitive permutation group. If a group G acts on a set V, then for any element $v \in V$ the subgroup of all permutations in G that fix v is the *stabilizer* of v; this subgroup is denoted by G_v.

13.1.1 Lemma. *Suppose G acts transitively on the set V and $1 \in V$. Then the number of orbitals of G is equal to the number of orbits of G_1 on V.*

Proof. We first sketch a combinatorial proof, and then offer an algebraic proof.

Suppose the rank of G is $r + 1$, and let $X_1, \ldots, X_r$ be the non-diagonal orbitals of G. Set $N_0(1) = \{1\}$ and for $i = 1, \ldots, r$, let $N_i(1)$ denote the set of out-neighbors of 1 in X_i. We claim that the sets $N_0(1), N_1(1), \ldots, N_r(1)$ are the orbits of G_1 on V. This immediately yields the result.

We turn to algebra. Burnside's Lemma tells us that the number of orbits of G on $V \times V$ is the average number of elements of $V \times V$ fixed by an element of G. If $\text{fix}(g)$ denotes the number of points in V fixed by $g \in G$, then the number of pairs in $V \times V$ fixed by g is $\text{fix}(g)^2$ and therefore the rank of G is

$$\frac{1}{|G|} \sum_{g \in G} \text{fix}(g)^2.$$

Now let ρ be the permutation character of G in its action on V. Then $\rho(g) = \text{fix}(g)$ and hence

$$\frac{1}{|G|} \sum_{g \in G} \text{fix}(g)^2 = \langle \rho, \rho \rangle_G.$$

We see that

$$\rho = \text{ind}_G (1_{G_1})$$

and so by Frobenius reciprocity

$$\langle \rho, \rho \rangle_G = \langle 1_{G_1}, \text{res}_{G_1} (\rho) \rangle_{G_1}$$

and (by Burnside's Lemma), this is equal to the number of orbits of G_1 on V. $\square$

The reason we have bothered with the algebraic proof here is that it provides a second expression for the permutation rank.

13.1.2 Lemma. *Suppose G acts transitively on V with permutation character ρ, and that we have the decomposition*

$$\rho = \sum_{i=0}^{k} m_i \phi_i$$

where $\phi_0, \ldots, \phi_k$ are irreducible characters of G. Then the permutation rank of G is $\sum_i m_i^2$.

Proof. We have

$$\langle \rho, \rho \rangle_G = \sum_{i=0}^{k} m_i^2$$

and the claim follows immediately.   □

13.2 Commutants of permutation groups

We first met the commutant Comm(G) of a permutation group G in Section 3.2. We saw there (in Lemma 3.2.1) that Comm(G) is a Schur-closed matrix algebra that contains I and J and is closed under taking transposes.

13.2.1 Theorem. *If G is a transitive permutation group on V, the adjacency matrices of its orbitals form a basis for Comm(G).*

Proof. Clearly the adjacency matrices of the orbitals are linearly independent and lie in Comm(G), so we only need to show that they span Comm(G).

Any Schur-closed vector space of matrices has a basis of 01-matrices (this is Exercise 3.2). If A belongs to such a basis, then A is the adjacency matrix of a directed graph X with G contained in its automorphism group. Therefore X is a union of orbitals of G, and it follows that A, and hence all of Comm(G), is contained in the span of the adjacency matrices of the orbitals.   □

13.2.2 Corollary. *If G is a transitive permutation group, the dimension of Comm(G) is equal to the number of orbitals of G.*   □

As we noted in Section 3.2, Comm(G) is the Bose-Mesner algebra of an association scheme if and only if it is commutative. In the following two sections we investigate when this happens.

13.3 Generously transitive groups

We present a simple condition on a permutation group that guarantees that its commutant is commutative, but first we need to consider some properties of orbitals.

If G is transitive on V and $i, j \in V$, then either the pairs (i, j) and (j, i) lie in the same orbital, or they are in distinct orbitals. In the first case, the orbital that contains both (i, j) and (j, i) must be a graph. In the second case, the two orbitals are directed graphs with at most one arc joining each pair of vertices. If A is the adjacency matrix of one of these orbitals, then A^T is the adjacency matrix of the other orbital. The orbital of G that contains (i, j) is said to be *paired* with the orbital containing (j, i). An orbital that is paired with itself is called *self-paired* (since orbitals are orbits they are either equal or disjoint, so this is well defined).

If the orbital that contains (i, j) is self-paired, then there is a permutation in G that swaps i and j. This permutation must have even order, but it need not be an involution. It follows that if $|G|$ is odd, then no non-diagonal orbital is self-paired. (The converse is also true, but we leave the proof to the reader.)

A permutation group on the set V is *generously transitive* if for each pair of points x and y in V, there is an element of G that swaps them.

13.3.1 Lemma. *A permutation group G is generously transitive if and only if each orbital is self-paired.* □

13.3.2 Lemma. *If G is a generously transitive permutation group, then the commutant of G is commutative.*

Proof. We use the interesting fact that if A and B are symmetric $v \times v$ matrices, then AB is symmetric if and only if A and B commute.

If G is generously transitive, then each of its orbitals is self-paired and so the adjacency matrices of its orbitals are symmetric. Therefore Comm(G) has a basis consisting of symmetric matrices, which implies that all the matrices in Comm(G) are symmetric. If A and B are members of Comm(G), then AB also lies in Comm(G) (since it is an algebra); thus all three of these matrices are symmetric, and from the fact just stated, A and B commute. □

13.3.3 Corollary. *The commutant of a generously transitive permutation group is the Bose-Mesner algebra of an association scheme.* □

To see an example of this, consider the group

$$(\mathrm{Sym}(2))^n = \mathrm{Sym}(2) \times \mathrm{Sym}(2) \times \cdots \times \mathrm{Sym}(2).$$

Let $x = (x_1, x_2, \ldots, x_n)$ be a length-n binary sequence. Any $g \in (\mathrm{Sym}(2))^n$ can be expressed as $g = (g_1, g_2, \ldots, g_n)$ and we can define an action on the binary sequences by

$$g(x) = (g_1(x_1), g_2(x_2), \ldots, g_n(x_n)).$$

It is not hard to see that this group action is generously transitive on the binary length-n sequences.

Let (x, y) be a pair of binary sequences. If these sequences agree in the positions $i_1, i_2, \ldots, i_k$, then the pair (gx, gy) will also agree in the positions $i_1, i_2, \ldots, i_k$. Conversely, if two pairs of sequences, say (x, y) and (x', y'), both agree in exactly the same positions, then there is a $g \in (\mathrm{Sym}(2))^n$ such that $(gx, gy) = (x', y')$. From this we can see that there are 2^n orbitals under this action, one for each subset of the n positions, and these orbits are composed of the pairs of sequences that agree in the positions in the subset.

13.4 Multiplicity-free representations

Suppose Ψ is a representation of the group G, with irreducible decomposition

$$\Psi = \sum_{i=1}^{m} m_i \Psi_i.$$

The representation Ψ is called *multiplicity-free* if each m_i is equal to 1. The *commutant* of Ψ is the set of matrices that commute with $\Psi(g)$ for all g in G; it is a matrix algebra but is not, in general, Schur-closed. Schur's Lemma implies that if Ψ is irreducible, then $\mathrm{Comm}(\Psi)$ consists only of scalar matrices.

13.4.1 Lemma. *Let G be a group and Ψ a representation of G. Then* $\mathrm{Comm}(\Psi)$ *is commutative if and only if Ψ is multiplicity-free.*

Proof. Assume that

$$\Psi = \sum_{i=1}^{m} m_i \Psi_i,$$

is the decomposition of Ψ into irreducible representations. From Corollary 11.7.4 the algebra spanned by $\Psi(g)$ for g in G is isomorphic to a direct sum of algebras of the form

$$I_{m_i} \otimes M_{d_i},$$

for $i = 1, \ldots, m$, where $M_{d_i} = \mathrm{Mat}_{d_i \times d_i}(\mathbb{C})$ is the full algebra of $d_i \times d_i$ matrices over $\mathbb{C}$ and d_i is the dimension of the irreducible representation Ψ_i.

The commutant of Ψ is then isomorphic to a direct sum of algebras of the form

$$N_{m_i} \otimes I_{d_i}$$

where $N_{m_i} = \text{Mat}_{m_i \times m_i}(\mathbb{C})$ is the full matrix algebra of dimension m_i. This is commutative if and only if $m_i = 1$ for all i. $\square$

Using the fact that the rank of a permutation group is equal to $\sum_i m_i^2$, it is not hard to show that any transitive permutation group with rank at most 4 is multiplicity-free. We leave the proof of this as an exercise.

There is one very important class of examples.

13.4.2 Theorem. *Let G be a group and let D be the diagonal subgroup of $G \times G$. The permutation representation of $G \times G$ on the cosets of D is multiplicity-free.*

Proof. The cosets of D in $G \times G$ are $(\text{id}, g)D$, where $g \in G$, so for each coset, there is a corresponding element of G.

The key observation is that in any orbital, the out-neighborhood of the coset $(\text{id}, \text{id})D$ consists of cosets $(\text{id}, g)D$ where g are all the elements in a conjugacy class of G. So the commutant of this representation is the Bose-Mesner algebra of the conjugacy class scheme. Since this algebra is commutative, the earlier lemma yields that the representation is multiplicity-free. $\square$

13.5 Multiplicity-free representations of the symmetric group

From Lemma 12.5.2 we have that

$$\text{ind}_{\text{Sym}(n)}\left(1_{\text{Sym}([n-k,k])}\right) = \sum_{i=0}^{k} \Phi_{[n-i,i]}.$$

This is an example of a permutation representation that is multiplicity free. Further, $\text{ind}_{\text{Sym}(n)}\left(1_{\text{Sym}([n-k,k])}\right)$ is the permutation representation from the action of $\text{Sym}(n)$ on the set of cosets $\text{Sym}(n)/\text{Sym}([n-k,k])$ (Theorem 12.1.1). Each of these cosets can be represented by a k-subset from $\{1, \ldots, n\}$, and the group $\text{Sym}(n)$ acts transitively on these k-sets.

The orbitals of $\text{Sym}(n)$ acting on these cosets are easy to determine. For each $i \in \{0, \ldots, k\}$ there is exactly one orbital that consists of all the pairs of k-sets whose intersection has size exactly i. In particular the kth orbital is the diagonal. The adjacency matrices of these orbitals span the commutant of $\text{ind}_{\text{Sym}(n)}\left(1_{\text{Sym}([n-k,k])}\right)$ and are exactly the graphs in the Johnson scheme. From Lemma 13.4.1 we know that these matrices commute and that the Johnson scheme is an association scheme (which should be no surprise at this point!)

In this example, the representation is an induced representation of the form $\operatorname{ind}_G (1_H)$. Since this representation is multiplicity-free, the adjacency matrices of the orbitals of G acting on the cosets G/H form an association scheme (this follows easily from Lemma 13.4.1 and Theorem 3.2.2).

13.5.1 Corollary. *If G and H are groups with $H \leq G$ such that $\operatorname{ind}_G (1_H)$ is a multiplicity-free representation of G, then the adjacency matrices of the orbitals of G on the cosets G/H form an association scheme.* □

For a fixed group G, it is interesting to consider the subgroups H which have the property that $\operatorname{ind}_G (1_H)$ is a multiplicity-free representation, since for each of these, the adjacency matrices of the orbitals form an association scheme. In the case that G is the symmetric group, all such subgroups are known. The list of all such subgroups is long and has many special cases (see [86] for a complete list). We will just note three infinite families of such subgroups that are of interest to us:

(a) $\operatorname{Sym}(n - k) \times \operatorname{Sym}(k)$,
(b) $\operatorname{Sym}(2) \wr \operatorname{Sym}(k)$
(c) $\operatorname{Sym}(k) \wr \operatorname{Sym}(2)$.

We have already seen that both the representations

$$\operatorname{ind}_{\operatorname{Sym}(n)} (1_{\operatorname{Sym}(2) \wr \operatorname{Sym}(k)}), \qquad \operatorname{ind}_{\operatorname{Sym}(n)} (1_{\operatorname{Sym}(k) \wr \operatorname{Sym}(2)})$$

are multiplicity-free (Lemma 12.8.1 and 12.8.3). Since each coset of $\operatorname{Sym}(2) \wr \operatorname{Sym}(k)$ in $\operatorname{Sym}(2k)$ corresponds to a perfect matching in the complete graph on $2k$ vertices, we can conclude that there is an association scheme on the perfect matchings where the relations are defined by the orbitals of $\operatorname{Sym}(2k)$. We study this association scheme in more detail in Section 15.2.

13.6 An equitable partition

Suppose X is an orbital for a transitive permutation group G and $A = A(X)$. If P is a permutation matrix representing an element of G, then P commutes with A, and, since E is a polynomial in A, this also implies that P also commutes with E. Consequently $\operatorname{col}(E)$ is fixed by P, and each eigenspace of X is a G-module. It follows that an eigenspace of X is a sum of irreducible G-modules, where the irreducible modules involved are the irreducible constituents of the permutation representation of G. Therefore each irreducible module of G appearing in the permutation representation is associated with an eigenvalue of X.

As an example we consider (once again) the Johnson graph $J(n, k)$. Here we may take G to be $\operatorname{Sym}(n)$ acting on the cosets of $\operatorname{Sym}(n - k) \times \operatorname{Sym}(k)$. Since

$\mathrm{Sym}(n - k) \times \mathrm{Sym}(k)$ is a Young subgroup, we have that the following is the decomposition into irreducibles of the permutation representation:

$$\mathrm{ind}_{\mathrm{Sym}(n)} \left(1_{\mathrm{Sym}([n-k,k])}\right) = \sum_{i=0}^{k} \phi_{[n-i,i]}.$$

(See Section 12.5 for more details.) We deduce immediately that $J(n, k)$ has at most $k + 1$ distinct eigenvalues. On the other hand, the diameter of $J(n, k)$ is k and therefore $J(n, k)$ has at least $k + 1$ distinct eigenvalues. Consequently it has exactly $k + 1$ eigenvalues and the associated eigenspaces are irreducible modules corresponding to the characters $\phi_{[n-i,i]}$ (for $i = 0, \ldots, k$).

Although the information just obtained is useful, there are important questions left unanswered. For example, to apply the ratio bound, we need the least eigenvalue of X, and nothing at hand indicates which irreducible modules lie in the associated eigenspace. Further, in cases of interest to us, X is a graph in an association scheme, and we would like to know the matrix of eigenvalues of this scheme. We do not have a useful approach for this problem in general, but for multiplicity-free representations we do.

Suppose $H \le G$ such that the permutation representation of G on cosets of H is multiplicity-free. We call the association scheme formed by the orbitals of the action of G on the cosets G/H the *orbital scheme*. Next we will show how to derive the matrix idempotents for the orbital scheme of G from the characters of G, by realizing the orbital scheme as a quotient of the conjugacy class scheme. The first step is to recognize an equitable partition of the vertices of the conjugacy class scheme.

Let $\mathcal{A} = \{A_0, \ldots, A_d\}$ be an association scheme with vertex set V and let π be a partition of V. We say that π is *equitable relative to* $\mathcal{A}$ if it is an equitable partition relative to each graph in the association scheme. If $\mathcal{A}$ is the conjugacy class scheme for G (so $V = G$), then each subgroup H of G determines two partitions. The first (which we will not need) is the partition given by the left cosets of H; here each cell is an orbit of H, The second partition is that given by the right cosets Hg of G (for g in G); we call this the *block partition* relative to H.

If $g \in G$, define the map $\lambda_g : G \to G$ by

$$\lambda_g(x) = xg^{-1}.$$

Then λ_g is a permutation of G and the map $g \mapsto \lambda_g$ is the right regular permutation representation of G. Each permutation λ_g is an automorphism of $\mathcal{A}$, and so G acts as a group of automorphisms of $\mathcal{A}$. If $H \le G$, then the block partition is G-invariant; equivalently, the right cosets of H form a partition of G into blocks of imprimitivity.

13.6.1 Lemma. *If X is a normal Cayley graph for G and $H \leq G$, then the block partition relative to H is equitable.*

Proof. Assume $\mathcal{C}$ is a union of conjugacy classes of G and $X = X(G, \mathcal{C})$. If $h \in H$, then

$$|\mathcal{C}h \cap Hx| = |h\mathcal{C} \cap Hx| = |h(\mathcal{C} \cap Hx)| = |\mathcal{C} \cap Hx|$$

and so the number of out-neighbors of h in Hx is the same for all h in H. Since $\mathcal{C}h = h\mathcal{C}$, the number of in-neighbors of h in Hx is the same for all h in H.

Finally, G acts transitively on the right cosets of H, and it follows that the block partition is equitable. $\qquad\square$

13.7 A homomorphism

Assume now that $H \leq G$ and β is the block partition of H. We construct a homomorphism from the conjugacy class scheme of G into the commutant of G acting on cosets of H.

Let S be the characteristic matrix for β. Then $S^T S = |H| I$ and $S S^T$ lies in the commutant of the regular representation of G. For M a $|G| \times |G|$ matrix, define

$$\overline{M} = \frac{1}{|H|} S^T M S.$$

This definition implies that

$$S \overline{M} = M S.$$

13.7.1 Lemma. *If M belongs to the Bose-Mesner algebra of the conjugacy class scheme of G, then $\overline{M}$ lies in the commutant of G acting on cosets of H, and the map $M \mapsto \overline{M}$ is a homomorphism.*

Proof. Suppose P is the right regular permutation representation of an element from G. Since the right cosets of H are blocks of imprimitivity for G, the action of P permutes the right cosets of H. This implies that there is a permutation matrix Q such that

$$PS = SQ,$$

and since $S^T S = |H| I$, we see that

$$Q = \frac{1}{|H|} S^T P S.$$

You may also verify that the map that sends P to Q is a homomorphism from G, acting regularly, to G, acting on the right cosets of H.

If M and N lie in the Bose-Mesner algebra of the conjugacy class scheme, they commute with SS^T and consequently

$$\overline{M}\,\overline{N} = \frac{1}{|H|^2} S^T M S S^T N S = \frac{1}{|H|^2} S^T S S^T M N S = \overline{MN}.$$

Similarly, we can prove that $PS\overline{M} = SQ\overline{M}$. Thus, if M lies in the Bose-Mesner algebra of the conjugacy class scheme of G, then

$$S\overline{M}Q = MSQ = MPS = PMS = PS\overline{M} = SQ\overline{M}.$$

As the columns of S are linearly independent, this implies that Q and $\overline{M}$ commute, and therefore $\overline{M}$ lies in the commutant of G on the cosets of H. $\square$

13.8 Eigenvalues of the orbital scheme

With the homomorphism given in the previous section we can determine the idempotents and the eigenvalues for the orbital scheme of H in G when H is a multiplicity-free subgroup of G.

13.8.1 Lemma. *Let G be a group and let Ω be the set of irreducible representations of G. Suppose $\{E_\phi : \phi \in \Omega\}$ are the matrix idempotents for the conjugacy class scheme of G and S is the characteristic matrix of the partition of G into right cosets of H. If H is multiplicity-free, the matrix idempotents for the orbital scheme on cosets of H are the nonzero matrices of the form $|H|^{-1} S^T E_\phi S$ where $\phi \in \Omega$.*

Proof. The matrices $|H|^{-1} S^T E_\phi S$, where $\phi \in \Omega$, are pairwise orthogonal idempotents that sum to I and lie in the commutant. So we need only show that the number of the matrices that are not zero is equal to the permutation rank of G on the cosets of H.

Let ϕ be an irreducible representation of G. From Section 11.11 we know that if $g, h \in G$, then

$$(E_\phi)_{g,h} = \frac{1}{\phi(1)} \phi(hg^{-1}).$$

Since $S^T E_\phi S$ belongs to the orbital scheme for G on H, its diagonal is constant and $S^T E_\phi S = 0$ if and only if $(S^T E_\phi S)_{1,1} = 0$. We can calculate this entry to be

$$(S^T E_\phi S)_{1,1} = \sum_{g,h \in H} \phi(hg^{-1}) = |H| \sum_{h \in H} \phi(h) = |H|^2 \langle 1_H, \mathrm{res}_H(\phi) \rangle.$$

By Frobenius reciprocity,

$$\langle 1_H, \text{res}_H(\phi) \rangle = \langle \text{ind}_G(1_H), \phi \rangle,$$

which shows us that $S^T E_\phi S \neq 0$ if and only if ϕ is a constituent of the permutation representation of G on the cosets of H. $\square$

13.8.2 Corollary. *Let H be a multiplicity-free subgroup of G. The matrix idempotents for the orbital scheme on cosets of H in G are E_ϕ where*

$$(E_\phi)_{x_i H, x_j H} = \frac{\phi(1)}{|G|} \sum_{\sigma \in H} \phi(x_i \sigma x_j^{-1})$$

and ϕ is an irreducible representation in the decomposition of $\text{ind}_G(1_H)$.

Proof. From the previous lemma, the idempotents of the orbital scheme are $|H|^{-1} S^T E_\phi S$ where ϕ is an irreducible representation of G that is in the decomposition of $\text{ind}_G(1_H)$. By (11.11.1), the $(x_i H, x_j H)$-entry of $|H|^{-1} S^T E_\phi S$ is

$$\frac{1}{|H|} \frac{\phi(1)}{|G|} \sum_{g \in x_i H} \sum_{h \in x_j H} \phi(hg^{-1}) = \frac{1}{|H|} \frac{\phi(1)}{|G|} \sum_{\sigma_1 \in H} \sum_{\sigma_2 \in H} \phi(x_j \sigma_2 \sigma_1^{-1} x_i^{-1})$$

$$= \frac{\phi(1)}{|G|} \sum_{\sigma \in H} \phi(x_j \sigma x_i^{-1}). \qquad \square$$

Note that these entries are well defined since the value of

$$\sum_{h \in H} \phi_\lambda(x_i h x_j^{-1})$$

does not depend on which coset representatives x_i, x_j are used.

Now that we have the idempotents for the orbital scheme, we can also produce a formula for the eigenvalues of the orbital scheme. The *degree of an orbital* is the number of times an element appears as the first in a pair of elements in the orbital. This is equal to the degree of the corresponding adjacency matrix of the orbital.

13.8.3 Lemma. *Let H be a multiplicity-free subgroup of G. The ϕ-eigenvalue of the adjacency matrix for the ℓ-th orbital in the orbital scheme of H in G is*

$$\eta_\phi(A_\ell) = \frac{d_\ell}{|H|} \sum_{h \in H} \phi(x_\ell h),$$

where d_ℓ is the degree of the ℓ-th orbital and x_ℓ is any element in G such that $(1H, x_\ell H)$ is in the ℓ-th orbital.

Proof. Let $X = \{x_1 = id, x_2, \ldots, x_{[G:H]}\}$ be a set of distinct representatives of the cosets of H in G. Set a vector $v = (v_1, \ldots, v_{[G:H]})$ to be a column of the ϕ

idempotent; from Corollary 13.8.2 this vector has entries

$$v_i = \frac{\phi(1)}{|G|} \sum_{h \in H} \phi\left(x_i h x_j^{-1}\right).$$

The entries of the vector $A_\ell v$ are

$$[A_\ell v]_i = \frac{\phi(1)}{|G|} \sum_{x_k} \sum_{h \in H} \phi\left(x_k h x_j^{-1}\right)$$

where the first sum is taken over the coset representatives x_k with $(x_i H, x_k H)$ in the same orbital as $(H, x_\ell H)$. To list all x_k with this property, consider the set of all $g \in G$ that map $(H, x_\ell H)$ to $(x_i H, x_k H)$. Clearly $g = x_i h$ for some $h \in H$, and for every $h \in H$ the coset $x_i h x_\ell H$ is equal to one of the cosets $x_k H$ in the sum. In the list of cosets $x_i h x_\ell H$, for all $h \in H$, each of the distinct cosets $x_k H$ is included exactly $\frac{|H|}{d_\ell}$ times. Thus

$$\sum_{x_k} \sum_{h \in H} \phi(x_k h x_j^{-1}) = \frac{d_\ell}{|H|} \sum_{g \in H} \sum_{h \in H} \phi\left(x_i h x_\ell g x_j^{-1}\right)$$

$$= \frac{d_\ell}{|H|} \sum_{g \in H} \sum_{h \in H} \phi\left(x_j^{-1} x_i h x_\ell g\right)$$

$$= \frac{d_\ell}{|H|} \sum_{h \in H} \phi\left(x_j^{-1} x_i h\right) \sum_{g \in H} \phi(x_\ell g) \qquad (13.8.1)$$

$$= \frac{d_\ell}{|H|} \sum_{h \in H} \phi\left(x_i h x_j^{-1}\right) \sum_{g \in H} \phi(x_\ell g).$$

(The proof that (13.8.1) holds follows from the proof of Lemma 11.12.1.) $\square$

The formula given in Lemma 13.8.3 is often not effective since we do not have good ways of calculating the value of a character over a coset. In the next section we see how to use this approach to find some of the eigenvalues in the orbital scheme.

13.9 Eigenspaces and Young subgroups

We consider subgroups $G \leq \mathrm{Sym}(n)$ such that the representation of $\mathrm{Sym}(n)$ induced by the trivial representation on G is multiplicity-free. Rather than denoting this representation by $\mathrm{ind}_{\mathrm{Sym}(n)}(1_G)$, we simply write $\mathrm{ind}(1_G)$, and it is assumed that we are inducing the representation to the symmetric group. The Young subgroups can be used to find eigenvalues of the matrices in these orbital schemes.

Let A_ℓ denote the adjacency matrix for the ℓ-th orbital of the action of $\mathrm{Sym}(n)$ on $\mathrm{Sym}(n)/G$. The Young subgroup $\mathrm{Sym}(\lambda)$ acts on the cosets $\mathrm{Sym}(n)/G$ and forms an equitable partition of the vertices of this graph; denote this partition by $\pi(\lambda)$. Thus the eigenvalues of the quotient graph $A_\ell/\pi(\lambda)$ are also eigenvalues of A_ℓ.

For an example, consider $G = \mathrm{Sym}(k) \times \mathrm{Sym}(n - k)$. In this case, the orbital scheme is the Johnson scheme and the matrices are A_{k-i} (two k-sets are adjacent in A_{k-i} if their intersection has size i). Let $\lambda = [n - 1, 1]$; then the quotient graph $A_{k-i}/\pi(\lambda)$ has the form:

$$A_{k-i}/\pi(\lambda) = \begin{pmatrix} \binom{k-1}{i-1}\binom{n-k}{k-i} & \binom{k-1}{i}\binom{n-k}{k-i} \\ \binom{k}{i}\binom{n-k-1}{k-i-1} & \binom{k}{i}\binom{n-k-1}{k-i} \end{pmatrix}.$$

The eigenvalues of this matrix are

$$d_i = \binom{k}{i}\binom{n-k}{k-i}, \qquad \eta_i = \binom{n-k-1}{k-i}\binom{k}{i} - \binom{k-1}{i}\binom{n-k}{k-i}.$$

If $i = 0$, then A_{k-i} is the adjacency matrix for the Kneser graph and these eigenvalues are $\binom{n-k}{k}$ and $-\binom{n-k-1}{k-1}$. To determine eigenvalues for the Johnson graph, set $i = k - 1$; these equations give that $k(n - k)$ and $(n - k)k - n$ are eigenvalues. These eigenvalues belong to the $[n]$-module and the $[n - 1, 1]$-module.

Choosing the partition λ of n carefully allows us to find additional information about the action of $\mathrm{Sym}(\lambda)$.

13.9.1 Theorem. *Let G be a subgroup of $\mathrm{Sym}(n)$ such that $\mathrm{ind}_{\mathrm{Sym}(n)}(1_G)$ is multiplicity-free. Let λ be the largest partition in the dominance ordering with $\lambda < [n]$ that is a constituent of $\mathrm{ind}_{\mathrm{Sym}(n)}(1_G)$. Then the action of $\mathrm{Sym}(\lambda)$ on the cosets $\mathrm{Sym}(n)/G$ has exactly two orbits.*

Proof. Since $\langle \mathrm{ind}(1_G), \phi_\mu \rangle = 0$ for all partitions with $\lambda < \mu < [n]$, the number of orbits of $\mathrm{Sym}(\lambda)$ acting on $\mathrm{Sym}(n)/G$ is equal to

$$\langle \mathrm{ind}(1_G), \mathrm{ind}(1_{\mathrm{Sym}(\lambda)}) \rangle = \langle \mathrm{ind}(1_G), \phi_\lambda + \sum_{\mu > \lambda} K_{\mu,\lambda}\phi_\mu \rangle$$

$$= \langle \mathrm{ind}(1_G), \phi_\lambda \rangle + \langle \mathrm{ind}(1_G), K_{[n],\lambda}\,\phi_{[n]} \rangle$$

$$= \langle \mathrm{ind}(1_G), \phi_\lambda \rangle + 1.$$

The subgroup G is multiplicity-free and ϕ_λ is a constituent, so $\langle \mathrm{ind}(1_G), \phi_\lambda \rangle = 1$. Thus the Young subgroup $\mathrm{Sym}(\lambda)$ has exactly two orbits on $\mathrm{Sym}(n)/G$. $\square$

Assume G is as in the previous theorem, and let λ be the largest partition (in the dominance ordering) with $\lambda < [n]$ such that ϕ_λ is a constituent of $\text{ind}(1_G)$. Further assume that $\pi = \{S, V \setminus S\}$ are the two orbits of $\text{Sym}(\lambda)$ acting on $\text{Sym}(n)/G$ (where V is the set of cosets $\text{Sym}(n)/G$). Let A_ℓ be the adjacency matrix of an orbital for $\text{Sym}(n)$ acting on $\text{Sym}(n)/G$. Then the quotient graph A_ℓ/π has the form:

$$\begin{pmatrix} a & d - a \\ \frac{(d-a)|S|}{|V|-|S|} & \frac{d|V|+a|S|-2d|S|}{|V|-|S|} \end{pmatrix} \tag{13.9.1}$$

where d is the degree of the orbital. The eigenvalues of this quotient graph are

$$d, \quad -\frac{(d-a)|S|}{|V|-|S|}.$$

Since this partition is equitable, both of these eigenvalues are eigenvalues of A_ℓ. If A_ℓ is the adjacency matrix for a graph X, then the set S is a coclique in X if and only if $a = 0$ or, equivalently, if and only if $-\frac{d|S|}{|V|-|S|}$ is the eigenvalue for the λ-module.

For example, we have seen that if $G = \text{Sym}(k) \times \text{Sym}(n - k)$ then $\text{ind}(1_G)$ is multiplicity-free, and the matrices A_ℓ are the matrices in the Johnson scheme. The largest non-trivial partition that is a constituent of $\text{ind}(1_G)$ is the partition $\lambda = [n - 1, 1]$. Assuming that $\text{Sym}(\lambda)$ is the subgroup of $\text{Sym}(n)$ that fixes 1, the set S is the collection of all k-sets that contain 1. The matrix in (13.9.1) is the quotient graph $A_{k-i}/\pi(\lambda)$, so we see that $a = \binom{k-1}{i-1}\binom{n-k}{k-i}$ and $d = \binom{k}{i}\binom{n-k}{k-i}$. Here $a = 0$ if and only if $i = 0$; in this case A_k is the adjacency matrix for the Kneser graph. The set S is a coclique in this graph and equality holds in the ratio bound (Theorem 2.4.1).

The approach used in Sections 13.6 and 13.7 to determine the matrix idempotents of an orbital scheme is based on [83].

13.10 Exercises

13.1 Assume G is a transitive permutation group acting on the set V and let $X_1, \ldots, X_r$ be its non-diagonal orbitals. If $i \in V$, show that the out-neighborhoods of i in $X_1, \ldots, X_r$, are the orbits of G_i on V.

13.2 Show that the group $\text{Sym}(n)$ acts generously transitively on k-subsets from $\{1, \ldots, n\}$. Find a basis for the commutant of this action.

13.3 Show that the group $\text{Sym}(2n)$ is generously transitive on the perfect matchings of K_{2n}.

13.4 Show that the group $\text{Sym}(16)$ is not generously transitive on the set of all 4×4 partitions of $\{1, \ldots, 16\}$.

13.5 Set $G = \text{Sym}(3) \wr \text{Sym}(3)$. Show that $\text{ind}_{\text{Sym}(9)}(1_G)$ is multiplicity-free.

13.6 Show that any transitive permutation group with rank at most 4 is multiplicity-free.

13.7 Let $\lambda \vdash n$ and assume that both the groups G and $\text{Sym}(n)$ act on a set V. Assume that A is the adjacency matrix for an orbital of the action of G on V and that π is the orbit partition of the action of $\text{Sym}(\lambda)$ on V. Show that if η is an eigenvalue of the quotient graph A/π, then η belongs to some μ-module; where $\mu \geq \lambda$ in the dominance ordering.

13.8 Let $G, H \leq \text{Sym}(n)$. Assume that H is transitive on $\text{Sym}(n)/G$. Show that the only irreducible representation that is common to both $\text{ind}(1_G)$ and $\text{ind}(1_H)$ is the trivial representation.

13.9 Using quotients, find the eigenvalue for the Kneser graph that belongs to the $[n-1, 1]$-module and find a basis for the module. Again using quotients, find the eigenvalue for the $[n-2, 2]$-module. Next use the fact that the eigenvectors in this module are orthogonal to the vectors in the $[n-1, 1]$-module to find a basis for the $[n-2, 2]$-module. Using this method, recursively find all the eigenvectors for the Kneser graph.

Notes

The book by Bannai and Ito [14] provides an extensive treatment of the relation between representation theory and association schemes. In particular, details about the conjugacy class scheme are also given in Section 2.7 of this work.

Saxl [148] determined properties of subgroups G such that $\text{ind}_{\text{Sym}(n)}(1_G)$ is a multiplicity-free representation. The exact list of all such groups was independently determined in [86] and in [171]. There is a fair amount of interest in multiplicity-free representations since they are important in invariant theory and are related to Gelfand pairs. For more details about Gelfand pairs and their spherical functions see [43, Chapter 4].

If a group action is transitive (rather than generously transitive), then the orbitals form a homogeneous coherent configuration. More details about orbital schemes can be found in [31, Section 2.9] and [14, Section 2.8]. We also recommend Wielandt's *Finite Permutation Groups* [170] for information about orbital schemes.

14

Permutations

We have seen many EKR-type theorems for objects other than sets. In Chapter 9 we gave a version of the EKR Theorem for vector spaces; in Chapter 10 we proved an EKR-type theorem for integer sequences. In this chapter we consider how an EKR-type theorem can be established for permutations.

To start, we recall from Section 7.5 our definition of intersection for permutations: two permutations $\pi, \sigma \in \mathrm{Sym}(n)$ are said to be *intersecting* if $\pi(i) = \sigma(i)$ for some $i \in \{1, \ldots, n\}$. Such permutations are also said to *agree* on the point i. For an integer $t \geq 1$, a pair of permutations from $\mathrm{Sym}(n)$ are said to be *t-intersecting* if they agree on at least t points from $\{1, \ldots, n\}$. A set of permutations is called intersecting (or t-intersecting) if any two permutations in the set are intersecting (or t-intersecting).

An example of a set of intersecting permutations is the set of all permutations in $\mathrm{Sym}(n)$ that fix some point i (this is the *stabilizer* of the point i). The stabilizer of a point is a set of intersecting permutations of size $(n - 1)!$. Similarly, the pointwise stabilizer of any t distinct elements is a set of t-intersecting permutations with size $(n - t)!$. Further, any coset of these groups is also a set of t-intersecting permutations of size $(n - t)!$.

One generalization (perhaps the most natural) of the EKR Theorem for permutations is the assertion that the largest set of intersecting permutations has size $(n - 1)!$, and the only sets that meet this bound are the cosets of point stabilizers. Cameron and Ku [40] and Larose and Malvenuto [113] independently proved this result in 2006. Since then other proofs of this result have appeared (see [168, Section 2] for a particularly simple proof).

In this chapter we present a proof that uses the ratio bound for cocliques, and an analysis of the corresponding eigenspaces. This is the proof given in [85]. We will consider extending this result to sets of t-intersecting permutations. We will also look at the largest subsets of intersecting permutations from subgroups of the symmetric group.

14.1 The derangement graph

For a positive integer n, define the *derangement graph* Γ_n to be the graph whose vertex set is the set of all permutations of an n-set, and vertices π and ρ are adjacent if $\pi(i) \neq \rho(i)$ for all $i \in \{1, \ldots, n\}$. This graph is defined so that a set of intersecting permutations is a coclique. (This is the same as the definition given in Section 7.5.)

The set of all derangements of n is denoted by $D(n)$. The number of derangements of a set of size n is $|D(n)| = d(n)$ and is given by the following recursive formula:

$$d(n) = (n-1)(d(n-1) + d(n-2)) \qquad (14.1.1)$$

with $d(1) = 0$ and $d(2) = 1$. The graph Γ_n is the Cayley graph for $\mathrm{Sym}(n)$ with connection set $D(n)$ and therefore has degree $d(n)$. Since $D(n)$ is the union of conjugacy classes on $\mathrm{Sym}(n)$, the derangement graph is a normal Cayley graph.

Further, Γ_n is a union of graphs in the conjugacy class association scheme on $\mathrm{Sym}(n)$ (see Section 3.3). It is a vertex-transitive graph, so the clique-coclique bound (Corollary 2.1.2) holds. Using this bound, it is straightforward to show that the largest intersecting sets in $\mathrm{Sym}(n)$ have size $(n-1)!$ (this was first shown by Deza and Frankl [69] in 1977).

14.1.1 Theorem. *The size of a maximum clique in Γ_n is n.*

Proof. If $i \in \{1, \ldots, n\}$, the images of i under the permutations in a clique of Γ_n must be distinct. Therefore a clique can have no more than n vertices. Since each row of a Latin square of order n is a permutation in $\mathrm{Sym}(n)$, the set of all rows in a Latin square of order n is a clique of size n in Γ_n. A Latin square of order n exists for every n, so the theorem holds. $\qquad \square$

Using the clique-coclique bound we have that

$$\alpha(\Gamma_n) \leq \frac{n!}{n} = (n-1)!,$$

and it is straightforward to see that this bound is tight. For any $i, j \in \{1, \ldots, n\}$, define the set $S_{i,j}$ as follows:

$$S_{i,j} = \{\pi \in \mathrm{Sym}(n) : \pi(i) = j\} \qquad (14.1.2)$$

(each $S_{i,j}$ is the coset of the stabilizer of a point). Each $S_{i,j}$ is a coclique in Γ_n of size $(n-1)!$.

14.1.2 Theorem. *The size of a maximum coclique in* Γ_n *is* $(n-1)!$.   □

This result can be considered to be the bound in the EKR Theorem for permutations. In Section 14.4 we show that the sets $S_{i,j}$ are all the maximum cocliques in Γ_n. This is a much more difficult result, which gives the characterization of the maximum sets of intersecting permutations.

Theorem 14.1.1 shows that the cliques in the derangement graph correspond to Latin squares, which is a much stronger and more useful result than simply determining the size of the maximum clique in Γ_n. With this fact, it is possible to find cliques with a particular structure. For an example of this, we use the result of Alspach, Gavlas, Šajna and Verrall [12] that the complete digraph on n vertices can be decomposed into directed n-cycles. Other decompositions of the complete digraph on n vertices produce cliques with different specific structures, but the structure in the following lemma is the only one that we require.

14.1.3 Lemma. *If* $n > 6$, *then there exists a clique in* Γ_n *that consists of the identity permutation and* $n-1$ *cycles of length* n.

Proof. For $n > 6$ there is a decomposition of the complete digraph on n vertices into $n-1$ directed cycles [12]. Each of these directed cycles is a cycle of length n in Sym(n). Moreover, no two cycles in the decomposition share an edge, so any two of these cycles correspond to permutations that are adjacent vertices in Γ_n. Thus there exists an n-clique whose elements are the n-cycles in this decomposition together with the identity of Sym(n).   □

14.2 Eigenvalues of the derangement graph

Since Γ_n is a Cayley graph for Sym(n), each eigenspace of Γ_n is the direct sum of Sym(n)-modules. Therefore, each such Sym(n)-module has an associated eigenvalue of Γ_n, which can be calculated using the representation theory of the symmetric group. By Theorem 11.12.3, if $\lambda \vdash n$, the associated eigenvalue is

$$\eta_\lambda = \frac{1}{\chi_\lambda(1)} \sum_{x \in D(n)} \chi_\lambda(x). \qquad (14.2.1)$$

(Note that different modules could have the same eigenvalue.) Unfortunately, except for small values of n, we do not have an effective way to calculate η_λ directly for all λ. Even bounds on the eigenvalues are difficult to obtain, but in some cases, this formula can be simplified.

Each conjugacy class of Sym(n) can be represented by a partition of n (the parts in the partition are the lengths of the cycles of the elements in the class).

Thus we use C_μ, where $\mu \vdash n$, to denote the conjugacy classes in $\mathrm{Sym}(n)$. The permutations in the conjugacy class C_μ are derangements if and only if the partition μ has no parts of size 1.

Since the character χ_λ is constant on the conjugacy classes and $D(n)$ is the union of conjugacy classes, the formula in (14.2.1) can be written as

$$\eta_\lambda = \frac{1}{\chi_\lambda(1)} \sum_\mu |C_\mu| \, \chi_\lambda(c),$$

where the sum is taken over all partitions μ that have no parts of size 1 and c is any element in C_μ.

For example, if $\lambda = [n]$ then

$$\eta_{[n]} = \sum_\mu |C_\mu| = d(n),$$

again where the sum is taken over all partitions μ that correspond to a derangement. It is no surprise that this is the degree of the graph and the largest eigenvalue.

If $\lambda = [n - 1, 1]$, then for any $x \in \mathrm{Sym}(n)$ the value of $\chi_\lambda(x)$ is one less than the number of points fixed by x; since we are summing over derangements this is

$$\eta_{[n-1,1]} = \frac{1}{n-1} \sum_{x \in D(n)} (-1) = -\frac{d(n)}{n-1}.$$

By (14.1.1),

$$-\frac{d(n)}{n-1} = -(d(n-1) + d(n-2))$$

so this eigenvalue is an integer.

It was conjectured by Ku and Wong [112] that this eigenvalue is the least eigenvalue of the derangement graph; this conjecture was proven to be true by Renteln [146] in 2007. In 2012, Ellis [54] gave another, much simpler proof of this.

14.2.1 Theorem. *The least eigenvalue of Γ_n is $-d(n)/(n - 1)$, and the associated eigenspace is the $[n - 1, 1]$-module for $\mathrm{Sym}(n)$.* $\qquad\square$

Once we know the least eigenvalue of Γ_n we can apply the ratio bound to confirm the bound

$$\alpha(\Gamma_n) \le \frac{n!}{1 - \frac{-\frac{d(n)}{n-1}}{d(n)}} = (n - 1)!.$$

Since the ratio bound holds with equality, the characteristic vector of a coclique of size $(n - 1)!$ is, when shifted by an appropriate multiple of the all-ones vector, an eigenvector for the smallest eigenvalue (see Theorem 2.4.1). In other words, it follows from Theorem 14.2.1 that the characteristic vector for any coclique of size $(n-1)!$ lies in the sum of the $[n]$-module and the $[n-1, 1]$-module. In the next section, we provide a proof of this without using Renteln's result. Further, it will be proven that the characteristic vectors for the sets $S_{i,j}$ form a spanning set for the sum of these two modules. With these two facts we conclude that any characteristic vector of a maximum coclique in Γ_n is a linear combination of the characteristic vectors of the sets $S_{i,j}$. These facts are used in the next two sections to characterize all maximum cocliques in Γ_n.

14.3 An eigenspace of the derangement graph

Let S be a coclique in Γ_n of size $(n-1)!$ and let v_S be the characteristic vector of S. We prove that $v_S - \frac{1}{n}\mathbf{1}$ is in the $[n-1, 1]$-module by showing that for all partitions λ, other than $[n]$ or $[n-1, 1]$, the projection of $v_S - \frac{1}{n}\mathbf{1}$ onto the λ-module is equal to zero.

To do this we use the fact that if S is a maximum coclique and T a maximum clique in Γ_n, then at most one of the projections of v_S and v_T onto a module is not zero (see Corollary 3.8.5). We construct a clique T in the derangement graph such that these projections are all not equal to zero.

For $\lambda \vdash n$ define the matrix E_λ as

$$(E_\lambda)_{\sigma,\pi} = \frac{\lambda(1)}{n!}\chi_\lambda(\pi\sigma^{-1}). \tag{14.3.1}$$

This matrix gives the projection onto the λ-module (see the comments following Lemma 11.11.1 to see why this is the projection). For T a maximum clique in Γ_n, define

$$\chi_\lambda(T) = \sum_{x \in T} \chi_\lambda(x).$$

If $\chi_\lambda(T) \neq 0$, then by (14.3.1), $E_\lambda v_T \neq 0$.

We need the following result about the values of characters of $\mathrm{Sym}(n)$. It is a special case of the Murnaghan-Nakayama rule, which provides a recursive formula for the character values of the symmetric group. For a proof see [147, Lemma 4.10.3], or any standard text on the representation theory of the symmetric group.

14.3.1 Lemma. *Suppose $\lambda \vdash n$ and σ is an n-cycle in $\mathrm{Sym}(n)$. If λ is the hook $[r, 1^{n-r}]$, then $\chi_\lambda(\sigma) = (-1)^{n-r}$. If λ is not a hook, then $\chi_\lambda(\sigma) = 0$.* □

The next proof holds only for $n > 6$. For $n \leq 6$ the results can be verified by hand or by an exhaustive computer search. This is true for several of the results and proofs given in this section.

14.3.2 Lemma. *If $n > 6$, then for every $\lambda \neq [n]$ or $[n-1, 1]$ there is a clique T in the derangement graph Γ_n such that $E_\lambda v_T \neq 0$.*

Proof. Let T be an n-clique in Γ_n whose elements are the identity and n-cycles in $\text{Sym}(n)$ (such a clique exists by Lemma 14.1.3). Since every $x \in T$, except the identity, is an n-cycle, for every $\lambda \vdash n$ the value of $\chi_\lambda(x)$ is the same. Thus

$$\chi_\lambda(T) = \sum_{x \in T} \chi_\lambda(x)$$

$$= \chi_\lambda(1) + (n-1)\chi_\lambda(x),$$

where x an n-cycle. By Lemma 14.3.1, $\chi_\lambda(x) \in \{-1, 0, 1\}$ for every character χ_λ. If $\chi_\lambda(T) = 0$, then, since $\chi_\lambda(1)$ is positive, it follows that $\chi_\lambda(x) = -1$ and, in turn, this implies that $\chi_\lambda(1) = n - 1$.

For $n > 6$, the only partitions of n with $\chi_\lambda(1) = n - 1$ are $[n-1, 1]$ and $[2, 1^{n-2}]$, so if λ is any other partition, $\chi_\lambda(T) \neq 0$. Further, if n is even, then for x an n-cycle $\chi_{[2,1^{n-2}]}(x) = 1$ (again by Lemma 14.3.1), so $\chi_\lambda(T) = 2(n-1) \neq 0$. Thus $E_\lambda v_T \neq 0$ for all partitions λ, except $\lambda = [2, 1^{n-2}]$ when n is odd.

To complete this theorem, we construct another specific clique T' with $\chi_{[2,1^{n-2}]}(T') \neq 0$, where n is odd. Consider an $n \times n$ Latin square with the first row $(1, 2, \ldots, n)$ and the second row $(2, 1, n, 3, 4, \ldots, n-1)$. Such a Latin square exists since any Latin rectangle can be extended to a Latin square [92]. The rows of this Latin square are the permutations in our clique T'. The first row corresponds to the identity permutation, and the second to an odd permutation. So, the first row contributes $n - 1$ to the sum, and the second row contributes 1. Each of the last $n - 2$ permutations contributes no less than -1 to the sum, so $\chi_{[2,1^{n-2}]}(T') \neq 0$. $\square$

14.3.3 Lemma. *Let n be an integer with $n > 6$. Let S be a maximum coclique in Γ_n and v_S be the characteristic vector of S. Then the vector $v_S - \frac{1}{n}\mathbf{1}$ is in the $[n-1, 1]$-module.*

Proof. The vector $v_s - \frac{1}{n}\mathbf{1}$ is balanced, so it is orthogonal to the $[n]$-module.

By Lemma 14.3.2, for every partition $\lambda \neq [n-1, 1]$ or $[n]$, there is a maximum clique T such that $E_\lambda v_T \neq 0$. By Corollary 3.8.5 this implies that $E_\lambda v_S = 0$, which in turn implies that $E_\lambda(v_S - \frac{1}{n}\mathbf{1}) = 0$ (since $\lambda \neq [n]$, we have that $E_\lambda \mathbf{1} = 0$). Thus the vector $v_S - \frac{1}{n}\mathbf{1}$ is orthogonal to the λ-module for every partition $\lambda \neq [n-1, 1]$, and must be in the $[n-1, 1]$-module. $\square$

The next step is to use the vectors from the previous lemma to find a basis for the $[n-1, 1]$-module. Let H be the $n! \times (n-1)^2$ matrix defined as follows: the rows are indexed by the elements of $\text{Sym}(n)$; the columns are indexed by the ordered pairs of elements in $\{1, \ldots, n-1\}$; and the $(\pi, (i, j))$-entry is 1 if $\pi(i) = j$ and 0 otherwise. Note that the columns of H are the characteristic vectors of the maximum cocliques $S_{i,j}$. We denote these vectors by $v_{i,j}$.

14.3.4 Lemma. *The vectors*

$$\left\{ v_{i,j} - \frac{1}{n}\mathbf{1} : 1 \leq i, j \leq n-1 \right\}$$

form a basis for the $[n-1, 1]$*-module of* Γ_n.

Proof. From Lemma 14.3.3, the vectors $v_{i,j} - \frac{1}{n}\mathbf{1}$ are in the $[n-1, 1]$-module of Γ_n. The dimension of the $[n-1, 1]$-module is $(n-1)^2$ (from Corollary 11.12.2), so we only need to show that these vectors are linearly independent. Since

$$\mathbf{1} \notin \mathrm{span}\{v_{i,j} : 1 \leq i, j \leq n-1\},$$

it is enough to show that the vectors $v_{i,j}$ are linearly independent.

We need to show that H has rank $(n-1)^2$. To do this, we show that $H^T H$ has rank $(n-1)^2$ by proving that $H^T H$ does not have 0 as an eigenvalue.

The adjacency matrix of the complete graph on n vertices is $A(K_n) = J_n - I_n$. If we order the columns of H such that the pair (i, j) occurs before (k, ℓ) whenever either $i < k$, or if $i = k$ and $j < \ell$, then it is not hard to see that

$$H^T H = (n-1)!\, I_{(n-1)^2} + (n-2)!\, (A(K_{n-1}) \otimes A(K_{n-1})).$$

Since 0 is not an eigenvalue of this matrix, $H^T H$ has rank $(n-1)^2$. □

14.4 Cocliques in the derangement graphs

Lemma 14.3.3 and Lemma 14.3.4 imply that the characteristic vector of any maximum coclique S of Γ_n is in the span of

$$\{v_{i,j} : 1 \leq i, j \leq n-1\} \cup \{\mathbf{1}\}.$$

Note that, using the matrix H defined in the previous section, this is equivalent to $v_S \in \mathrm{span}(\mathrm{col}(H) \cup \{\mathbf{1}\})$. Further, since

$$\mathbf{1} - \sum_{i=1}^{n-1} v_{i,i} = v_{n,n},$$

this span is equal to the span of $\{\mathrm{col}(H) \cup \{v_{n,n}\}\}$. Define H to be the $n! \times (n^2 - 2n + 2)$ matrix in which the first n columns are the vectors $v_{1,1}, \ldots, v_{n,n}$. The next $(n-1)(n-2)$ columns are $v_{i,j}$ with $1 \leq i, j \leq n-1$ and $i \neq j$ (in arbitrary order). The columns of H are exactly the columns of H with the addition of $v_{n,n}$. The fact that the vector $v_S - \frac{1}{n}\mathbf{1}$ is a linear combination of the columns of H' will be used to show that S must be one of the cocliques $S_{i,j}$.

Let id be the identity permutation in $\mathrm{Sym}(n)$. Then the set of all derangements in $\mathrm{Sym}(n)$ are the neighbors of id in Γ_n. Consider the following three submatrices of H':

(a) N, the submatrix whose rows are the derangements of $\mathrm{Sym}(n)$;
(b) M, the submatrix of N with columns indexed by the pairs (i, j) with $i, j \in \{1, \ldots, n-1\}$ and $i \neq j$; and
(c) W, the submatrix of H' with columns indexed by the pairs (i, i) with $i \in \{1, \ldots, n\}$.

Assume that the first row of H' corresponds to the identity permutation and the next $d(n)$ rows correspond to the neighbors of the identity (these are exactly the derangements). Then H' has the following block structure:

$$H' = \left(\begin{array}{c|c} 1 & 0 \\ \hline 0 & M \\ \hline H_1 & H_2 \end{array} \right).$$

Further, we consider the following two submatrices of H':

$$N = \left(\, 0 \, \middle| M \, \right), \qquad W = \left(\begin{array}{c} 1 \\ \hline 0 \\ \hline H_1 \end{array} \right).$$

14.4.1 Lemma. *For all n the rank of M is $(n-1)(n-2)$.* $\quad\square$

We omit the details of this proof and only claim that M contains $K_{n-1} \otimes I_{n-2}$ as a submatrix and refer the reader to [85].

14.4.2 Lemma. *If y is in the kernel of N, then $H'y$ lies in the column space of W.*

Proof. Assume y is in the kernel of N. Let y_M denote the vector of length $(n-1)(n-2)$ formed by taking the final $(n-1)(n-2)$ entries of y. Then

$$0 = Ny = [0|M]y = My_M.$$

Since M has rank $(n-1)(n-2)$, the last $(n-1)(n-2)$ entries of y are all 0. Thus $H'y$ is in the column space of W. $\quad\square$

We now have all the tools to prove the EKR Theorem for permutations. The result is true for all n, but we have only proven it for $n > 6$. For smaller values the result can be checked by hand or computer.

14.4.3 Theorem. *Let $n > 6$. If $S \subseteq \mathrm{Sym}(n)$ is an intersecting family of permutations, then:*

(a) $|S| \leq (n-1)!$; and,

(b) if $|S| = (n-1)!$, then S is a coset of a stabilizer of a point.

Proof. Statement (a) follows directly from Theorem 14.1.2.

Let S be a coclique of size $(n-1)!$ in Γ_n, and let v_S be the characteristic vector of S. Without loss generality, we may assume that the identity permutation, id, is in S. By Lemma 14.3.3, v_S is in the column space of H'. Since S is a coclique, no neighbors of id can be in S, and so $Ny = 0$. This implies, by Lemma 14.4.2, that $v_S = Wx$ for some vector x.

Suppose that x_i is nonzero for some $i \in \{1, \ldots, n\}$. As $n \geq 3$, there is a permutation π with $\pi(i) = i$ and no other fixed points. This means that the entry of v_S corresponding to π must be equal to the ith entry of x. Since v_S is a 01-vector, we can conclude that x must also be a 01-vector.

Further, since $n \geq 4$, for every pair of distinct $i, j \in \{1, \ldots, n\}$ there is a permutation π that fixes i and j but no other points. If the ith and jth entries of x are both nonzero, then the π-entry of v_S is 2. Since v_S is a 01-vector, we conclude that there is only one nonzero entry in x. Thus v_S is equal to one of the columns of W and $S = S_{i,i}$ for some $i \in \{1, \ldots, n\}$. $\square$

14.5 t-Intersecting permutations

Deza and Frankl [69] conjectured that the EKR Theorem can also be extended to sets of permutations in $\mathrm{Sym}(n)$ in which any two permutations agree on at least some fixed number of points (so are t-intersecting). This conjecture was proved by Ellis, Friedgut and Pilpel [56].

14.5.1 Theorem. *For n sufficiently large relative to t, the size of a maximum set of permutations of an n-set in which every pair of permutations agree on at least t elements is $(n-t)!$.* $\square$

Ellis, Friedgut and Pilpel also proved that if a set of t-intersecting permutations has size $(n-t)!$ (still assuming that n is sufficiently large), then all permutations in the set must agree on a set of t elements. This simply means that the only sets of permutations that meet this bound are the cosets of the pointwise stabilizer of a t-subset of elements.

The proof requires that n be sufficiently large with respect to t, and this condition cannot be dropped. It is conjectured, provided that t is at least 4, that the exact lower bound on n is $2t+1$ and for $n \leq 2t$ there are t-intersecting sets of permutations that are larger than the pointwise stabilizer of a t-subset. The construction of such sets is derived from the set systems in Section 1.2.

Let S_i be the set of all permutations that fix at least $t + i$ elements from the set $\{1, \ldots, t + 2i\}$ (this requires that $n \geq 2i + t$). For any $i \in \{0, \ldots, \lfloor(n - t)/2\rfloor\}$, any two permutations in the set S_i agree on at least t elements. Also, the set S_0 is the pointwise stabilizer of a set of size t. If $n \leq 2t$ and $n \geq t + 4$, the size of the set S_1 is strictly larger than S_0.

The smallest value of n and t that satisfy both $n \leq 2t$ and $n \geq t + 4$ is $t = 4$ and $n = 8$. In this case the pointwise stabilizer of 4 points in Sym(8) has size 24. The set S_1 is the set of all permutations that fix at least 5 of the points $\{1, \ldots, 6\}$. Thus the size of S_1 is

$$|S_1| = \binom{6}{6} 2! + \binom{6}{5}(3! - 2!) = 26.$$

We conclude with a conjecture by Ellis [54, Conjecture 4.1]. This conjecture is the natural version of Theorem 1.3.1 extended to permutations.

14.5.2 Conjecture. *For all n and t, the largest set of permutations of an n-set that has the property that any two permutations agree on at least t elements is one of the sets S_i.*

14.6 Transitive permutation groups

For any permutation group $G \leq \text{Sym}(n)$, we can define the *derangement graph of G* to be the graph with vertex set G, and $g_1, g_2 \in G$ are adjacent if and only if $g_1^{-1} g_2$ is a derangement. This graph is denoted by Γ_G; it is the normal Cayley graph $\Gamma(G, G \cap D(n))$. Just as in the case for the derangement graph, a coclique in Γ_G is a set of intersecting permutations in G. The question we now ask is, what is the size and what is the structure of the maximum cocliques in Γ_G? Specifically, for which groups are the cosets of point stabilizers in G the maximum cocliques in Γ_G?

We say that a permutation group G has the *EKR property* if the size of the largest intersecting set of permutations in G is the same as the size of the largest stabilizer of a point in G. Further, G has the *strict EKR property* if the only sets of intersecting permutations in G of maximum size are the cosets of point stabilizers in G. These cosets are the collections of all permutations in G that map some i to j; again, they will be denoted by $S_{i,j}$. Following the standard notation, we denote the stabilizer in G of the point i by G_i.

Theorem 14.4.3 is equivalent to the symmetric group having the strict EKR property. Other groups are known to have the strict EKR property. For example, Ku and Wong [112] proved that the alternating group has the strict EKR property. If q is a prime power then PSL$(2, q)$, PGL$(2, q)$ and PGL$(3, q)$ have the EKR property, and for PGL$(2, q)$ the strict EKR property also holds [131, 132].

Many groups can be shown to have the EKR property simply by showing that the derangement graph for the group has a clique of the appropriate size. A subgroup of Sym(n) is *regular* if for every pair $(i, j) \in \{1, \ldots, n\}$, there is exactly one element in the subgroup that maps i to j. These subgroups are also known as *sharply transitive* subgroups.

14.6.1 Lemma. *If a permutation group $G \leq$ Sym(n) contains a regular subgroup, then G has the EKR property.*

Proof. For any group G, the derangement graph Γ_G is vertex transitive, so the clique-coclique bound (Corollary 2.1.2) can be applied. The elements of the regular subgroup form a clique of size n in Γ_G; thus the size of a coclique is no more than $|G|/n$. Since the regular subgroup is transitive, G must also be transitive; the size of the stabilizer of a point is $|G|/n$. $\square$

The following result, due to Pablo Spiga, shows that if a subgroup of Sym(n) contains a transitive subgroup that has the EKR property, then the larger subgroup also has the EKR property. This motivates finding minimal transitive subgroups with the EKR property.

14.6.2 Theorem. *Let G be a transitive subgroup of* Sym(n) *and let H be a transitive subgroup of G. If H has the EKR property, then G also has the EKR property.*

Proof. Since H has the EKR property and is transitive, the size of the maximum coclique is $|H|/n$. Since the graph Γ_H is vertex transitive, its fractional chromatic number is n (Lemma 2.12.1).

The embedding $\Gamma_H \to \Gamma_G$ is a homomorphism, so the fractional chromatic number of Γ_H is a lower bound on the fractional chromatic number of Γ_G (Lemma 2.13.2). Since Γ_G is vertex transitive this means that

$$n \leq \frac{|G|}{\alpha(\Gamma_G)},$$

and thus $\alpha(\Gamma_G) \leq \frac{|G|}{n}$. Finally, since G is transitive the stabilizer of a point achieves this bound. $\square$

If $H \leq G$ and Γ_H and Γ_G have the same fractional chromatic number, then by Lemma 2.13.3, if S is a maximum coclique in Γ_G, then $S \cap H$ is a maximum coclique in Γ_H. In the next section, this fact is used to show that if H is a 2-transitive group and every maximum coclique in H is the coset of the stabilizer of the point, then the same is true for every maximum coclique in G.

It is interesting to note that there are infinite families of transitive groups that do not have the EKR property. In Section 14.5 it was shown that there is subset of Sym(8) that is 4-intersecting and larger than the pointwise stabilizer of a set of 4 distinct elements. The group Sym(8) acts transitively on the set of all ordered 4-sets from $\{1, \ldots, 8\}$. Under this action Sym(8) is a transitive subgroup of Sym(1680) (since $8 \times 7 \times 6 \times 5 = 1680$); call this subgroup G. The size of the stabilizer of a point in G, under this action, is the same as the size of the pointwise stabilizer in Sym(8) of a set of 4 distinct elements. Further, the set of all permutations from Sym(8) that fix at least 5 points from $\{1, \ldots, 6\}$ is isomorphic to an intersecting set of size 26 in G. This means that G is a transitive group that does not have the EKR property. Provided that $t \geq 4$, for every value of n with $t + 2 \leq n \leq 2t$ it is possible to use this construction to find groups that do not have the EKR property.

Note that the derangement graphs of these two isomorphic groups are not isomorphic, as the actions of the groups are different. In fact the derangement graph for Sym(8) is a subgraph of the derangement graph of G. When considering the EKR property for groups, it is important to be clear about which group action is being considered.

14.7 2-Transitive subgroups

The 2-transitive subgroups are a particularly interesting family of groups in terms of EKR-type results. If G is a 2-transitive subgroup of Sym(n), then the action of G on pairs from $\{1, \ldots, n\}$ has exactly two orbits. From this we can deduce that the permutation character of G is the sum of two irreducible characters (see Exercise 11.9), namely the trivial character and an irreducible character that we denote by ψ. This irreducible character is the *standard character*.

The value of ψ on a derangement of G is -1 and the degree of ψ is $n - 1$. Using (14.2.1), the eigenvalue of the derangement graph Γ_G arising from ψ is $-d/(n - 1)$, where d is the valency of Γ_G. The multiplicity of this eigenvalue is at least $(n - 1)^2$. The module corresponding to ψ will be called the *standard module*.

If this eigenvalue is indeed the least eigenvalue of Γ_G, then by Theorem 2.4.1, we have that the size of a coclique is no bigger than $|G|/n$, which is the size of the stabilizer of a point in G (since G must also be transitive). There are examples of 2-transitive groups for which this eigenvalue is not the least eigenvalue of the derangement graph (see [6] for examples).

Determining if a 2-transitive group of Sym(n) has the strict EKR property is difficult, and there are 2-transitive groups that have the EKR property but do not have the strict EKR property (see [131]). The next results show that many of the results we used in Section 14.4 to show that the symmetric group has the strict EKR property hold more generally for all 2-transitive groups. The next two lemmas show that both Lemma 14.3.3 and Lemma 14.3.4 hold for all 2-transitive groups.

14.7.1 Lemma. *Let G be a 2-transitive subgroup of* $\text{Sym}(n)$ *and* $S_{i,j}$ *the canonical cocliques of G. Then for all $i, j \in \{1, \ldots, n\}$ the vector* $v_{i,j} - \frac{1}{n}\mathbf{1}$ *lies in the standard module.*

Proof. Let ψ be the standard character; we will show that

$$E_\psi \left(v_{i,j} - \frac{1}{n}\mathbf{1} \right) = v_{i,j} - \frac{1}{n}\mathbf{1}.$$

Since G is 2-transitive, it is also transitive, so we can assume, without loss of generality, that $i = j = n$. Since ψ is orthogonal to the trivial character,

$$E_\psi \left(v_{n,n} - \frac{1}{n}\mathbf{1} \right) = E_\psi(v_{n,n}).$$

The entries of E_ψ are given in by (11.11.1). We use $[E_\psi]_g$ to denote the row of E_ψ corresponding to $g \in G$.

If $g(n) = n$, then

$$[E_\psi]_g \cdot v_{n,n} = \frac{n-1}{|G|} \sum_{\substack{h(n)=n \\ h \in G}} \psi(hg^{-1}) = \frac{n-1}{|G|} \sum_{\substack{h(n)=n \\ h \in G}} \psi(h).$$

Let $\text{fix}'(g)$ denote the number of fixed points an element g from G_n (the stabilizer of n in G) has on the set $\{1, \ldots, n - 1\}$. Then

$$[E_\psi]_g \cdot v_{n,n} = \frac{n-1}{|G|} \sum_{g \in G_n} \text{fix}'(g) = \frac{n-1}{|G|} \left(\frac{|G|}{n} \right) = \frac{n-1}{n}.$$

Note that the first equality holds by the orbit counting lemma and the fact that G_n is transitive.

Since $\sum_{g \in G} \psi(g) = 0$, this implies that

$$\sum_{i=1}^{n-1} \sum_{g(n)=i} \psi(g) = -\frac{|G|}{n}.$$

Since G is 2-transitive, for any $i \neq n$

$$\sum_{g(n)=i} \psi(g) = -\frac{|G|}{n(n-1)}.$$

We can apply this in the case that $g(n) \neq n$ to get that

$$[E_\psi]_g \cdot v_{n,n} = \frac{n-1}{|G|} \sum_{\substack{h(n)=i \\ h \in G}} \psi(hg^{-1}) = \frac{n-1}{|G|} \left(-\frac{|G|}{n(n-1)} \right) = -\frac{1}{n}. \quad \square$$

14.7.2 Lemma. *Let G be a 2-transitive group. The set*

$$B = \{v_{i,j} - \frac{1}{n}\mathbf{1} \; : \; 1 \le i, j \le n - 1\}\}$$

is a basis for the standard module of G.

Proof. According to Lemma 14.7.1, it suffices to show that B is linearly independent. Note also that since $\mathbf{1}$ is not in the span of the $v_{i,j}$ for $i, j \in \{1, \ldots, n - 1\}$, it is enough to prove that the set

$$\{v_{i,j} \; : \; 1 \le i, j \le n - 1\}$$

is linearly independent.

Define H to be the matrix with the vectors $v_{i,j}$, with $i, j \in \{1, \ldots, n - 1\}$, as its columns. Then the rows of H are indexed by the elements of G and the columns are indexed by the ordered pairs (i, j). If we assume that the ordered pairs are listed in lexicographic order, then it is easy to see that

$$H^\top H = \frac{(n - 1)!}{2} I_{(n-1)^2} + \frac{(n - 2)!}{2} (A(K_{n-1}) \otimes A(K_{n-1})).$$

Just as the case in Lemma 14.3.4, this matrix does not have 0 as an eigenvalue, so the set B is linearly independent. $\qquad\square$

Following what we did in Section 14.4, define the $|G| \times (n^2 - 2n + 2)$ matrix H' to be the matrix whose columns are the vectors $v_{i,j}$, for all $i, j \in \{1, \ldots, n - 1\}$, and the vector $v_{n,n}$. The column space of this matrix is exactly the sum of the standard module and the trivial module. Just as in the case for the symmetric group, the matrix H' has the following block structure:

$$H' = \begin{pmatrix} \begin{array}{c|c} 1 & 0 \\ \hline 0 & M \\ \hline H_1 & H_2 \end{array} \end{pmatrix}. \tag{14.7.1}$$

The next proposition, which appears in [7], shows that the submatrix H_1 in H' contains an $n \times n$ identity matrix.

14.7.3 Proposition. *Let $n > 2$ and $G \le \mathrm{Sym}(n)$ be 2-transitive. For any $x \in \{1, \ldots, n\}$, there is an element in G that has x as its only fixed point.*

Proof. Since G is transitive, we only need to show that the stabilizer of n in G, which we denote by G_n, contains a derangement. Suppose that $|\mathrm{fix}(g)| \ge 1$ for every element $g \in G_n$. Then

$$\frac{1}{|G_n|} \sum_{g \in G_n} |\operatorname{fix}(g)| \geq \frac{(n-1) + |G_n| - 1}{|G_n|} = \frac{(n-2+|G_n|)}{|G_n|}.$$

Provided that $n > 2$, this greater than 1. But since G_n acts transitively on $\{1, \ldots, n-1\}$, this contradicts the orbit counting lemma. □

The final step in the proof that $\operatorname{Sym}(n)$ has the strict EKR property is to show that the submatrix M in H' has full rank. This step cannot be generalized to all 2-transitive groups; in fact there are many such groups for which the corresponding matrix M does not have full rank.

These results give a convenient method to test if a 2-transitive group has the strict EKR property. First check if the group has the EKR property (either compute the eigenvalues of the derangement graph or find a clique in the graph of size n). Then calculate the rank of the matrix M; if this matrix has full rank then the group has the strict EKR property. Unfortunately, this method cannot be used to determine if the group does not have the strict EKR property. Indeed, if M does not have full rank, then this test is inconclusive.

Theorem 14.6.2 shows that if a group contains a subgroup that has the EKR property and is 1-transitive, then the group also has the EKR property. Next we will show that if a 2-transitive subgroup of $\operatorname{Sym}(n)$ has the strict EKR property, then any other subgroup of $\operatorname{Sym}(n)$ that contains the subgroup also has the strict EKR property. (This result is also due to Pablo Spiga.)

14.7.4 Theorem. *Let G be a transitive subgroup of* $\operatorname{Sym}(n)$ *and let H be a 2-transitive subgroup of G. If H has the strict EKR property, then G also has the strict EKR property.*

Proof. Assume that S is a coclique in Γ_G of size $|G|/n$ that contains the identity; we will prove that S is the stabilizer of a point.

Let $\{x_1 = \operatorname{id}, \ldots, x_{[G:H]}\}$ be a set of representatives of the cosets of H in G and set $S_i = S \cap x_i H$. For each i the set $x_i^{-1} S_i$ is a coclique in Γ_H with size $|H|/n$. Since H has the strict EKR property, each $x_i^{-1} S_i$ is the coset of a stabilizer of a point.

Since $\operatorname{id} \in S_1$, we have that S_1 is the stabilizer of a point, and we can assume that $S_1 = H_\alpha$ for some $\alpha \in \{1, \ldots, n\}$. We need to show that every permutation in S also fixes the point α. Assume that there is a $\pi \in S$ that does not fix α. Since S is intersecting, for every $\sigma \in S_1$ the permutation $\sigma \pi^{-1}$ fixes some element (but not α and not $\pi(\alpha)$). From this it follows that

$$H_\alpha \pi^{-1} = \bigcup_{\beta \neq \alpha, \pi(\alpha)} (G_\beta \cap H_\alpha \pi^{-1}).$$

Assume that

$$\sigma \pi^{-1} \in G_\beta \cap H_\alpha \pi^{-1}.$$

Then $\sigma \pi^{-1}$ fixes β, and σ fixes α. So the permutation $\sigma \pi^{-1}$ must map (α, β) to $(\pi^{-1}(\alpha), \beta)$. Since the group H is 2-transitive there are exactly $|H|/n(n-1)$ such permutations and we have that

$$|G_\beta \cap H_\alpha \pi^{-1}| = \frac{|H|}{n(n-1)}.$$

It follows that the size of $H_\alpha \pi^{-1}$ is

$$\sum_{\beta \neq \alpha, \pi(\alpha)} \frac{|H|}{n(n-1)} = (n-2)\frac{|H|}{n(n-1)}.$$

But since this is strictly less that $|H|/n$, this is a contradiction. $\qquad \square$

14.8 Direct products

In this section, we construct permutation groups that are not transitive and have the EKR property. Let G be a subgroup of $\mathrm{Sym}(n)$ and H a subgroup of $\mathrm{Sym}(m)$ (where the set of points on which G acts is disjoint from the set on which H acts). We consider the group $G \times H$ as a subgroup of $\mathrm{Sym}(n+m)$; this is the *internal direct product* of G and H. This group clearly acts on a set of size $n+m$, but this action is not transitive. The first step is to express the derangement graph $G \times H$ in terms of the derangement graphs of G and H; this result appeared in [8].

14.8.1 Lemma. *The derangement graph for $G \times H$ is $\Gamma_G \times \Gamma_H$.*

Proof. A vertex of $\Gamma_{G \times H}$ is the product of an element from G and an element from H; vertices $g_1 h_1$ and $g_2 h_2$ are adjacent if and only if $(g_1 h_1)(g_2 h_2)^{-1} = g_1 g_2^{-1} h_1 h_2^{-1}$ is a derangement. This happens exactly when both $g_1 g_2^{-1}$ and $h_1 h_2^{-1}$ are derangements on their respective sets. This is exactly the condition for $g_1 h_1$ and $g_2 h_2$ to be adjacent in the product graph $\Gamma_G \times \Gamma_H$. $\qquad \square$

By Theorem 10.10.2 we know that the size of the maximum coclique in $\Gamma_{G \times H}$ is

$$\max\{|G|\alpha(H), \alpha(G)|H|\}.$$

We can apply this to the direct product of any two groups that both have the EKR property.

14.8.2 Theorem. *If the groups G and H have the EKR property, then $G \times H$ also has the EKR property.*

Proof. Assume that $|G|\alpha(H) \leq \alpha(G)|H|$. Then the size of the maximum coclique in $\Gamma_{G \times H}$ is $\alpha(G)|H|$, which is the size of the stabilizer in $G \times H$ of some point from the domain of G. □

For a partition $\lambda \vdash n$, the Young subgroup $\text{Sym}(\lambda)$ is the direct product of groups that have the EKR property. If λ has a part of size 1, then any two permutations in $\text{Sym}(\lambda)$ are intersecting, so these Young subgroups trivially have the strict EKR property. By the previous theorem and Theorem 14.1.2, the group $\text{Sym}(m) \times \text{Sym}(n)$ has the EKR property, and repeatedly applying the previous theorem yields the result for all Young subgroups.

14.8.3 Corollary. *Every Young subgroup has the EKR property.* □

Ku and Wong [112] have determined exactly which Young subgroups have the strict EKR property; these are given in the next theorem. Their proof uses compression and is similar to the proof that $\text{Sym}(n)$ has the strict EKR property given by Ku and Cameron [40].

14.8.4 Theorem. *Let $\lambda = [n_1, n_2, \ldots, n_q]$ be a partition of n with all parts larger than 1. The Young subgroup $\text{Sym}(\lambda)$ has the strict EKR property unless one of the following holds:*

(a) $n_p = 3$ and $n_{p+1} = \cdots = n_q = 2$;

(b) $n_q = n_{q-1} = 3$; or,

(c) $n_q = n_{q-1} = n_{q-2} = 2$. □

14.9 Exercises

14.1 Show that the derangement graph Γ_n is connected except when $n > 3$.

14.2 Show that $-d(n)/(n-1)$ is the least eigenvalue of the derangement graph Γ_n. (Hint: use Exercise 12.7.2 and follow the proof of Theorem 15.5.2.)

14.3 Let π be the orbit partition from the action of $\text{Sym}(1) \times \text{Sym}(n-1)$ on the vertices of Γ_n. Find the quotient graph Γ_n/π and give the eigenvalues of the quotient graph.

14.4 Let H be the matrix in Lemma 14.3.4. Prove that

$$H^T H = (n-1)! I_{(n-1)^2} + (n-2)! (K_{n-1} \otimes K_{n-1}),$$

and find all the eigenvalues of $H^T H$ with their multiplicities.

14.5 Show that any subgroup of $\text{Sym}(n)$ that contains an n-cycle has the EKR property.

14.6 Let $G = \text{PGL}(3, q)$ where q is a prime power. The group G acts on the projective space $\mathbb{P}_q^3$. Show that both the stabilizer of a point and the stabilizer of a hyperplane in $\mathbb{P}_q^3$ form maximum intersecting sets of permutations from G. Conclude that G does not have the strict EKR property.

14.7 Prove that the Young subgroups listed in Theorem 14.8.4 do not have the strict EKR property.

14.8 Show that any cyclic group has the strict EKR property.

14.9 Let $G \leq \text{Sym}(n)$. Show that G is a Frobenius group if and only if the derangement graph of G is the disjoint union of n-cliques.

14.10 Show that the dihedral groups have the strict EKR property.

14.11 Consider a sequence of permutation groups $G_1 \leq \text{Sym}(n_1), \ldots, G_k \leq \text{Sym}(n_k)$. Their *external direct product* is defined to be the group $G = G_1 \times \cdots \times G_k$. The elements of this group are $(g_1, \ldots, g_k)$, where $g_i \in G_i$, for $1 \leq i \leq k$, and the group operation is component wise multiplication. This group has a natural action on the set

$$\Omega = \{1, \ldots, n_1\} \times \cdots \times \{1, \ldots, n_k\},$$

namely

$$(x_1, \ldots, x_k)^{(g_1, \ldots, g_k)} = (x_1^{g_1}, \ldots, x_k^{g_k}).$$

Describe the derangement graph for G with this action.

14.12 Prove that if each $G_i \leq \text{Sym}(n_i)$ for $i = 1, \ldots, k$ has the (strict) EKR property, then the external direct product of the groups G_i also has the (strict) EKR property.

14.13 Show that the 2-transitive Mathieu groups have the strict EKR property.

Notes

There are many definitions of intersection for permutations. For example, in [108] *cycle-intersecting* permutations are considered, and in [55] two permutations are *t-setwise intersecting* if they both map a set of size t to the same set. In both of these papers a version of the EKR Theorem is shown to hold for the given type of intersection. The proof for t-setwise intersecting permutations by Ellis uses the representation theory of the symmetric group and requires that the permutations come from $\text{Sym}(n)$ where n is large relative to t, but the author conjectures that all that is required is that $n \geq 2t + 1$.

Renteln [146] first determined the least eigenvalue of the derangement graph. Further details about the eigenvalues of this graph, based on Renteln's work, are given by Ku and Wales [109] and by Ku and Wong [111]. Exercise 14.2 is based on Ellis's proof of the result that is given in Exercise 12.9.

In her master's thesis [141], Purdy modified Ahlswede and Khachatrian's proof to prove Conjecture 14.5.2 with the additional assumption that the permutation systems are *compressed*.

We know the complete character table for the group PGL(2, q), so using Corollary 2.1.2 it is easy to show that this group has the EKR property. (Similarly we can show PSL(2, q) has the EKR property.) As with most of these problems, the difficulty lies in showing that the group also has the strict EKR property. Specifically, it is much more difficult to prove that the result analogous to Lemma 14.4.1 holds in this case. For complete details see [131].

From Exercise 14.6, it is clear that the group PGL(3, q) does not have the strict EKR property. In this case, the result analogous to Lemma 14.4.1 does not hold. But, the kernel of the matrix, analogous to the matrix in the lemma, can be found. Using this result, all the maximum cocliques in $\Gamma_{\mathrm{PGL}(3,q)}$ can be determined. See [132] for details.

Wang [169] uses an eigenvalue technique similar to that used by Ellis, Friedgut and Pilpel [56] to prove that irreducible imprimitive reflection groups have the strict EKR property.

Both Exercise 14.12 and Exercise 14.13 are solved in [7].

In response to a question posed by Peter Cameron, Pablo Spiga pointed out that there are groups that do not have the EKR property. Spiga observed that when $q \equiv 1 \pmod 4$ the stabilizer in PSL(2, q), acting on unordered pair of points, is a 2-intersecting set. This set has size $q - 1$, which is twice the size of the stabilizer of an ordered pair of points.

15

Partitions

A *uniform k-partition of an n-set* is a partition $P = \{P_1, \ldots, P_k\}$ of the set $\{1, \ldots, n\}$ into k cells of size n/k. In order for such a partition to exist, k must divide n, and throughout this chapter we assume that n and k are integers such that $n = \ell k$ for some integer ℓ. The number of uniform k-partitions of an n-set is

$$U(n, k) = \frac{1}{k!} \binom{n}{\ell} \binom{n-\ell}{\ell} \cdots \binom{\ell}{\ell} = \frac{1}{k!} \prod_{i=0}^{k-1} \binom{n - i\ell}{\ell}.$$

In this chapter we consider the EKR problem for uniform partitions. First we need to specify which partitions we consider to be intersecting.

A partition of $\{1, \ldots, n\}$ is a set of k cells (each of which is an ℓ-set). We say that two partitions are intersecting if they contain a common cell. This is a natural generalization of intersection for sets, as a partition is a set of sets and we require that the partitions share a common element. An EKR-type theorem for partitions does not follow directly from the EKR Theorem for sets. This is because there are restrictions on which set of cells can be used to form a partition, namely the cells must be all disjoint.

However, since each cell of a partition is also a set, we could define two partitions to be intersecting if they each contain cells that intersect non-trivially. We will consider the problem of finding the largest set of uniform k-partitions such that any two contain cells that intersect in at least m points. If $m = \ell$, then this is equivalent to the partitions having a cell in common. For each value of m, there are obvious candidates for the largest intersecting sets of uniform k-partitions. If S is a fixed subset of $\{1, \ldots, n\}$ of size m, then the family of all partitions that contain a cell that contains S has the desired property.

In this chapter we consider the cases where these families are either known or conjectured to be maximum, and our focus is on maximum intersecting sets of uniform k-partitions of $\{1, \ldots, k^2\}$ where $m = 2$. In this case, two partitions are not intersecting if and only if any cell from one partition intersects any cell from the other partition in exactly one point. We say such a pair of partitions are *skew*.

To avoid trivial situations, we assume that both k and ℓ are greater than 1. Further, if $k = 2$ a uniform 2-partition of a 2ℓ set is an ℓ-subset of 2ℓ and its complement. This case is essentially subsets and it is covered in Chapter 6, thus we do not specifically discuss it in this chapter.

15.1 Intersecting partitions

In this section, we consider two partitions to be intersecting if they contain a common cell. More precisely, if P and Q are uniform k-partitions with

$$P = \{P_1, \ldots, P_k\}, \qquad Q = \{Q_1, \ldots, Q_k\},$$

then P and Q are intersecting if $P_i = Q_j$ for some $i, j \in \{1, \ldots, k\}$. In this section we show that this definition of intersection leads to a natural version of the EKR Theorem for partitions.

As we have done in the previous chapters, we can define a graph in such a way that the cocliques in the graph are sets of intersecting partitions. Denote this graph by $P(n, k)$. The vertex set of $P(n, k)$ is the set of all uniform k-partitions of an n-set, and vertices are adjacent if and only if they contain no common cells. By considering the action of $\mathrm{Sym}(n)$ on the vertices, it can easily be seen that this graph is vertex transitive. Further reflection reveals that this graph is neither arc nor edge transitive.

A clique in $P(n, k)$ is a collection of uniform partitions in which no cell occurs in more than one partition. Since the total number of possible cells is $\binom{n}{\ell}$, the maximum number of such partitions is $\frac{1}{k}\binom{n}{\ell}$; this gives an upper bound on the size of clique in $P(n, k)$. To show that this bound can be met, we point out that a 1-factorization of the complete uniform hypergraph $K_n^{(\ell)}$ is a clique in $P(n, k)$ of this size. Since such a 1-factorization exists for all n and k (see [16] for a proof of this fact), we have that

$$\omega(P(n, k)) = \frac{1}{k}\binom{n}{\ell} = \binom{n-1}{\ell-1}.$$

By the clique-coclique bound, a coclique in $P(n, k)$ has size no larger than

$$\frac{\frac{1}{k!}\prod_{i=0}^{k-1}\binom{n-i\ell}{\ell}}{\binom{n-1}{\ell-1}} = \frac{1}{(k-1)!}\prod_{i=1}^{k-1}\binom{n-i\ell}{\ell}.$$

If we fix a subset S of size ℓ from $\{1, \ldots, n\}$, then the set of all uniform k-partitions that contain S as a cell forms a coclique in $P(n, k)$ of this size. Thus we have that

$$\alpha(P(n, k)) = \frac{1}{(k-1)!} \prod_{i=1}^{k-1} \binom{n - i\ell}{\ell}. \tag{15.1.1}$$

We define the *canonical cocliques* in $P(n, k)$ to be the cocliques that consist of all partitions that contain a fixed cell. We have shown that these canonical cocliques are of maximum size. The next step in deriving an EKR-type theorem for partitions is to prove that these canonical cocliques are the only maximum cocliques. This can be done using a version of the *kernel method* introduced by Hajnal and Rothschild [91]. The proof we give requires that the cells in the partitions have size larger than 2 (so $\ell > 2$); a version of the argument works for $\ell = 2$, but it requires a less straightforward application of the kernel method. See [128] for the complete details in the case when $\ell = 2$.

15.1.1 Theorem. *Let $k, \ell > 2$, and $n = k\ell$. If S is a coclique in $P(n, k)$, then*

$$|S| \leq \frac{1}{(k-1)!} \prod_{i=1}^{k-1} \binom{n - i\ell}{\ell}.$$

Moreover, this bound is tight if and only if S is a canonical coclique in $P(n, k)$.

Proof. The bound follows from (15.1.1), so all that needs to be shown is that if a coclique meets this bound, then all partitions in the coclique contain a common fixed ℓ-set. Let S be a coclique in $P(n, k)$ and assume that there is no ℓ-subset that is a cell of every partition of S. We now show that S is not a maximum coclique.

Let $P = \{P_1, \ldots, P_k\}$ be any partition in S. Since the set S is a coclique, every partition in S must contain at least one of the cells P_i. So we can divide up the partitions in S according to which cell from P they contain. Let S_i denote the set of all partitions in S that contain the cell P_i.

No single cell is contained in every partition in S. So for every i, there is a partition $Q \in S$ that does not contain the cell P_i. Every partition in S_i must intersect Q, and this fact can be used to bound the size of S_i. There are at most $k - 2$ cells in Q that do not contain any of the elements that are in the cell P_i. Each partition in S_i must contain at least one of these $k - 2$ cells. By counting the number of partitions that can contain both the cell P_i, and one of the $k - 2$

cells from Q disjoint from P_i, we that have for any i

$$|\mathcal{S}_i| \le (k-2)\frac{1}{(k-2)!}\prod_{i=2}^{k-1}\binom{n-i\ell}{\ell}.$$

Every partition in $\mathcal{S}$ is in at least one of the sets $\mathcal{S}_i$, so it follows that

$$|\mathcal{S}| \le k(k-2)\frac{1}{(k-2)!}\prod_{i=2}^{k-1}\binom{n-i\ell}{\ell}.$$

We have assumed that $k > 2$ and $\ell > 2$, so the inequality $k(k-2) < \binom{n-\ell-1}{\ell-1}$ holds and this bound becomes

$$|\mathcal{S}| < \binom{n-\ell-1}{\ell-1}\frac{1}{(k-2)!}\prod_{i=2}^{k-1}\binom{n-i\ell}{\ell} < \frac{1}{(k-1)!}\prod_{i=1}^{k-1}\binom{n-i\ell}{\ell}.$$

Thus any coclique $\mathcal{S}$ in which the partitions do not all contain some common cell is strictly smaller than the maximum size of a coclique in $P(n, k)$.   □

Algebraic methods do not seem to work well for this graph. For example, the ratio bound is not tight in general. To see this, consider the graph $P(9, 3)$. The degree of $P(9, 3)$ is 252 and the least eigenvalue is -12. (The eigenvalues of $P(9, 3)$ are exactly the values found by summing the last three columns in Table 15.2.) The ratio bound for cocliques gives

$$\alpha(P(9, 3)) \le \frac{280}{1 - \frac{252}{-12}} \approx 12.73,$$

and by the previous theorem $\alpha(P(9, 3)) = 10$. We will see $P(9, 3)$ again in Section 15.11.

15.2 Perfect matchings

In the previous section, we noted that our proof of Theorem 15.1.1 can be extended so that it applies when $\ell = 2$ (in this case, the uniform partitions are perfect matchings) but we omitted the details. This case is particularly interesting because the algebraic methods, which do not work well for general partitions, work for perfect matchings. Throughout this section we assume that $\ell = 2$ and that our partitions are *perfect matchings* in the complete graph K_{2k}. The *intersection of two matchings* is the set of edges that are in both matchings, and two matchings are *intersecting* if they contain at least one common edge (this agrees with the definitions given in Section 7.3). In this special case, we call the graph $P(n, k)$ the *perfect matching graph* and denote it by $M(2k)$ (this graph is also defined in Section 7.4).

The number of vertices in $M(2k)$ is

$$U(2k, 2) = \frac{1}{k!} \prod_{i=0}^{k-1} \binom{2k - 2i}{2} = (2k - 1)!!$$

(this also follows from (7.4.1)). The degree of $M(2k)$ is denoted by $d(2k)$, or simply d if the value of k is clear. This is the number of perfect matchings that do not contain any the edges from some fixed perfect matching. This number can be calculated using the principle of inclusion-exclusion:

$$d(2k) = \sum_{i=0}^{k-1} (-1)^i \binom{k}{i} (2k - 2i - 1)!!. \tag{15.2.1}$$

In practice, this formula is tricky, but we will make use of the following simple lower bound on $d(2k)$.

15.2.1 Lemma. *For any k*

$$d(2k) > (2k - 1)!! - \binom{k}{1}(2k - 3)!!.$$

Proof. For any $i \in \{0, \ldots k - 1\}$

$$\binom{k}{i}(2k - 2i - 1)!! = \frac{i+1}{k-i}\binom{k}{i+1}(2k - 2i - 1)(2k - 2(i+1) - 1)!!$$

$$> \binom{k}{i+1}(2k - 2(i + 1) - 1)!!$$

(since $\frac{i+1}{k-i}(2k - 2i - 1) > 1$ for these values of i). Thus the terms in (15.2.1) are strictly decreasing and, since it is an alternating sequence, the first two terms give a lower bound on $d(2k)$. □

A maximum clique in $M(2k)$ is a 1-factorization of the complete graph on $2k$ vertices, and thus has size exactly $2k - 1$. By the ratio bound, a coclique in $M(2k)$ can be no larger than

$$\frac{(2k - 1)!!}{2k - 1} = (2k - 3)!!.$$

As in the case for partitions, it is easy to construct cocliques of this size. For any edge e in K_{2k}, let S_e be the collection of all matchings that contain e. Then,

$$|S_e| = \frac{1}{(k - 1)!} \prod_{i=1}^{k-1} \binom{2k - 2i}{2} = \prod_{i=1}^{k-1} 2k - 2i - 1 = (2k - 3)!!. \tag{15.2.2}$$

This is an easy proof of the bound in the EKR Theorem for perfect matchings. The characterization of the sets that meet this bound is also known (see the comments preceding Theorem 15.1.1), but the proof is not quite so simple. In the following sections, we give another proof of the characterization.

15.3 Eigenvalues of the perfect matching graph

It is interesting to note that it is straightforward to find an eigenvalue of the matching graph that will give equality in the ratio bound – what is not straightforward is to show that this eigenvalue is the least eigenvalue of the graph. For any edge e in K_{2k}, the partition $\pi = \{S_e, V(M(2k))/S_e\}$ is an equitable partition. In fact it is the orbit partition formed by the stabilizer of the edge e in $\mathrm{Sym}(2k)$ (this subgroup is isomorphic to $\mathrm{Sym}(2) \times \mathrm{Sym}(2k - 2)$) acting on the set of all vertices of $M(2k)$. The quotient graph of $M(2k)$ with respect to this partition is

$$M(2k)/\pi = \begin{pmatrix} 0 & d(2k) \\ \frac{1}{2k-2}\, d(2k) & \frac{2k-3}{2k-2}\, d(2k) \end{pmatrix},$$

where $d(2k)$ is the degree of the perfect matching graph $M(2k)$. The eigenvalues for the quotient graph $M(2k)/\pi$ are

$$d(2k), \quad -\frac{d(2k)}{2k - 2}.$$

Clearly $d(2k)$ is the largest eigenvalue for $M(2k)$, and we show in the following sections that $-d(2k)/(2k - 2)$ is the smallest eigenvalue. But first, we find a lower bound on the multiplicity of this eigenvalue.

15.3.1 Lemma. *The multiplicity of* $-d(2k)/(2k - 2)$ *as an eigenvalue of* $M(2k)$ *is at least* $2k^2 - 3k$.

Proof. For any edge e from K_{2k}, let v_e be the characteristic vector of S_e. Direct calculation shows that $v_e - \frac{1}{2k-1}\mathbf{1}$ is a $-d(2k)/(2k - 2)$-eigenvector of $M(2k)$. (This uses the fact that each vertex not in S_e is adjacent to $d(2k)/(2k - 2)$ vertices in S_e.)

To complete the lemma, we need to prove that the span of the vectors $v_e - \frac{1}{2k-1}\mathbf{1}$ for all edges e from K_{2k} has dimension $2k^2 - 3k$. Let H be the matrix with the vectors v_e as its columns. Since $\mathbf{1}$ is orthogonal to the vectors $v_e - \frac{1}{2k-1}\mathbf{1}$, it is sufficient to prove that the rank of H is $2k^2 - 3k + 1$.

Let A be the adjacency matrix of the Kneser graph $K(2k, 2)$. Then

$$H^T H = (2k - 5)!!\,((2k - 3)I + A).$$

By Corollary 6.6.1, 0 is an eigenvalue of this matrix with multiplicity $2k - 1$. Thus the rank of $H^T H$ (and hence H) is exactly $\binom{2k}{2} - (2k - 1) = 2k^2 - 3k + 1$. $\qquad\square$

In Section 15.5, we show that this lemma gives the exact multiplicity of the eigenvalue $-d(2k)/(2k - 2)$ and that it is the least eigenvalue of the matching graph. This result is used in Section 15.6 to prove the EKR Theorem for perfect matchings. But first, we show that the perfect matching graph is a graph in an orbital association scheme.

15.4 The perfect matching association scheme

The union of any two perfect matchings of K_{2k} is a union of disjoint even cycles in K_{2k} (where a pair of parallel edges are considered to be a 2-cycle). Thus the union of any two perfect matchings has a corresponding integer partition of $2k$ where each part is the length of a cycle in the union – so each part is an even integer. We denote such partitions by $2\lambda = (2\lambda_1, \ldots, 2\lambda_i)$, where $\lambda = (\lambda_1, \ldots, \lambda_i)$ is a partition of k.

This way of representing the union of two perfect matchings can be used to define the classes of an association scheme on the perfect matchings called the *perfect matching association scheme*. The vertices of the perfect matching association scheme are all the perfect matchings of K_{2k}, and the classes of the association scheme are defined by the cycle structure of the union of two matchings. For each partition 2λ with $\lambda \vdash k$, there is a class in the perfect matching association scheme labeled with 2λ. Two perfect matchings are adjacent in the class 2λ if and only if their union corresponds the partition 2λ.

We still need to show that these classes do indeed define an association scheme on the perfect matchings. To do this we could prove that the five conditions for an association scheme hold (see Section 3.1). Of these, only condition (d) is difficult. But, rather than doing this, we will see that the adjacency matrix for each class is also the adjacency matrix of an orbital of a group action, and then we can apply Corollary 13.5.1.

There is a one-to-one correspondence between the perfect matchings of K_{2k} and the cosets $\mathrm{Sym}(2k)/(\mathrm{Sym}(2) \wr \mathrm{Sym}(k))$. With this correspondence, the set of orbitals from the action of $\mathrm{Sym}(2k)$ on the cosets $\mathrm{Sym}(2k)/(\mathrm{Sym}(2) \wr \mathrm{Sym}(k))$ are the classes in the matching association scheme. To see this, it is only necessary to realize that for any $2\lambda \vdash 2k$, the group $\mathrm{Sym}(2k)$ acts transitively on pairs of matchings whose union corresponds to that partition 2λ (the details of this are left as an exercise).

By Lemma 13.4.1, the orbitals of the action of $\mathrm{Sym}(2k)$ on the cosets $\mathrm{Sym}(2k)/(\mathrm{Sym}(2) \wr \mathrm{Sym}(k))$ define an association scheme if and only if the induced representation $\text{ind}_{\text{Sym}(2k)}(1_{\text{Sym}(2)\wr\text{Sym}(k)})$ is multiplicity-free. The decomposition of the permutation representation of this action is known by Lemma 12.8.1 to be

$$\text{ind}_{\text{Sym}(2k)}(1_{\text{Sym}(2)\wr\text{Sym}(k)}) = \sum_{\lambda\vdash k} \Phi_{2\lambda}.$$

Clearly this is a multiplicity-free representation, and hence the adjacency matrices of the orbitals of $\text{Sym}(2k)$ form an association scheme on the perfect matchings in K_{2k}.

The graph $M(2k)$ is the union of all the classes in this association scheme in which the corresponding partition contains no part of size 2. This means that each eigenspace of $M(2k)$ is the union of 2λ-modules where $\lambda \vdash k$. (So if ξ is an eigenvalue of $M(2k)$, and its eigenspace includes the 2λ-module, then we say that ξ is the eigenvalue belonging to the 2λ-module.) Further, Corollary 13.8.2 describes the projections to the modules, and Lemma 13.8.3 describes a formula to calculate the eigenvalue belonging to the 2λ-module using the character $\phi_{2\lambda}$. This gives considerable information about the eigenvalues of $M(2k)$.

15.5 Modules of the perfect matching graph

In this section, we determine the eigenvalues that belong to two of the modules of the perfecting matching graph. We show that the eigenvalue belonging to any other module will be smaller, in absolute value, than these two eigenvalues.

15.5.1 Lemma. *The eigenvalue belonging to the $[2k]$-module is $d(2k)$, and the eigenvalue belonging to the $[2k-2,2]$-module is $-d(2k)/(2k-2)$.*

Proof. The first statement is clear using the formula in Lemma 13.8.3.

To prove the second statement, we consider the equitable partition defined in Section 15.3. The partition π is the orbit partition of $\text{Sym}(2k-2) \times \text{Sym}(2)$ acting on the perfect matchings. Let $H = \text{Sym}(2k-2) \times \text{Sym}(2)$ and denote the cosets of H in $\text{Sym}(2k)$ by $\{x_0 H = H, x_1 H, \ldots, x_{2k^2-k-1}H\}$.

The $-d(2k)/(2k-2)$-eigenvector of $M(2k)/\pi$ lifts to an eigenvector v of $M(2k)$. The entries of v are $1 - \frac{1}{2k-1}$ or $-\frac{1}{2k-1}$, depending on whether the index of the entry is in S_e, or not. This means that $v = v_e - \frac{1}{2k-1}\mathbf{1}$, where v_e is the characteristic vector of S_e (this is the eigenvector given in Lemma 15.3.1).

The group $\text{Sym}(2k)$ acts on the edges of K_{2k}, and we can define

$$v^\sigma = v_{e^\sigma} - \frac{1}{2k-1}\mathbf{1}.$$

The vector v is fixed by any $\sigma \in H$ under this action. If we define

$$V = \text{span}\{v^\sigma : \sigma \in \text{Sym}(2k)\},$$

then V is a subspace of the $-d(2k)/(2k-2)$-eigenspace. Moreover, V is invariant under the action of $\text{Sym}(2k)$, so it is also a $\text{Sym}(2k)$-module. To prove this lemma we show that V is isomorphic to the $[2k-2, 2]$-module.

Let W be the $\text{Sym}(2k)$-module for the induced representation $\text{ind}_{\text{Sym}(2k)}(1_H)$. By Lemma 12.5.2, W is the sum of irreducible modules of $\text{Sym}(2k)$ that are isomorphic to $M_{[2k]}$, $M_{[2k-1,1]}$ and $M_{[2k-2,2]}$. By the comments in Section 11.13, W is isomorphic to the vector space of functions $f \in L(\text{Sym}(2k))$ that are constant on H. For each coset xH, the function $\delta_{xH}(\sigma)$ is equal to 1 if σ is in xH and 0 otherwise. The functions δ_{xH} form a basis for W.

Define the map f so that

$$f(v^\sigma) = \delta_{\sigma H} - \frac{1}{2k-1} \sum_{i=0}^{2k^2-k-1} \delta_{x_i H}.$$

Since $v^\sigma = v^\pi$ if and only if $\sigma H = \pi H$, this function is well defined. Further, it is a $\text{Sym}(2k)$-module homomorphism. Thus V is isomorphic to a submodule of W. Since V is not trivial, it must be the $[2k-2, 2]$-module, since it is the only module (other than the trivial) that is common to both $\text{ind}_{\text{Sym}(2k)}(1_H)$ and $\text{ind}_{\text{Sym}(2k)}(1_{\text{Sym}(2)\wr\text{Sym}(k)})$. $\square$

Next we bound the size of the other eigenvalues. This bound follows from the straightforward fact that if A is the adjacency matrix of a graph, then the trace of the square of A is equal both to the sum of the squares of the eigenvalues of A, and to twice the number of edges in the graph. The proof of this result closely follows the proof the least eigenvalue of the derangement graph of the symmetric group by Ellis [54].

15.5.2 Theorem. *For $\lambda \vdash k$, the absolute value of the eigenvalue of $M(2k)$ belonging to the (2λ)-module is strictly less than $d(2k)/(2k-2)$, unless $\lambda = [2k]$ or $\lambda = [2k-2, 2]$.*

Proof. Let A be the adjacency matrix of $M(2k)$ and use $\xi_{2\lambda}$ to denote the eigenvalue for the 2λ-module.

The sum of the eigenvalues of A^2 is twice the number of edges in $M(2k)$, that is

$$\sum_{\lambda \vdash k} \phi_{2\lambda}(1)\xi_{2\lambda}^2 = (2k-1)!!\, d(2k).$$

From Lemma 15.5.1 we know the eigenvalues for two of the modules, so this bound can be expressed as

$$\sum_{\substack{\lambda\vdash k \\ \lambda\neq[k],[k-1,1]}} \phi_{2\lambda}(1)\xi_{2\lambda}^2 = (2k-1)!!\, d(2k) - d(2k)^2 - (2k^2-3k)\left(\frac{d(2k)}{2k-2}\right)^2.$$

Since all the terms in the left-hand side of this summation are positive, any single term is less than the sum. Thus

$$\phi_{2\lambda}(1)\xi_{2\lambda}^2 \le (2k-1)!!\, d(2k) - d(2k)^2 - (2k^2-3k)\left(\frac{d(2k)}{2k-2}\right)^2$$

(where $\lambda \vdash k$ and $\lambda \neq [k], [k-1,1]$). If $|\xi_{2\lambda}| \ge d(2k)/(2k-2)$, then this reduces to

$$\phi_{2\lambda}(1) \le \frac{(2k-1)!!(2k-2)^2}{d(2k)} - 6k^2 + 11k - 4.$$

Using the bound in Lemma 15.2.1, this implies that

$$\phi_{2\lambda}(1) < 2k^2 - k = \frac{(2k)^2 - (2k)}{2}.$$

If $|\xi_{2\lambda}| \le d(2k)/(2k-2)$, then 2λ must be one of the eight irreducible representations in Lemma 12.7.3. Thus 2λ must be one of $[2k]$ or $[2k-2,2]$, which proves the result. $\qquad\square$

We restate this result in terms of the least eigenvalue of the matching graph, noting that only the $[2k-2,2]$-module has $-d(2k)/(2k-2)$ as its eigenvalue.

15.5.3 Corollary. *The smallest eigenvalue of $M(2k)$ is $-d(2k)/(2k-2)$ and the multiplicity of this eigenvalue is $2k^2 - 3k$.*

Finally, we also have a spanning set for the $[2k-2,2]$-module.

15.5.4 Corollary. *The balanced characteristic vectors of the canonical cocliques span the $[2k-2,2]$-module.*

Proof. From the proof of Theorem 15.5.2, the balanced characteristic vectors of the canonical cocliques are all in the $[2k-2,2]$-module. In Lemma 15.3.1, it is shown that the span of these vectors has dimension $2k^2 - 3k$. This is exactly the dimension of the $[2k-2,2]$-module (simply using the hook length formula), so these vectors are a spanning set for the module. $\qquad\square$

15.6 EKR Theorem for perfect matchings

Applying Hoffman's ratio bound with the fact that $-d(2k)/(2k-2)$ is the least eigenvalue of $M(2k)$ proves that the cocliques S_e are ratio tight, since

$$\frac{|V(M(2k)|}{1 - \frac{d}{\tau}} = \frac{(2k-1)!!}{1 - \frac{d(2k)}{-\frac{d(2k)}{2k-2}}} = (2k-3)!!.$$

The ratio bound further implies that the balanced characteristic vector for any maximum coclique in $M(2k)$ is an eigenvector for the least eigenvalue. Since the $[2k-2, 2]$-module is the only module for which the corresponding eigenvalue is the least, this means that the balanced characteristic vector for any maximum coclique must be in the $[2k-2, 2]$-module. This fact can be used to determine the structure of any maximum coclique of $M(2k)$.

15.6.1 Theorem. *The largest coclique in $M(2k)$ has size $(2k-3)!!$. The only cocliques that meet this bound are the canonically intersecting sets of perfect matchings.*

Proof. Let S be a maximum coclique in $M(2k)$ and let v_S be the characteristic vector of S. Then $|S| = (2k-3)!!$, by Hoffman's ratio bound and Corollary 15.5.3. Hoffman's ratio bound, along with Theorem 15.5.2, further implies that the vector $v_S - \frac{1}{2k-1}\mathbf{1}$ is in the $[2k-2, 2]$-module.

By Corollary 15.5.4, $v_S - \frac{1}{2k-1}\mathbf{1}$ is a linear combination of the balanced characteristic vectors of the canonical cocliques. Similar to the case in Section 14.4 for permutations, this also implies that v_S is a linear combination of the characteristic vectors of the canonical cocliques.

Recall that in the proof of Lemma 15.3.1 we defined the matrix H in which the columns are the characteristic vectors of the canonical cocliques in $M(2k)$. So there exists a vector y such that $Hy = v_S$. Finally, by Lemma 7.4.2, this implies that S is a canonical coclique of $M(2k)$. $\square$

15.7 Skew partitions

We now turn our attention to a different type of intersection between partitions. Two partitions are *skew* if every cell from the first partition has intersection of size exactly 1 with every cell of the second partition. It is clear that two partitions can be skew only if they are both uniform k-partitions of a k^2-set, so throughout this and the following sections, we consider only partitions of this type.

Just as with the partitions and matchings in the previous sections, we can define a graph on the set of partitions using the property of being skew. The *skew-partition graph* is the graph whose vertex set is the set of all uniform k-partitions of a k^2-set where two partitions are adjacent if and only if the partitions are skew. This graph is denoted by $P(k)$.

15.7.1 Lemma. *The graph $P(k)$ is $(k!)^{k-1}$-regular.*

Proof. For any partition $P = \{P_1, \ldots, P_k\}$, we construct all $(k!)^{k-1}$ partitions that are skew to P. First place the k-elements from P_1 into k distinct empty cells. Next place the elements from P_i for $i = 2, \ldots, k$ into these cells in such a way that no two elements from P_i are in the same cell. There are $k!$ ways to do this for each P_i for a total of $(k!)^{k-1}$ ways to construct a uniform k-partition that is skew to P. Since each of the partitions constructed this way is unique and every partition that is skew to P arises in this way, this exactly gives the degree of the skew-partition graph. $\square$

The graph $P(k)$ is connected. Rather than proving this fact, we show the following stronger result that originally appeared in [133].

15.7.2 Lemma. *For all k the diameter of $P(k)$ is 2.*

Proof. For any pair of partitions $P = \{P_1, \ldots, P_k\}$ and $Q = \{Q_1, \ldots, Q_k\}$, we show how to construct a third partition R that is skew to both P and Q. To do this we construct an auxiliary multi-graph G as follows.

The multi-graph has $2k$ vertices, one for each cell P_i of P and one for each cell Q_i of Q. Between the vertices corresponding to P_i and Q_j we add $|P_i \cap Q_j|$ parallel edges. We label these edges with the elements of $P_i \cap Q_j$. Clearly G is bipartite, k-regular and has exactly k^2 edges.

It is an easy consequence of König's Theorem that the edges of every k-regular bipartite multi-graph can be partitioned into k disjoint perfect matchings. Let $M_1, \ldots, M_k$ be a partition of the edges of G into perfect matchings. For each $1 \le i \le k$, let R_i be the set of labels on the edges in M_i. We leave it as an exercise to show that $R = \{R_1, \ldots, R_k\}$ is a uniform k-partition that is skew to both P and Q. $\square$

15.7.3 Lemma. *For all k, the graph $P(k)$ is both vertex and arc transitive.*

Proof. The group $\mathrm{Sym}(k^2)$ acts transitively on the uniform k-partitions of a k^2-set and it is not hard to see that, under this action, the symmetric group is a subgroup of the automorphism group of the skew-partition graph.

To see that this graph is also arc transitive, let

$$P = \{P_1, \ldots, P_k\}, \quad Q = \{Q_1, \ldots, Q_k\}$$

be one pair of skew partitions, and let

$$R = \{R_1, \ldots, R_k\}, \quad S = \{S_1, \ldots, S_k\}$$

be a second pair of skew partitions. Since P and Q are skew, for each pair (i, j) with $i, j \in \{1, \ldots, k\}$, there is exactly one $a_{i,j} \in \{1, \ldots, k^2\}$ such that $a_{i,j} \in P_i \cap Q_j$. Similarly, there is also a unique element $b_{i,j} \in R_i \cap S_j$ for all $i, j \in \{1, \ldots, k\}$. The permutation in $\mathrm{Sym}(k^2)$ that maps $a_{i,j}$ to $b_{i,j}$ for all $i, j \in \{1, \ldots, k\}$ is an automorphism of the skew-partition graph that maps the arc (P, Q) to the arc (R, S). $\quad\square$

15.8 Cliques and cocliques in the skew-partition graph

In the previous section, we defined the skew-partition graph so that the cliques in the graph are sets of partitions that are pairwise skew. One interesting feature of the skew-partition graph is that its cliques are equivalent to interesting combinatorial objects.

For example, any triangle in $P(k)$ can be used to construct a Latin square of order k. To see this, let P, Q and R be three partitions that form a triangle in $P(k)$; and let P_i, Q_i and R_i for $i = 1, \ldots, k$ be the cells of the partitions. To build the Latin square, place the entry h in row i and column j of the Latin square if and only if $P_h \cap Q_i \cap R_j$ is non-empty (the proof that this is a Latin square is left as an exercise). An appropriate reordering of the cells of the partitions produces a Latin square that is in standard form, and changing the order of the partitions gives the conjugate Latin squares.

Conversely, any Latin square of order k can be used to construct a triangle in the skew-partition graph. This means that a triangle in this graph is equivalent to a Latin square, so even just counting the number of triangles in $P(k)$ is as difficult as enumerating the Latin squares of order k. (See [127] for a short history of this problem.)

Similarly, an m-clique in the graph $P(k)$ is equivalent to an orthogonal array $OA(m, k)$. Orthogonal arrays are discussed in Section 5.5, and we noted there that if an $OA(m, k)$ exists, then $m \le k + 1$. Moreover, if equality holds, then the orthogonal array is equivalent to an affine plane of order k. Hence the size of clique in $P(k)$ is at most $k + 1$, and equality holds if and only if there is an affine plane of order k. Affine planes exist when k is a prime power. Of course the existence of an affine plane whose order is not a prime power is a famous open problem. Currently, it has been proved that there is no plane of order 6 or 10, and no examples are known when k is not a prime power.

Since the graph $P(k)$ is vertex transitive, the clique-coclique bound, Corollary 2.1.2, can be applied. If k is a prime power, then the size of the maximum clique is known and the size of a maximum coclique is no larger than

$$\frac{|V(P(k))|}{k+1} = \binom{k^2-2}{k-2} U(k^2 - k, k - 1). \tag{15.8.1}$$

If two partitions P and Q are not adjacent in the skew-partition graph, then it must be the case that a cell of P has intersection of size at least 2 with a cell of Q. In this sense, partitions that are not skew can be considered to have a larger intersection. Once again we can define an EKR-type problem. We define two partitions to be intersecting if they contain cells that have an intersection of at least size 2. Then we ask, what is the size of the largest set of intersecting partitions?

Similar to all the other instances of EKR-type problems that we have considered, there is an obvious candidate for the largest set of uniform partitions in which every pair of partitions has this intersection property. Let i, j be two distinct values in $\{1, \ldots, k^2\}$ and define the set $S_{i,j}$ to be the collection of all partitions in which i and j are in the same cell. Clearly, any two partitions in $S_{i,j}$ have cells that intersect in at least two points, so $S_{i,j}$ is a coclique in $P(k)$ with size

$$\binom{k^2-2}{k-2} U(k^2 - k, k - 1) = \binom{k^2-2}{k-2} \frac{1}{(k-1)!} \prod_{i=1}^{k-1} \binom{k^2 - ki}{k}.$$

This is exactly the bound given in (15.8.1).

15.8.1 Theorem. *If k is a prime power, then the maximum size of a coclique in $P(k)$ is*

$$\binom{k^2-2}{k-2} \frac{1}{(k-1)!} \prod_{i=1}^{k} \binom{k^2 - ki}{k}. \qquad \square$$

We conjecture that this theorem holds for all skew-partition graphs and, moreover, that the sets $S_{i,j}$, for distinct $i, j \in \{1, \ldots, k^2\}$, are all of the maximum cocliques in the skew-partition graph (see Conjecture 15.9.2). This conjecture can be considered to be a version of the EKR Theorem for skew partitions.

Since the skew-partition graph is a vertex-transitive graph and the sets $S_{i,j}$ are cocliques, the fractional chromatic number is bounded by

$$\chi^*(P(k)) \le \frac{U(k^2, k)}{\frac{1}{k+1} U(k^2, k)} = k + 1.$$

Note that this bound also follows from the existence of a simple homomorphism to a Kneser graph.

15.8.2 Lemma. *For all values of k, there is a homomorphism from $P(k)$ to the Kneser graph $K(k^2 - 1, k - 1)$.*

Proof. To prove this we simply construct the homomorphism. For each partition P, let P_k be the cell that contains the element k^2. Then map the partition P to the set $P_k \setminus \{k^2\}$. Clearly, this is a map of the vertices of $P(k)$ onto the vertices of $K(k^2 - 1, k - 1)$. This map is a homomorphism, since if P and Q are two skew partitions, then $P_k \cap Q_k = \{k^2\}$ and this implies that $(P_k \setminus \{k^2\}) \cap (Q_k \setminus \{k^2\}) = \emptyset$. $\qquad\square$

By Theorem 15.8.1, if k is a prime power, then $\chi^*(P(k)) = k + 1$. We conjecture that this is the correct fractional chromatic number for all k.

15.8.3 Conjecture. *For all k, the fractional chromatic number of $P(k)$ is $k + 1$.*

The sets S_{ij} can be further used to construct large cocliques in the neighborhood of a vertex. For example, any partition P in the vertex set of $P(k)$ is adjacent to exactly $(k!)^{k-1}$ other partitions. Let a, b be elements in $\{1, \ldots, k^2\}$ that are contained in different cells of P and consider all the partitions that are adjacent to P in which a and b are contained in the same cell. It is not hard to see that there are $\frac{1}{k}(k!)^{k-1}$ such partitions, and they form a coclique in the neighborhood of P. We conjecture that this is the largest such coclique.

15.8.4 Conjecture. *In the graph $P(k)$, the size of the largest coclique in the neighborhood of a vertex in $P(k)$ is $\frac{1}{k}(k!)^{k-1}$.*

If this conjecture is true, then $\alpha_1(P(k)) = \frac{1}{k}(k!)^{k-1}$ and equality would hold in Theorem 2.8.1. Further, this would prove that Theorem 15.8.1 holds for all skew-partition graphs.

15.9 Eigenvalues of the skew-partition graph

Since the graph $P(k)$ is arc transitive, we can apply the ratio bound for cliques, Theorem 2.6.1. If τ is the least eigenvalue for $P(k)$, then

$$\omega(P(k)) \leq 1 - \frac{k!^{k-1}}{\tau}.$$

In the case when k is a prime power, $\omega(P(k)) = k + 1$, which implies that $-(k!^{k-1})/k \leq \tau$. It is not difficult to directly show that $-(k!^{k-1})/k$ is an eigenvalue of $P(k)$ for all values of k. Let $v_{i,j}$ denote the characteristic vector of the set $S_{i,j}$ (defined in the previous section) for some distinct pair i, j from $\{1, \ldots, k^2\}$. We leave it as an exercise to show that $v_{i,j} - \frac{1}{k+1}\mathbf{1}$ is an eigenvector for $P(k)$ with eigenvalue $-(k!^{k-1})/k$.

Alternatively, we can show that $-(k!^{k-1})/k$ is an eigenvalue of $P(k)$ using a quotient graph. For any distinct $i, j \in \{1, \ldots, k^2\}$ consider the partition

$$\pi = \{S_{i,j}, V(P(k)) \setminus S_{i,j}\}$$

of the vertices in the skew-partition graph. This is the orbit partition from the action on the vertices of $P(k)$ by the stabilizer of the set $\{i, j\}$ in $\text{Sym}(k^2)$. Hence π is an equitable partition of the vertices, and the quotient matrix is

$$P(k)/\pi = \begin{pmatrix} 0 & k!^{k-1} \\ \frac{k!^{k-1}}{k} & k!^{k-1} - \frac{k!^{k-1}}{k} \end{pmatrix}.$$

The eigenvalues of this quotient graph are

$$k!^{k-1}, \quad -\frac{k!^{k-1}}{k}.$$

If k is a prime power, then $-(k!^{k-1})/k$ is the least eigenvalue of $P(k)$; we conjecture that this is the least eigenvalue for all k.

15.9.1 Conjecture. *For all integers k, the least eigenvalue of $P(k)$ is $-(k!^{k-1})/k$.*

If Conjecture 15.9.1 is true, then the ratio bound gives

$$\alpha(P(k)) \leq \frac{|V(P(k))|}{1 - \frac{k!^{k-1}}{-\frac{k!^{k-1}}{k}}} = \frac{U(k^2, k)}{1 + k}.$$

This implies that Theorem 15.8.1 holds for all values of k. So we conclude this section with one final conjecture.

15.9.2 Conjecture. *For all k, the maximum size of a coclique in $P(k)$ is*

$$\binom{k^2 - 2}{k - 2} \frac{1}{(k-1)!} \prod_{i=1}^{k} \left(\frac{k^2 - ki}{k} \right) = \frac{U(k^2, k)}{1 + k}.$$

Further, the only cocliques that meet this bound are $S_{i,j}$ for distinct $i, j \in \{1, \ldots, k^2\}$.

For the bound in this conjecture, the first open case is when $k = 6$. The graph $P(6)$ has

$$\frac{1}{6!} \binom{36}{6} \binom{30}{6} \cdots \binom{6}{6} \approx 3.7 \times 10^{21}$$

vertices, which makes it computationally difficult to find all the eigenvalues!

15.10 Eigenspaces of the skew-partition graph

The group $\mathrm{Sym}(k^2)$ acts on the set of uniform k-partitions, and this action produces a permutation representation of the symmetric group. Since the stabilizer of a uniform k-partition under this action is the subgroup $\mathrm{Sym}(k) \wr \mathrm{Sym}(k)$ (see Section 12.8 for a description of this group), by Theorem 12.1.1, this permutation representation is the representation induced by the trivial representation on $\mathrm{Sym}(k) \wr \mathrm{Sym}(k)$. We denote this representation by $\mathrm{ind}_{\mathrm{Sym}(k^2)}(1_{\mathrm{Sym}(k)\wr\mathrm{Sym}(k)})$ and note that just as in the case of the perfect matchings, each uniform k-partition corresponds to a coset of $\mathrm{Sym}(k) \wr \mathrm{Sym}(k)$ in $\mathrm{Sym}(k^2)$.

We saw in Lemma 15.7.3 that the action of $\mathrm{Sym}(k^2)$ is transitive on the arcs of the skew-partition graph. This means that all pairs of partitions that are adjacent in the skew-partition graph are in the same orbital under this action. Thus, the adjacency matrix of the skew-partition graph is the characteristic matrix of an orbital of the action of $\mathrm{Sym}(k^2)$ on the cosets of $\mathrm{Sym}(k) \wr \mathrm{Sym}(k)$. From Lemma 13.8.1, each eigenspace of the skew-partition graph is a union of modules in the representation $\mathrm{ind}_{\mathrm{Sym}(k^2)}(1_{\mathrm{Sym}(k)\wr\mathrm{Sym}(k)})$ – so to obtain information about the eigenspaces, we look to the modules.

Using orbit counting, it is possible to find a partial decomposition of this representation into irreducible representations. The irreducible constituents of $\mathrm{ind}_{\mathrm{Sym}(k^2)}(1_{\mathrm{Sym}(k)\wr\mathrm{Sym}(k)})$ include Φ_λ for $\lambda = [k^2]$, $[k^2 - 2, 2]$, $[k^2 - 3, 3]$ and $[k^2 - 4, 4]$ (with multiplicities $1, 1, 1, 2$, respectively), as well as other $\lambda <[k^2 - 4, 4]$.

The multiplicity of $\Phi_{[k^2]}$ as a constituent is 1 and the dimension of the $[k^2]$-module is 1. Since $\Phi_{[k^2]}$ is the trivial representation, the all-ones vector is a basis for the $[k^2]$-module. From Lemma 13.8.3, this is a subspace of the $(k!)^{k-1}$-eigenspace. Since $P(k)$ is a connected $(k!)^{k-1}$-regular graph, we know that the $(k!)^{k-1}$-eigenspace has dimension 1 and that this eigenspace is exactly the $[k^2]$-module.

It follows from Exercise 13.7 applied to the quotient graph in the previous section that the $[k^2 - 2, 2]$-module is contained in the $-(k!^{k-1})/k$-eigenspace. We cannot conclude that this module is the entire eigenspace, since another module may also have the same eigenvalue (although we have no example of a skew-partition graph where this occurs). The multiplicity of $\Phi_{[k^2-2,2]}$ in the representation $\mathrm{ind}_{\mathrm{Sym}(k^2)}(1_{\mathrm{Sym}(k)\wr\mathrm{Sym}(k)})$ is 1 and from (12.6.1) the dimension of this irreducible module is

$$\binom{k^2}{2} - \binom{k^2}{1} = \binom{k^2 - 1}{2} - 1.$$

We can find a spanning set for the $[k^2 - 2, 2]$-module using the cocliques $S_{i,j}$ of $P(k)$ from Section 15.8. Recall that for distinct $i, j \in \{1, \ldots, k^2\}$, we let

$v_{i,j}$ be the characteristic vector of the set of all uniform k-partitions of a k^2-set in which i and j are in the same cell. The following result is due to Mike Newman and originally appears in [138].

15.10.1 Theorem. *For i, j distinct values in $\{1, \ldots, k^2\}$, let $v_{i,j}$ be the characteristic vector of the set $S_{i,j}$. The set*

$$\left\{ v_{i,j} - \frac{1}{k+1}\mathbf{1} \ : \ i, j \in \{1, \ldots, k^2\}, \ i \neq j \right\}$$

is a spanning set for the $[k^2 - 2, 2]$-module.

Proof. As stated before the theorem, it follows from Exercise 13.7 applied to the quotient graph from the previous section that each $v_{i,j} - \frac{1}{k}\mathbf{1}$ is in the $[k^2 - 2, 2]$-module.

We now show that these vectors span a subspace with dimension $\binom{k^2-1}{2} - 1$, which is exactly the dimension of the $[k^2 - 2, 2]$-module. To do this, we prove that the vectors $v_{i,j}$ span a subspace of dimension $\binom{k^2-1}{2}$. Since the sum of all the vectors $v_{i,j}$ is a multiple of the all-ones vector, this is sufficient.

Let H be the matrix whose columns are the characteristic vectors $v_{i,j}$ for all distinct i and j in $\{1, \ldots, k^2\}$. We show that the rank of H is $\binom{k^2-1}{2}$ by considering the matrix $H^T H$. Each entry in this matrix is the dot product of two columns of H (i.e., the dot product of two vectors $v_{i,j}$ and $v_{k,\ell}$). The value of this dot product depends only on the size of $\{i, j\} \cap \{k, \ell\}$ and can be calculated directly. For $\{i, j\}$ and $\{k, \ell\}$ with $|\{i, j\} \cap \{k, \ell\}| = m$ we let $v_{i,j} \cdot v_{k,\ell} = s_{2-m}$. If $m = 2$, then s_0 can be calculated as

$$s_0 = \binom{k^2 - 2}{k - 2} \frac{1}{(k-1)!} \prod_{i=1}^{k-1} \binom{k(k-i)}{k} = \frac{|V(P(k))|}{k+1}.$$

Likewise, s_1 can be calculated and expressed in terms of s_0 as

$$s_1 = \binom{k^2 - 3}{k - 3} \frac{1}{(k-1)!} \prod_{i=1}^{k-1} \binom{k(k-i)}{k} = \frac{k - 2}{k^2 - 2} s_0.$$

Finally, we calculate s_2 as

$$s_2 = \binom{k^2 - 4}{k - 4} \frac{1}{(k-1)!} \prod_{i=1}^{k-1} \binom{k(k-i)}{k}$$

$$+ \binom{k^2 - 4}{k - 2}\binom{k^2 - k - 2}{k - 2} \frac{1}{(k-2)!} \prod_{i=2}^{k-1} \binom{k(k-i)}{k}$$

$$= \frac{(k-2)(k-3) + k(k-1)^2}{(k^2 - 3)(k^2 - 2)} s_0.$$

Table 15.1 *The eigenvalues of $H^T H$.*

Eigenvalue	multiplicity
$\lambda_1 = s_0 + s_1(2k^2 - 4) + s_2\binom{k^2-2}{2}$	1
$\lambda_2 = s_0 + s_1(k^2 - 4) - s_2(k^2 - 3)$	$k^2 - 1$
$\lambda_3 = s_0 - 2s_1 + s_2$	$\frac{k^4 - 3k^2}{2}$

Since the values of the entries in $H^T H$ depend only on the size of the intersection of $\{i, j\}$ and $\{k, \ell\}$, it is possible to express the matrix $H^T H$ as a linear combination of the matrices in the Johnson scheme $J(k^2, 2)$. Let A_0 be the identity matrix, A_1 be the adjacency matrix of the Johnson graph $J(k^2, 2, 1)$, and A_2 be the adjacency matrix for the Kneser graph $K(k^2, 2)$ (the complement of $J(k^2, 2, 1)$). Then

$$H^T H = s_0 A_0 + s_1 A_1 + s_2 A_2.$$

The eigenvalues of these matrices are known (see Theorem 6.5.2), so the eigenvalues (with their multiplicities) of $H^T H$ are straight forward to calculate. They are given in Table 15.1.

It is not difficult to determine that only $\lambda_2 = 0$. Thus the nullity of $H^T H$ is $k^2 - 1$, and the rank of $H^T H$, and hence H, is

$$\binom{k^2}{2} - (k^2 - 1) = \binom{k^2 - 1}{2}. \qquad \square$$

A very similar proof may be used to obtain a basis for the direct sum of the first two modules.

15.10.2 Corollary. *The vectors $v_{i,j}$ for distinct $i, j \in \{1, \ldots, k^2 - 1\}$ form a basis for the sum of the $[k^2]$-module and the $[k^2 - 2, 2]$-module.* $\qquad \square$

Finally we conclude with a conjecture. If this conjecture is true, then the only module that has $-(k!^{k-1})/k$ as its eigenvalue is the $[k^2 - 2, 2]$-module. This would mean that the characteristic vector of any maximum coclique would be in the sum of the $[k^2]$-module and the $[k^2 - 2, 2]$-module, and thus would be a linear combination of the vectors $v_{i,j}$. The hope is that this could be used to show that the only cocliques in the skew-partition graph of size $\frac{1}{k+1} U(k^2, k)$ are the sets $S_{i,j}$. Indeed, we see in the following section that this is the case for $k = 3$.

15.10.3 Conjecture. *The least eigenvalue of $P(k)$ is $-(k!^{k-1})/k$ and it has multiplicity $\binom{k^2}{2} - \binom{k^2}{1}$.*

This conjecture is known to be true for $k = 3$ and $k = 4$.

Table 15.2 *Modified matrix of eigenvalues for the association scheme on 3×3 partitions.*

Representation	Dimension	A_0	A_1	A_2	A_3	A_4
$\Phi_{[9]}$	1	1	27	162	54	36
$\Phi_{[7,2]}$	27	1	11	-6	6	-12
$\Phi_{[6,3]}$	48	1	6	-6	-9	8
$\Phi_{[5,2,2]}$	84	1	-3	12	-6	-4
$\Phi_{[4,4,1]}$	120	1	-3	-6	6	2

15.11 3×3 partitions

The graph $P(3)$ is an interesting example of a skew-partition graph. The vertices of this graph are the 280 partitions of a set of size 9 into three cells, each of size 3. Three is a prime number, so there exists an orthogonal array $OA(4, 3)$ and the size of the maximum clique in $P(3)$ is 4. The size of a maximum coclique is 70 and the sets $S_{i,j}$ from the previous section give examples of maximum cocliques. Thus the clique-coclique bound holds with equality in this graph. Further, each vertex has degree 36 and is contained in exactly 12 edge-disjoint cliques of size 4.

The group Sym(9) acts on the 280 partitions, and this action produces a representation of Sym(9), namely $\text{ind}_{\text{Sym}(9)}(1_{\text{Sym}(3)\wr\text{Sym}(3)})$. The decomposition of $\text{ind}_{\text{Sym}(9)}(1_{\text{Sym}(3)\wr\text{Sym}(3)})$ can be easily calculated (either by orbit counting or by using a computer algebra package) to be

$$\Phi_{[9]} + \Phi_{[7,2]} + \Phi_{[6,3]} + \Phi_{[5,2,2]} + \Phi_{[4,4,1]}. \tag{15.11.1}$$

This decomposition is multiplicity-free, so by Lemma 13.4.1 the commutant of this representation is commutative. Further, there is a basis of the commutant of this representation that forms an association scheme whose vertices are the 280 uniform 3-partitions of a 9-set.

Mathon and Rosa first published the table of eigenvalues for this association scheme [124] and we present these values in Table 15.2. The columns in the table give the eigenvalues of the five matrices A_0, A_1, A_2, A_3, A_4 in the association scheme. The eigenvalues are ordered according to the eigenspace they belong to (labeled by the corresponding constituent of the representation $\text{ind}_{\text{Sym}(9)}(1_{\text{Sym}(3)\wr\text{Sym}(3)})$).

We see that the matrix A_3 has exactly three distinct eigenvalues and hence is the adjacency matrix for a strongly regular graph. This is the graph that interested Mathon and Rosa. Our interest lies mainly in the final matrix A_4, as this is the adjacency matrix for the graph $P(3)$.

We can give a nice description of the matrices in this association scheme, as the classes can be defined using the intersections of the 3×3 partitions. We have labeled the matrices in the scheme by A_i for $i = 0, \ldots, 4$, and we will say two partitions are i-related if the corresponding entry in A_i is 1. Two partitions are 4-related if and only if they are skew. If two partitions are skew, then there are nine pairs of cells from the two partitions that have non-trivial intersection. This way of viewing the intersection of two partitions can be used to describe the other classes in the association scheme.

For example, if two partitions have only three pairs of cells that have non-trivial intersection, then the partitions must be identical – in this case the partitions are 0-related. The other three classes of the association scheme can be similarly defined. Two partitions are 1-related if and only if they have exactly five pairs of cells that have non-empty intersection, 2-related if and only if there are seven such pairs and 3-related if and only if there are six. This association scheme and these relations are discussed further in Section 15.14.

From Table 15.2, the largest eigenvalue of $P(3)$ is (not surprisingly) 36 and the least is -12. With these values, it is easy to see that both the ratio bound for cocliques and the ratio bound for cliques hold with equality. Further, the (-12)-eigenspace is exactly the module corresponding to the representation $\Phi_{[7,2]}$, since no other module has -12 as its eigenvalue. By Theorem 15.10.1, the vectors $v_{i,j} - \frac{1}{4}\mathbf{1}$, for distinct $i, j \in \{1, \ldots, 9\}$, span this module. From Theorem 2.4.1, it follows that if S is a maximum coclique in $P(3)$ of size 70 and v_S is the characteristic vector for S, then $v_S - \frac{1}{4}\mathbf{1}$ is a (-12)-eigenvector. Thus v_S is a linear combination of the vectors $v_{i,j} - \frac{1}{4}\mathbf{1}$. In [84] this fact was used to show that S is actually equal to some $S_{i,j}$ for some pair of $i, j \in \{1, \ldots, 9\}$. In the following two sections we present this proof, excepting one step that depends on a computer search.

15.12 Inner distributions of cliques

Suppose S is a subset of the vertices of an association scheme with d classes, and let x be its characteristic vector. The *inner distribution* of S is the sequence

$$\left(\frac{x^T A_i x}{|S|} \right),$$

where $i = 0, \ldots, d$. In this section we show that all maximum cocliques in $P(3)$ have the same *inner distribution*. In particular, any maximum coclique in $P(3)$ has the same inner distribution as the set $S_{i,j}$. To start we consider the projection to the Bose-Mesner algebra of a matrix based on the characteristic vector of S.

15.12.1 Lemma. *Let x be the characteristic vector for a coclique of size 70 in $P(3)$ and let $M = xx^T$. Define $\widehat{M}$ to be the projection of M onto the Bose-Mesner algebra of the association scheme. Then*

$$\widehat{M} = \frac{1}{4}A_0 + \frac{5}{36}A_1 + \frac{1}{18}A_2 + \frac{1}{12}A_3. \qquad (15.12.1)$$

Proof. We can calculate $\widehat{M}$ using the idempotents of the association scheme. We denote these idempotents by E_i with $i = 0, \ldots, 4$, where E_i gives the projection onto the eigenspace corresponding to the $(i + 1)$-st row of Table 15.2. The matrices E_i also form an orthonormal basis for the algebra. We calculate the projection to be

$$\widehat{M} = \sum_{i=0}^{4} \frac{\langle M, E_i \rangle}{\langle E_i, E_i \rangle} = \sum_{i=0}^{4} \frac{x^T E_i x}{m_i} E_i.$$

This equation can be simplified. First, since E_0 is a multiple of the all-ones matrix,

$$x E_0 x = \frac{70^2}{280} = \frac{70}{4}.$$

Next, E_1 is the projection to the (-12)-eigenspace, so we know that

$$E_1 \left(x - \frac{1}{4}\mathbf{1} \right) = x - \frac{1}{4}\mathbf{1}.$$

Since the idempotents are orthogonal $E_1 \mathbf{1} = 0$, which implies that $E_1 x = x - \frac{1}{4}\mathbf{1}$. So we can calculate

$$\frac{x^T E_1 x}{m_1} = \frac{1}{27} 70 \left(1 - \frac{1}{4} \right) = \frac{70}{36}.$$

Finally, since x is in the sum of two eigenspaces (the (36)-eigenspace and the (-12)-eigenspace), it is orthogonal to the remaining three eigenspaces. Thus $E_i x = 0$ for $i = 2, 3, 4$. Now we have that the projection of M onto the Bose-Mesner algebra is simply

$$\widehat{M} = \frac{70}{4} E_0 + \frac{70}{36} E_1.$$

The matrix of eigenvalues for the association scheme is a change of basis matrix from the matrices A_i to the idempotents E_i; similarly, the matrix of dual eigenvalues (simply the inverse of the matrix of eigenvalues) is the reverse change of basis matrix. Using the matrix of dual eigenvalues for the association scheme, we can calculate the values of $x^T A_i x$ for all i and find that

$$\widehat{M} = \frac{1}{4}A_0 + \frac{5}{36}A_1 + \frac{1}{18}A_2 + \frac{1}{12}A_3. \qquad \square$$

Note that the matrix $\widehat{M}$ does not depend on which coclique we use for x. From (15.12.1) it is possible to glean information about the structure of any maximum coclique in $P(3)$.

15.12.2 Corollary. *If S is a coclique in $P(3)$ with size 70, then for a fixed element $a \in S$ there are:*

(a) 15 *elements in S that are 1-related to a;*
(b) 36 *elements that are 2-related to a;*
(c) 18 *elements that are 3-related to a.*

Proof. As in the previous lemma, let x be the characteristic vector for S and $\widehat{M}$ the projection of xx^T on the Bose-Mesner algebra.

The matrices A_i are an orthonormal basis of the algebra, and the projection of M into this algebra can be calculated to be

$$\widehat{M} = \sum_{i=0}^{4} \frac{\langle M, A_i \rangle}{\langle A_i, A_i \rangle} A_i = \sum_{i=0}^{4} \frac{x^T A_i x}{vv_i} A_i.$$

In particular, from Lemma 15.12.1

$$\frac{x^T A_1 x}{vv_1} = \frac{5}{36},$$

since $v = 280$ and from Table 15.2 $v_1 = 27$, this means that $x^T A_1 x = 1050$. So there are 1050 pairs of partitions in S that are 1-related, since these are vertices in a vertex-transitive graph and $|S| = 70$, each vertex is 1-related to exactly 15 other vertices in S. A similar calculation works for the number of elements that are 2-related and 3-related. $\qquad\square$

15.13 Characterizing cocliques

Finally, we show that Conjecture 15.9.2 holds for $P(3)$. Only an outline of the proof is given; more details can be found in [84].

15.13.1 Theorem. *If S is a coclique of size 70 in $P(3)$, then $S = S_{i,j}$ for some pair of distinct $i, j \in \{1, \ldots, 9\}$.*

Proof. By Corollary 15.12.2, we can assume that there are two partitions that are 3-related (these partitions have seven pairs of cells that have non-empty intersection). Thus we can assume without loss of generality that the partitions

$$P = \{\{123\}, \{456\}, \{789\}\}, \qquad Q = \{\{127\}, \{458\}, \{369\}\}$$

are both in S.

Let x be the characteristic vector of S, and H be the 280×28 matrix whose columns are the characteristic vectors of the sets $S_{i,j}$ where i, j are distinct elements in $\{1, \ldots, 8\}$. The columns of H form a basis for the sum of the (36)-eigenspace and the (-12)-eigenspace (from Corollary 15.10.2). Thus there is a vector y such that $Hy = x$. We will see that the vector y has exactly one entry equal to 1, and all other entries equal to 0; this implies that $S = S_{i,j}$ for some $i, j \in \{1, \ldots, 8\}$.

Let M be the submatrix of H that is composed of the rows in H which represent the neighborhoods of P and Q. Since none of these partitions are in the coclique, it must be that $My = 0$. Using the facts that $My = 0$ and that Hy is a 01-vector, a computer search of all possible such vectors shows that y must have one entry equal to 1 and all other entries equal to 0. In fact, it must be either the characteristic vector of $S_{1,2}$ or $S_{4,5}$. $\qquad\square$

There are several barriers that prevent us from generalizing this result to larger values of k. First, $-(k!^{k-1})/k$ is the least eigenvalue for $P(k)$ when k is a prime power, but we do not have a proof of this when k is not a prime power. Also, when k is a prime power, we do not know the dimension of $-(k!^{k-1})/k$-eigenspace. Specifically, we cannot prove that the $[k^2 - 2, 2]$-module is the only module to have $-(k!^{k-1})/k$ as its eigenvalue. Finally, in the proof of Theorem 15.13.1, a computer search is necessary. This search is impractical for general k, and even for $k = 4$ this search has not been completed.

15.14 Meet tables

In this chapter we have defined two families of graphs on the set of all uniform k-partitions of an n-set where adjacencies are defined by intersection. All of these graphs are vertex transitive, which can be seen from the fact that the group $\mathrm{Sym}(n)$ acts transitively on the set of partitions and this action preserves the intersection patterns of the partitions.

Further, since $\mathrm{Sym}(n)$ acts on the partitions, there is a permutation representation of $\mathrm{Sym}(n)$ from this action. It is not hard to see that this representation is a familiar representation. By Theorem 12.1.1 it is the representation induced on $\mathrm{Sym}(n)$ by the trivial representation of $\mathrm{Sym}(\ell) \wr \mathrm{Sym}(k)$.

The commutant of a permutation representation of a group is the set of all matrices that commute with the matrix representation of every element in the group. By Lemma 3.2.1, the commutant of a permutation representation is a Schur-closed algebra; and by Theorem 13.2.1, the adjacency matrices of the orbitals from the action of $\mathrm{Sym}(k\ell)$ on the cosets of $\mathrm{Sym}(\ell) \wr \mathrm{Sym}(k)$ form a basis for this commutant. In this section, we see that the adjacency matrix

Table 15.3 *All meet tables for uniform 3-partitions of a 9-set.*

$$\begin{pmatrix} 3 & 0 & 0 \\ 0 & 3 & 0 \\ 0 & 0 & 3 \end{pmatrix}, \begin{pmatrix} 3 & 0 & 0 \\ 0 & 2 & 1 \\ 0 & 1 & 2 \end{pmatrix}, \begin{pmatrix} 2 & 0 & 1 \\ 0 & 2 & 1 \\ 1 & 1 & 1 \end{pmatrix}, \begin{pmatrix} 2 & 1 & 0 \\ 0 & 2 & 1 \\ 1 & 0 & 2 \end{pmatrix}, \begin{pmatrix} 1 & 1 & 1 \\ 1 & 1 & 1 \\ 1 & 1 & 1 \end{pmatrix}$$

of each graph that we have considered in this chapter is the sum of adjacency matrices of these orbitals. To prove this, we give a simple combinatorial description of the adjacency matrices of the orbitals.

Let $P = \{P_1, \ldots, P_k\}$ and $Q = \{Q_1, \ldots, Q_k\}$ be two uniform k-partitions of an n-set. Define the *meet table*, $M(P, Q)$, of P and Q to be the $k \times k$ array with the (i, j)-entry equal to $|P_i \cap Q_j|$. Two meet tables are *isomorphic* if one table can be obtained by permuting the rows and columns of the other table. Changing the order of the cells in either P or Q may produce different, but isomorphic, meet tables. A meet table $M(P, Q)$ may not be isomorphic to the meet table $M(Q, P)$, but it will be isomorphic to the transpose of $M(Q, P)$. For example, the 3-partitions of a 9-set have exactly five non-isomorphic meet tables; these tables are given in Table 15.3.

For a given k and n, let $\mathcal{MT}(n, k)$ be the set of all non-isomorphic meet tables for the uniform k-partitions of an n-set. For each meet table $M_i \in \mathcal{MT}(n, k)$, define a matrix A_i with the rows and columns indexed by the uniform k-partitions of the n-set. For partitions P and Q, set the (P, Q)-entry of A_i to be 1 if the the meet table $M(P, Q)$ is isomorphic to the meet table M_i, and 0 otherwise.

For any values of n and k there will be a meet table in $\mathcal{MT}(n, k)$ isomorphic to ℓ times the $k \times k$ identity matrix; the meet table for partitions P and Q is isomorphic to this table if and only if $P = Q$. We denote this meet table by M_0, and the corresponding matrix A_0 is the identity matrix. Since the meet tables $M(P, Q)$ and $M(Q, P)$ may not be isomorphic, the matrices A_i may not be symmetric. But, $(A_i)^T = A_j$ for some j, since $M(Q, P)$ will be one of the meet tables in $\mathcal{MT}(n, k)$.

15.14.1 Theorem. *Assume k, ℓ, n are integers with $n = k\ell$. The set of matrices A_i defined earlier in this section form a basis for the commutant of the representation* $\mathrm{ind}_{\mathrm{Sym}(n)} (1_{\mathrm{Sym}(\ell) \wr \mathrm{Sym}(k)})$.

Proof. We only need to show that matrices A_i are the adjacency matrices of the orbitals of the action of $\mathrm{Sym}(n)$ on the uniform k-partitions.

Assume that P, Q, P' and Q' are uniform k-partitions with $M(P, Q)$ isomorphic to $M(P', Q')$. Order the cells of P' and Q' so that the meet tables

$M(P, Q)$ and $M(P', Q')$ are equal. Then

$$|P_i \cap Q_j| = |P'_i \cap Q'_j|,$$

for all $i, j \in \{1, \ldots, k\}$. Clearly, it is possible to define a one-to-one map from $P_i \cap Q_j$ to $P'_i \cap Q'_j$. Doing this for every (i, j) will define a permutation of $\{1, \ldots, n\}$ that maps P to P', and Q to Q'. Thus (P, Q) and (P', Q') are in the same orbital under the action of $\mathrm{Sym}(n)$.

Conversely, assume that (P, Q) and (P', Q') are in the same orbital. Then there is a permutation σ in $\mathrm{Sym}(n)$ such that $\sigma(P) = P'$ and $\sigma(Q) = Q'$. Order the cells of P' and Q' so that $\sigma(P_i) = P'_i$ and $\sigma(Q_i) = Q'_i$. Then $\sigma(P_i \cap Q_j) = P'_i \cap Q'_j$ so clearly

$$|P_i \cap Q_j| = |P'_i \cap Q'_j|.$$

Thus the meet tables $M(P, Q)$ and $M(P', Q')$ are isomorphic. $\square$

With this characterization, it is easy to see that for each of the graphs that we have defined in this chapter, the adjacency matrix is a sum of the matrices A_i. For example, two uniform k-partitions of a k^2-set are adjacent in the skew-partition graph if and only if their meet table is the matrix with every entry equal to 1. Thus the adjacency matrix of the skew-partition graph is a single matrix in this basis.

The adjacency matrices of the graph $P(n, k)$, defined in Section 15.1, are the union of the matrices whose corresponding meet table has at least one entry of size ℓ (so two partitions are adjacent in this graph if and only if they have a cell in common). For the 3×3 partitions, the graph $P(n, k)$ is the second graph in the association scheme. The degree of this graph (from Table 15.2) is 27 and the least eigenvalue is -3. Since this graph is a single graph in an association scheme, the ratio bound for cliques applies and the size of a clique can be no more than 10. This is exactly the size of a canonical intersecting partition set for these parameters. For this special case, Theorem 15.1.1 can be shown to hold using eigenvalues (this cannot be done in general).

In Section 15.11, we saw that there is an association scheme on the 3×3 partitions. Each class in the association scheme corresponds to one of the meet tables given in Table 15.3 (the meet tables are given in the same order as the classes in Table 15.2). For example, the strongly regular graph in the association scheme is the graph in which the partitions are adjacent if and only if their meet table is isomorphic to the third matrix in Table 15.3. In this special case the commutant is commutative, and the representation of $\mathrm{Sym}(9)$ induced from the trivial representation on $\mathrm{Sym}(3) \wr \mathrm{Sym}(3)$ is multiplicity-free. The representation $\mathrm{ind}_{\mathrm{Sym}(n)}(1_{\mathrm{Sym}(\ell)\wr\mathrm{Sym}(k)})$ is not multiplicity-free in general.

In fact, it is multiplicity-free if and only if either $k = 2$, or $\ell = 2$, or (k, n) is one of $\{(3, 3), (3, 4), (3, 5)\}$ [86].

15.15 Exercises

15.1 Show that the chromatic number of the perfect matching graph $M(2k)$ is $2k - 1$. What is the core of $M(2k)$?

15.2 Show that the set of orbitals from the action of $\mathrm{Sym}(2k)$ on the cosets of $\mathrm{Sym}(2) \wr \mathrm{Sym}(k)$ in $\mathrm{Sym}(2k)$ are the classes in the matching association scheme.

15.3 Determine if the perfect matching association scheme metric. Is it cometric?

15.4 Let $H \leq \mathrm{Sym}(n)$ and let X be a graph in the orbital scheme of $\mathrm{Sym}(n)$ on $\mathrm{Sym}(n)/H$. Let $K \leq \mathrm{Sym}(n)$ and let π be the equitable partition of the vertices of X from the action of K on the cosets $\mathrm{Sym}(n)/H$. Show that if ξ is an eigenvalue of the quotient graph X/π, then ξ belongs to a module that is common to $\mathrm{ind}_{\mathrm{Sym}(n)}(1_K)$ and $\mathrm{ind}_{\mathrm{Sym}(n)}(1_H)$. This is a generalization of the proof of Lemma 15.5.1.

15.5 Prove that the construction given in Section 15.7 produces a Latin square from a triangle in $P(k)$. Next, prove that an m-clique in $P(k)$ is equivalent to an $OA(m, k)$.

15.6 Show that the graph $P(k)$ is $\binom{k+1}{2}$-colorable (it is conjectured that this is the chromatic number).

15.7 Prove that the vector $v_{i,j} - \frac{1}{k+1}\mathbf{1}$ is an eigenvector for the graph $P(k)$ with eigenvalue $-(k!^{k-1})/k$ by multiplying the adjacency matrix of $P(k)$ and the vector.

15.8 Using Theorem 15.13.1, prove that the graph $P(3)$ is a core. Hint: this is similar to the proof that the Kneser graphs are cores [87, Section 7.9].

15.9 Show that the orbitals of $\mathrm{Sym}(16)$ acting on $\mathrm{Sym}(4) \wr \mathrm{Sym}(4)$ do not form an association scheme. Find a 4×4 meet table that is not isomorphic to its transpose.

15.10 Let $n = k\ell$. Consider the decomposition of

$$\mathrm{ind}_{\mathrm{Sym}(n)}\left(1_{\mathrm{Sym}(\ell) \wr \mathrm{Sym}(k)}\right)$$

into irreducible representations.

(a) Show that this decomposition includes the representations

$$\Phi_{[n]}, \ \Phi_{[n-2,2]}, \ \Phi_{[n-3,3]}, \ \ldots, \ \Phi_{[n-\ell,\ell]}.$$

(b) Prove that if $k > 3$, then the representation is not multiplicity-free.

(c) If $n = k^2$, find the multiplicity and corresponding eigenvalue of each of the following representations:

$$\Phi_{[k^2]}, \quad \Phi_{[k^2-2,2]}, \quad \Phi_{[k^2-3,3]}, \quad \Phi_{[k^2-4,4]}.$$

15.11 Show that the graph $P(n, k)$ is not edge transitive.

15.12 Define a graph $P(n, k, t)$ on the set of all uniform k-partitions of an n-set (where $n = k\ell$ for some integer ℓ). Two vertices

$$P = \{P_1, \ldots, P_k\}, \quad Q = \{Q_1, \ldots, Q_k\}$$

are adjacent in this graph if and only if for all $i, j \in \{1, \ldots, k\}$

$$|P_i \cap Q_j| < t.$$

(a) Show that $P(n, k, t)$ is neither the empty graph nor the complete graph if and only if

$$\max\{1, \lceil \ell/k \rceil\} < t \le \ell.$$

(b) Show that a set of partitions is a equivalent to *resolvable* t-$(n, \ell, 1)$ *design* if and only if it is a clique in $P(n, k, t)$ of size

$$\frac{(n-1)!(\ell-t)!}{(\ell-1)!(n-t)!}.$$

(c) Show that there is a coclique in $P(n, k, t)$ of size

$$\frac{1}{(k-1)!} \binom{n-t}{\ell-t} \prod_{i=1}^{k-1} \binom{n-i\ell}{\ell},$$

and if a resolvable t-$(n, \ell, 1)$ design exists, then this coclique is of maximum size.

15.13 Show that the graph $P(n, k, t)$ is connected and find its diameter.

15.14 Let k and λ be positive integers. Define a graph $OA_\lambda(k)$ to have the set of all uniform k-partitions of a λk^2 set as its vertices. Two partitions

$$P = \{P_1, \ldots, P_k\}, \quad Q = \{Q_1, \ldots, Q_k\}$$

are adjacent if and only if $|P_i \cap Q_j| = \lambda$ for all i, j. Show, using the method described in Section 15.9, that the largest clique in this graph is no larger than $(\lambda k^2 - 1)/(k - 1)$. The cliques in this graph correspond to orthogonal arrays and this bound is the *Bose-Bush bound*.

Notes

Expanding on the notion of intersecting partitions from Section 15.1, a set of partitions could be considered to be *t-intersecting* if any two partitions in the set have at least t cells in common. Theorem 15.1.1, with a very similar proof, can be generalized to a set of partitions in which any two partitions are t-intersecting with $t > 1$ [128].

Peter Erdős and László Székely also looked at maximum sets of intersecting partitions, but without the restriction that the cells of the partition all have the same size. In this case, provided that the base set is sufficiently large, a version of the EKR Theorem holds. They also consider two other definitions of intersection for partitions; see [59] for complete details.

Ku and Renshaw in [108] also prove a version of the EKR Theorem for set partitions, but without the requirement that each partition have the same number of cells.

Muzychuk [136] investigated the perfect matching association scheme and gave all the eigenvalues for this association scheme for some small values of k. He was able to construct these tables using some heavy computation with Jack symmetric functions.

Finally, Lindzey [115] has managed to prove the EKR Theorem for perfect matchings using a variant of the method we presented for derangements in the previous chapter.

16

Open problems

Throughout this book we have presented many open problems. In this final chapter, we outline these problems, as well as some new ones, and give some conjectures. The sections in this chapter roughly follow the order of the material presented in the previous chapters.

16.1 Generalize the ratio bound for cliques

In Chapter 2, several versions of the ratio bound are given. For example, we show that the ratio bound for cocliques holds for all graphs that are regular (Theorem 2.4.1). We prove that the ratio bound for cliques can be be applied to regular graphs for which there is a set of maximum cliques that forms a regular point/block structure that covers all the edges in the graph (Theorem 2.6.1).

In Chapter 3, we prove that a ratio bound for cliques holds for the graphs in an association scheme (Corollary 3.7.2) and that there is a ratio bound for T-cocliques in an association scheme (Corollary 3.8.3). There is also a version of the ratio bound for cross cocliques (Theorem 2.7.1).

Despite all of these variants on the ratio bound, we have encountered examples of graphs that meet the bound given in the ratio bound for cliques, but do not satisfy the conditions of Theorem 2.6.1 or Corollary 3.7.2. One example is the perfect matching graph defined in Section 15.2, $M(2k)$. It is easy to determine the exact size of the largest clique in $M(2k)$, and it is also easy to show that

$$\frac{|V(M(2k))|}{1 - \omega(M(2k))}$$

is an eigenvalue of $M(2k)$.

308

What we suspect is that the ratio bound for cliques holds more generally than as stated in Theorem 2.6.1, and that it can be applied to a more general family of graphs that includes the perfect matching graph.

16.2 Extend the Katona cycle proof

In Section 2.14, we give a proof of the EKR Theorem that is based on a graph homomorphism. This proof is also known as the Katona cycle proof, and it is probably the most popular of the proofs of the standard EKR Theorem. Katona wrote a paper [106] entitled "The cycle method and its limits" that described some of the different applications of this method and outlined the difficulties in generalizing his method to t-intersecting sets. Katona did not view his method using graph homomorphisms, so this perspective may give new insight.

To see how this method could be applied to t-intersecting sets, consider the *generalized Kneser graph*. This graph is denoted by $K_t(n, k)$. It has the k-sets of $\{1, \ldots, n\}$ as its vertices, and two sets are adjacent if and only if their intersection has size no more than t. So $K(n, k) = K_0(n, k)$ and a coclique in $K_{t-1}(n, k)$ is a t-intersecting set. The EKR Theorem says that if $n > (t + 1)(k - t + 1)$, then any maximum coclique in $K_{t-1}(n, k)$ consists of all the sets that contain a fixed t-set. To generalize the cycle proof to this situation, we need to find a graph that will play the role of cyclic interval graph.

16.2.1 Problem. *Does there exist a graph H that satisfies the following conditions:*

(a) *there is a homomorphism* $\psi : H \to K_t(n, k)$; *and,*

(b) $\frac{V(H)}{\alpha(H)} = \frac{\binom{n}{k}}{\binom{n-t}{k-t}}$?

If such an H exists, then by Lemma 2.13.2 the size of a maximum coclique in $K_{t-1}(n, k)$ will be exactly $\binom{n-t}{k-t}$. Katona [106] notes that if a t-$(n, k, 1)$ design exists, then the subgraph of $K_{t-1}(n, k)$ induced by the sets in the design will satisfy the conditions for H (in this case H would be a complete graph).

Lemma 2.13.3 gives a way to apply the homomorphism method in other situations. For any set of objects for which there is a concept of intersection, we can define the *intersection graph*; this graph has the objects as vertices and two vertices are adjacent if and only if they do not intersect. If this is graph is vertex transitive, then we can apply Lemma 2.13.3. The challenge is to find a subgraph of the intersection graph that has the same fractional chromatic number as the intersection graph and also has the property that the maximum cocliques are easily identified. For example, the proof by Kamat and Misra [102] that the EKR Theorem holds for perfect matchings uses this method.

An unpublished proof by Mohammad Mahdian uses a version of the Katona's cycle proof to show that the bound in the EKR Theorem holds for multisets. In this case, the intersection graph is not regular, so it is not so straightforward to determine its fractional chromatic number. Mahdian's proof uses a subgraph of the intersection graph that has a weighting on the vertices. The vertices of intersection graph for multisets is then uniformly covered with copies of this weighted graph. We suspect a similar approach can be used for other intersection graphs that are not regular.

16.3 Prove the EKR Theorem for the block graph of a design

In Section 5.3 we introduce the block graph of a 2-$(n, m, 1)$ design. Theorem 5.3.2 states that the set of all blocks that contain a fixed element is an example of a maximum clique in the block graph (these are the canonical cliques). This result is equivalent to the bound in the EKR Theorem for blocks in a design. For any design where n is sufficiently large relative to m, the canonical cliques are the only maximum cliques (this is Corollary 5.3.5). This gives a large number of designs for which the bound in the natural generalization of the EKR Theorem holds, but for which we do not know if the characterization holds.

16.3.1 Problem. *Determine a characterization of the 2-$(n, m, 1)$ designs, based only on the parameters of the design, for which the only maximum cliques in the block graph are the canonical cliques.*

In Section 5.3, there is an example of a design with a block graph that has maximum cliques which are not canonical cliques. These maximum cliques have an interesting structure – namely, they form a subdesign isomorphic to the Fano plane. It is not clear if this a result of a wider phenomenon.

16.3.2 Problem. *When the block graph of a design has maximum cliques that are not canonical, are the non-canonical cliques isomorphic to smaller designs?*

In general, the problem is to determine all the maximum cliques in the block graph for any t-(n, m, λ) design.

16.4 Prove the EKR Theorem for the orthogonal array graph

In Section 5.5 we define the orthogonal array graph for any orthogonal array. We show that the set of all columns in an orthogonal array that have the same entry in a fixed row form a clique of maximum size in the orthogonal array graph. Not surprisingly, we called these cliques the canonical cliques of the orthogonal array graph. We also show that if $OA(m, n)$ is an orthogonal array and n is sufficiently large relative to m, then all the maximal cliques in the orthogonal array graph are canonical cliques (this is Corollary 5.5.3). But this result is nowhere near to the result we really want, which is a characterization of the maximum cliques in an orthogonal array graph.

16.4.1 Problem. *Find a characterization of the orthogonal arrays, based only on the parameters of the array, for which all of the maximum cliques in the orthogonal array graph are canonical cliques.*

Section 5.5 includes an example of an orthogonal array which has a maximum clique that is not a canonical clique. It is interesting to note that this clique forms a smaller orthogonal array.

16.4.2 Problem. *Assume that $OA(m, (m - 1)^2)$ is an orthogonal array and its orthogonal array graph has non-canonical cliques of size $(m - 1)^2$. Do these non-canonical cliques form subarrays that are isomorphic to orthogonal arrays with entries from $\{1, \ldots, m - 1\}$?*

The most general problem is to determine all the maximum cliques in the orthogonal array graph for any orthogonal array.

16.5 Find an algebraic proof of the EKR Theorem for the Paley graphs

The Paley graph $P(q)$ is defined in Section 5.8, where q is the order of a finite field with $q \equiv 1 \pmod 4$. In Section 5.9 we prove that if q is a perfect square, then the size of a maximum clique in $P(q)$ (and hence also the size of a maximum coclique) is exactly $\sqrt{q}$. Blokhuis [20] proved that in such a Paley graph every maximum clique is a square translate of the set of squares in the finite field. We defined these cliques to be the canonical cliques of the Paley graph. In Section 5.9, a possible alternate proof of Blokhuis's result is presented. Solving the next two problems would fill in the details of this alternate proof (the θ-eigenspace is defined in Section 5.9).

16.5.1 Problem. *Show that the θ-eigenspace of $P(q^2)$ is spanned by the balanced characteristic vectors of the canonical cliques.*

16.5.2 Problem. *Prove that the only balanced characteristic vectors of sets of size q, in the eigenspace belonging to θ of P(q²), are the balanced characteristic vectors of the canonical cliques.*

16.6 Determine the chromatic number of the Johnson graphs

In Section 6.1, we describe two families of cliques in the Johnson graph. These give the following lower bound on the chromatic number of any Johnson graph:

$$\max\{n - k + 1, k + 1\} \le \chi(J(n, k)).$$

It is an old result that for all n and k the bound $\chi(J(n, k)) \le n$ holds (see [89]). Further, it is not hard to determine that

$$\chi(J(n, 2)) = \begin{cases} n - 1, & \text{if } n \text{ is even;} \\ n, & \text{if } n \text{ is odd.} \end{cases}$$

The value of $\chi(J(n, 3))$ is known for all n with $n \not\equiv 5 \pmod 6$:

$$\chi(J(n, 3)) = \begin{cases} n - 2, & \text{if } n \ge 7 \text{ and } n \equiv 1 \text{ or } 3 \pmod 6; \\ n - 1, & \text{if } n > 7 \text{ and } n \equiv 0 \text{ or } 2 \pmod 6; \\ n, & \text{if } n \equiv 4 \pmod 6. \end{cases}$$

These values were initially determined in a series of papers by Lu [119, 120] and are nicely summarized by Etzion and Bitan in [60]. In the Etzion and Bitan paper, the chromatic number and the size of maximum cocliques in the Johnson graph are considered. Specifically, they give constructions of proper colorings of some Johnson graphs $J(n, k)$ that use fewer than n colors. A more recent paper by Brouwer and Etzion [35] provides a table summarizing of the current known values for the chromatic number of the Johnson graphs. There are many interesting open questions concerning the chromatic number of the Johnson graphs, but we state just the most general of these problems.

16.6.1 Problem. *Find the chromatic number for the Johnson graph J(n, k) for all values of n and k.*

Naturally, a related problem is the size of the largest coclique in $J(n, k)$.

16.6.2 Problem. *What is the size of the largest coclique in the Johnson graph J(n, k) for all n and k?*

These problems are of a wider interest since they are related to bounds on codes. A coclique in the Johnson graph $J(n, k)$ is a code with distance at least 4; thus the chromatic number of $J(n, k)$ is the smallest number of such codes into which the vertices can be partitioned.

16.7 Prove the EKR Theorem for 2-transitive groups

In Section 14.7 we consider 2-transitive subgroups of $\mathrm{Sym}(n)$. These groups are interesting since it is easy to see that the derangement graph has both d (the number of derangements in the group) and $-d/(n-1)$ as eigenvalues. In general, the eigenvalues can be found from the irreducible representations of the group. The eigenvalue d comes from the trivial character, and $-d/(n-1)$ comes from the standard character. If $-d/(n-1)$ is the least eigenvalue, then the group must have the EKR property (this follows easily by the ratio bound).

16.7.1 Problem. *Are there properties of 2-transitive subgroups of $\mathrm{Sym}(n)$ that can be used to determine if the value of the least eigenvalue of the derangement graph for the group is $-d/(n-1)$ (where d is the number of derangements in the group)?*

The group $\mathrm{PGL}(2, q)$ has $-d/(n-1)$ as its least eigenvalue. But there are two more irreducible characters, other than the standard character, which also give the same eigenvalue. In this case, it is possible that the characteristic vector of a maximum coclique of the derangement graph is in the modules corresponding to these other irreducible characters. If this were to happen, it would be more difficult to characterize the maximum cocliques. But, it can be shown that these other two characters are constant on the subgroup $\mathrm{PSL}(2, q)$. This fact can be used to show that the characteristic vectors of the maximum cocliques are orthogonal to the modules corresponding to these other irreducible characters. Hence, the characteristic vector of any maximum coclique is in the sum of the trivial module and the standard module. From this it is possible to give a characterization of the maximum cocliques (see [131] for details).

16.7.2 Problem. *Assume that G is a 2-transitive subgroup of $\mathrm{Sym}(n)$ with d derangements. Further, assume that G has an irreducible representation χ for which the corresponding eigenvalue is smaller than $-d/(n-1)$. Is it possible to prove that the characteristic vector for any maximum coclique is orthogonal to the χ module?*

In Section 14.7, we show that for any 2-transitive group, the canonical cocliques in the derangement graph lie in the span of the trivial and the standard module. In fact, for all the 2-transitive groups we have considered, the characteristic vector of any maximum coclique is always in the direct sum of the trivial module and the standard module (see [7] for details on small 2-transitive groups). But there is no clear reason why this should hold for every 2-transitive subgroup.

For any group for which this holds, the characteristic vector of any maximum coclique is a linear combination of the characteristic vectors of canonical cocliques. In this case, the question of determining if a 2-transitive group has the strict EKR property becomes a question of whether or not there is a 01-vector in the span of the characteristic vectors of canonical cocliques with the correct weight.

16.7.3 Problem. *Are there examples of 2-transitive groups for which there is a χ-module, where χ is an irreducible representation other than the standard representation, such that the span of the χ-module and the trivial module contains the characteristic vector of a maximum coclique?*

In Section 14.7 we outline a method to prove that a 2-transitive group has the strict EKR property. In this method, we consider a matrix M for the subgroup, defined in (14.7.1) in Section 14.7. A key step in our proof of Theorem 14.4.3 was to show that M for Sym(n) has full rank (this is Lemma 14.4.1). In order to use the same proof to show that a 2-transitive subgroup has the strict EKR property, we would need to show that M for the subgroup has full rank. We do not have an effective general way to do this.

16.7.4 Problem. *Let G be a subgroup of Sym(n). Define a matrix M where the rows are indexed by the derangements in G and the columns by the distinct ordered pairs from $\{1, \ldots, n\}$. The $(\sigma, (i, j))$-entry of M is 1 if $\sigma(i) = j$ and 0 otherwise. Are there properties of the group G that can be used to determine if M has full rank?*

16.8 Determine cocliques in groups that do not have the EKR property

Our focus has been on groups that have the EKR property, but there are many groups that do not have the EKR property. For these groups, it is still interesting to consider the size and structure of the maximum cocliques in the derangement graph. The general problem is to characterize all the maximum cocliques in the derangement graph for a group or for a family of groups.

A simple example is the family of Frobenius groups. For these groups the derangement graph is the union of complete graphs, so it is trivial to determine the maximum cocliques in the derangement graph. For many families of groups this problem is hard, and this is a project that is wide open. For example, in [131] the following conjecture is made.

16.8.1 Conjecture. *Every maximum coclique in the derangement graph for PGL(n, q) is either the coset of the stabilizer of a point or the coset of the stabilizer of a plane.*

It is known that any maximum intersecting set in PGL$(3, q)$ is a coset of the stabilizer of either a point or a plane [132].

16.9 Prove the EKR property holds for other group actions

In Chapter 14, we considered the EKR property for a permutation group. We focused on the natural action on $\{1, \ldots, n\}$, but there are many other choices for a group action. For any group G that acts on any set Ω, the group action (as opposed to the group itself) is defined to have the EKR property if the largest set of intersecting elements from G is no larger than the size of the largest point stabilizer. Moreover, the group action is defined to have the strict EKR property if the only maximum intersecting sets of elements from G are the cosets of point stabilizers. Every group action of a group G is equivalent to the action of G on the cosets of H in G, for some subgroup $H \leq G$. This gives a convenient way to consider all possible group actions.

16.9.1 Problem. *Let G and H be groups with $H \leq G$. Find conditions on the groups G and H such that G acting on G/H has the EKR property or the strict EKR property.*

Bardestani and Mallahi-Karai [17] give a very interesting alternate definition of the EKR property. They define a group G to have the *weak EKR property* if for every subgroup $H \leq G$, the group G acting on G/H has the EKR property (as we define it). They show that every nilpotent group has the weak EKR property and that any finite group with the weak EKR property is solvable.

16.10 Calculate the eigenvalues of the perfect matching scheme

Let H be the wreath product Sym$(2) \wr$ Sym(k). Then H is a subgroup of Sym$(2k)$ and the orbitals of the action of Sym$(2k)$ on the cosets of H in Sym$(2k)$ produces an association scheme called the perfect matching scheme. The eigenvalues of the graphs in this scheme can be calculated from the irreducible representations of Sym$(2k)$ that occur in the decomposition of ind$_{\text{Sym}(2k)}(1_H)$. These are exactly the representations whose corresponding partitions of $2k$ have all parts even (see Section 12.8 for details).

Our focus is on the graph in this association scheme that we call the perfect matching graph (see Section 15.2 for the definition of this graph). It is easy to see that the eigenvalue corresponding to the module $[2k]$ is the largest eigenvalue for the perfect matching graph. We have proved in Corollary 15.5.3 that the next largest eigenvalue, in absolute value, corresponds to the module $[2k - 2, 2]$.

Muzychuk [136] calculated the entire spectrum of $M(10)$. These calculations are complicated and involve the Jack symmetric functions rather than the Schur functions used by Renteln. Using GAP [76] we also have computed the eigenvalues of the perfect matching graph $M(2k)$ for $k \leq 5$. Ideally, we would like to have a closed form for all the eigenvalues of all the perfect matching graphs, but this seems to be a very difficult problem. Toward this goal, it would useful to know more of the eigenvalues of the perfect matching graph. There are two possible approaches to this.

16.10.1 Problem. *Calculate all of the eigenvalues for specific perfect matching graphs $M(2k)$ with $k \geq 6$.*

16.10.2 Problem. *Calculate the eigenvalue of the perfect matching graph $M(2k)$ corresponding to a partition of $2k$ in which all the parts are even (for the partitions $[2k]$ and $[2k-2, 2]$ this is straightforward) for all k.*

Ku and Wong [109] (using Renteln's recurrence relation) showed that there are interesting relations between the eigenvalues of the derangement graph. For example, let $\lambda = (\lambda_1, \ldots, \lambda_i)$ be a partition of n, and ξ_λ be the eigenvalue of the λ-module for the derangement graph. Ku and Wong proved that the sign of ξ_λ is equal to the sign of $(-1)^{\lambda_1}$. A similar pattern holds for the perfect matching graph $M(2k)$ for $k = 2, 3, 4, 5$, and we conjecture that this holds for all k.

16.10.3 Conjecture. *Assume that $\lambda = (\lambda_1, \ldots, \lambda_i)$ is a partition of $2k$ in which every part is even. If ξ_λ is the eigenvalue of the perfect matching graph corresponding to λ, then the sign of ξ_λ is $(-1)^{(2k-\lambda_1)/2}$.*

There are more general questions about the relative values of the eigenvalues of the perfect matching graph based on properties of their corresponding partition. For example, if α is fixed, out of all the partitions $\lambda = (\alpha, \lambda_2, \ldots, \lambda_\ell)$ of $2k$, is it possible to predict which partition will give an eigenvalue with the largest absolute value?

16.11 Prove the EKR Theorem for partitions

In Section 15.7, we defined two uniform k-partitions of a k^2-set to be skew if any two cells from the different partitions have non-trivial intersection. Conjecture 15.9.2 predicts the size and structure of the largest set of partitions in which no two are skew. This conjecture predicts the bound in the natural extension of the EKR Theorem to skew partitions.

16.11.1 Conjecture. *For all* k, *the maximum size of a set of uniform* k-*partitions of a* k^2-*set in which no two partitions are skew is*

$$\binom{k^2 - 2}{k - 2} \frac{1}{(k - 1)!} \prod_{i=1}^{k} \binom{k^2 - ki}{k}.$$

The only sets that meet this bound consist of all partitions that contain a pair of distinct $i, j \in \{1, \ldots, k^2\}$ *in the same cell.*

The bound in this result can be proven if k is a prime power (see Theorem 15.8.1), and we know that the characterization holds if $k = 3$. Strangely, this is the problem that initially motivated the work that has led up to this book; yet we really are no further ahead on it than we were when we began!

One route to proving this conjecture would be to apply the ratio bound to the skew-partition graph. This would require knowing the least eigenvalue of this graph. We believe that we know what the least eigenvalue is (and have given the value in Conjecture 15.9.1). But we have been unable, so far, to prove it. More generally we can ask the following.

16.11.2 Problem. *What are the eigenvalues of the skew-partition graph?*

Similar to the work done by Renteln [146] and Ku and Wong [111], it might be possible to find relations between the eigenvalues or relative bounds on the eigenvalues. The case of skew partitions may be more difficult; unlike the derangement graph and the matching graph, the skew partition graph is not a graph in an association scheme.

16.12 Prove EKR Theorems for other objects

Throughout this book, we have considered different types of objects for which a version of the EKR Theorem holds. In fact, for any object for which there is a notion of intersection, it is possible to consider if a version of the EKR Theorem holds. One approach for any such object is to define the *intersection graph*; the vertices are the objects, and two objects are adjacent if they are not intersecting. Now the goal becomes determining the maximum cocliques in the intersection graph.

For almost all the objects that we have considered in this book, the intersection graph is regular and vertex transitive; frequently we can find large subgroups in the automorphism group of the intersection graph. In such cases, we are able to use tools such as the clique-coclique bound and the ratio bound to get information about the cocliques.

There are, however, interesting objects whose intersection graph is neither vertex transitive nor regular. In these cases, we need to find new methods. One example is the multisets of an n-set. In this case, the natural analog of the EKR Theorem does hold. One proof of this result uses the fact that there is a homomorphism of the intersection graph for multisets to a Kneser graph [129]. This approach may be used for other objects, provided that there is a homomorphism to an intersection graph in which an EKR-type theorem is already known to hold.

Another possible approach is to consider the Laplacian matrix of an irregular intersection graph rather than the adjacency matrix. In [42, Lemma 2.4.5] it is shown that a version of the ratio bound holds for the Laplacian matrix. One example of where this works is for intersecting subsets (with no fixed size) from a set of size n. In this case, the ratio bound for the Laplacian matrix does give the correct number of maximum intersecting subsets, but this is a long route to a very simple result. At this point, we do not have any more interesting examples using the ratio bound for the Laplacian matrix.

16.12.1 Problem. *Can the ratio bound for the Laplacian matrix be used to prove EKR-type theorems for objects whose intersection graph is not regular?*

16.13 Find maximal cocliques in graphs from an orbital scheme

The comments at the end of the final section of Chapter 13 indicate that there is an interesting project in looking at the orbital schemes from the action of $\mathrm{Sym}(n)$ on the cosets of G in $\mathrm{Sym}(n)$ where $G \leq \mathrm{Sym}(n)$ and $\mathrm{ind}_{\mathrm{Sym}(n)}(1_G)$ is multiplicity free.

16.13.1 Problem. *Let $G \leq \mathrm{Sym}(n)$ with $\mathrm{ind}_{\mathrm{Sym}(n)}(1_G)$ multiplicity-free. Let $\lambda < [n]$ be the largest (in the dominance ordering) partition of n such that χ_λ is a constituent of $\mathrm{ind}_{\mathrm{Sym}(n)}(1_G)$. Let $\{S, V \setminus S\}$ be the partition formed by the action of $\mathrm{Sym}(\lambda)$ on the cosets of G in $\mathrm{Sym}(n)$.*

For this situation consider the following two questions:

   (a) Is there a graph A_ℓ in the orbital scheme such that S forms a coclique?

   (b) Is the eigenvalue corresponding to the λ-module the least eigenvalue for such a graph?

In [86] and in [171] a complete list of the subgroups G of $\mathrm{Sym}(n)$ that are multiplicity free is given. Each of these groups satisfies the conditions of Problem 16.13.1.

16.14 Develop other compression operations

Ahlswede and Khachatrian [3] found the largest collection of t-intersecting k-subsets of $\{1, \ldots, n\}$ when n is small relative to k and t. These are the collections of all k-sets that contain at least $t + i$ elements from a fixed set of $t + 2i$ elements. The value of i depends on the relative values of n, k and t. This result they called the "Complete Intersection Theorem" (see Section 1.3). Shortly after they published this proof, Ahlswede and Khachatrian produced a second, very similar proof of the result [5]. A key tool in both proofs is the compression or shifting of set systems.

"Complete" EKR Theorems are conjectured to hold for both permutations and perfect matchings. A barrier in proving these conjectures is that the natural generalization of the compression operation to collections of permutations and to collections of perfect matchings does not quite work.

A natural generalization of the compression operation to collections of permutations requires two operations [141]. The first operation is called *fixing*; this operation increases the number of fixed points in the permutations in the collection. When this fixing operation is done on a collection of t-intersecting permutations, not only is the resulting set of permutations t-intersecting, but any two permutations share at least t fixed points. We say that a set of permutations is *fixed* if applying, the fixing operation does not change the collection. The second operation (called *compression*) transforms a permutation so that the collection of fixed points of the new permutation is the set that would result from doing the standard compression on the fixed points of the original permutation (for details see [141]). The problem with this generalization is in the fixing operation. Specifically, the fixing operation can take two distinct permutations and transform them both into the same permutation.

If it is possible to show that every maximum t-intersecting set of permutations has the property that it is "fixed" (this is the case for $t = 1$; see [40]), then the following conjecture can be proven following the method of Ahlswede and Khachatrian in [3] (see [141] for complete details).

16.14.1 Conjecture. *If $n \geq 2t + 1$, a t-intersecting family of permutations will have size at most $(n - t)!$, and only canonically t-intersecting families achieve this bound.*

Another approach would be to define a different compression operation for permutations. In fact, Ku and Renshaw [108] defined a compression on collections of permutations that does not change the size of the collection. But this operation does not preserve the property of t-intersection for the collections of permutations. Rather, it preserves a different type of intersection, called *t-cycle intersection.* Using this compression with the Ahlswede and Khacha-trian method, the EKR Theorem, including the exact bound bound on n, can be proven for t-cycle intersecting permutations in Sym(n). See either of [18, 130] for complete details.

Similar to the case for permutations, it is possible to define two operations on a collection of perfect matchings that together act like the compression oper-ation for sets. But these two operations may reduce the size of the collection of perfect matchings. If it were true that for any maximum t-intersecting col-lection of perfect matchings the compression operation did not change size of the set, then (using the Ahlswede and Khachatrian method) we could prove the following conjectured "Complete" EKR-type theorem for perfect matchings.

16.14.2 Conjecture. *Let $\mathcal{M}$ be a t-intersecting system of $2k$-matchings. If $k \geq 3t/2 + 1$, then*

$$|\mathcal{M}| \leq (2k - 2t - 1)!!.$$

Further, equality holds if and only if $\mathcal{M}$ is a canonical t-intersecting system.

Glossary: Symbols

1_G trivial representation of G 210

1_λ trivial representation on the Young subgroup $\mathrm{Sym}(\lambda)$ 235

1   all-ones vector   25

$\alpha(X)$ size of largest coclique in X 24

$\mathbb{C}[\mathcal{A}]$ Bose-Mesner algebra of the association scheme $\mathcal{A}$ 51

$C(n, k)$ cyclic interval graph   44

$D(n)$ set of all derangements of a set of size n 261

$d(n)$ number of derangements of a set of size n 261

$\{E_0, \ldots, E_d\}$ basis of idempotents an association scheme   56

E_ψ idempotent of the group scheme   226

$\mathrm{End}(V)$ group of all complex valued invertible linear maps of V 210

$\mathcal{F}_i$ intersecting set systems   6

$\mathrm{fix}(g)$ number of elements fixed by a permutation g 221

Γ_n derangement graph for $\mathrm{Sym}(n)$ 261

Γ_G derangement graph for a group G 269

$G(n, k)$ Grassmann scheme   161

$H(n, q)$ Hamming scheme   184

$\mathrm{ind}_G(\Phi)$ representation of G induced by Φ 228

$\Phi \!\uparrow^n$ representation of $\mathrm{Sym}(n)$ induced by Φ 235

$\lambda = [\lambda_1, \ldots, \lambda_k]$ an integer partition of n 285

$\lambda \vdash n$ an integer partition of n 235

J all-ones matrix   25

$J_q(n, k)$ Grassmann graph   161

$\mathcal{J}(n, k)$ the Johnson scheme   112

$J(n, k, i)$ ith graph in the Johnson scheme   49

$K_{\mu\lambda}$ Kostka number for μ and λ 237

$K(n, k)$ Kneser graph   24

Glossary: Operations and relations

$\lambda \leq \mu$ partition μ is larger than λ in the dominance ordering   236

$(2m - 1)!!$ double factorial   131

$\partial_i(x)$ operation that removes the ith element from a word x 195

$\partial_i(\mathcal{A})$ operation that removes the ith element from all words in set of words $\mathcal{A}$ 195

$\langle M, N \rangle$ inner product of matrices M and N 59

$\langle \varphi, \psi \rangle_G$ inner product of representations φ and ψ of the group G 211

$\langle u, v \rangle$ inner product of vectors u and v 36

$\partial(S)$ boundary of a set of vertices   130

$\partial^g(\mathcal{A})$ gth shadow of a set system   11

$\partial(\mathcal{A})$ shadow of a set system   11

X^n direct product of n copies of a graph X 200

$\begin{bmatrix} n \\ k \end{bmatrix}_q$ q-binomial coefficients   162

X/π quotient graph of X 27

$M^{\circ(r)}$ rth Schur power of M 74

$A \circ B$ Schur product of A and B 50

References

[1] R. Ahlswede, H. Aydinian, and L. H. Khachatrian. The intersection theorem for direct products. *European J. Combin.*, 19(6):649–661, 1998.

[2] Rudolf Ahlswede and Vladimir Blinovsky. *Lectures on Advances in Combinatorics*. Universitext. Springer-Verlag, Berlin, 2008.

[3] Rudolf Ahlswede and Levon H. Khachatrian. The complete intersection theorem for systems of finite sets. *European J. Combin.*, 18(2):125–136, 1997.

[4] Rudolf Ahlswede and Levon H. Khachatrian. The diametric theorem in Hamming spaces – optimal anticodes. *Adv. in Appl. Math.*, 20(4):429–449, 1998.

[5] Rudolf Ahlswede and Levon H. Khachatrian. A pushing-pulling method: new proofs of intersection theorems. *Combinatorica*, 19(1):1–15, 1999.

[6] Bahman Ahmadi. *Maximum Intersecting Families of Permutations*. PhD thesis, University of Regina, Regina, 2013.

[7] Bahman Ahmadi and Karen Meagher. The Erdős–Ko–Rado property for some 2-transitive groups. *Annals of Combinatorics*, 20 pp., 2015.

[8] Bahman Ahmadi and Karen Meagher. The Erdős–Ko–Rado property for some permutation groups. *Australas. J. Combin.*, 61(2):23–41, 2015.

[9] Michael O. Albertson and Karen L. Collins. Homomorphisms of 3-chromatic graphs. *Discrete Math.*, 54(2):127–132, 1985.

[10] N. Alon, I. Dinur, E. Friedgut, and B. Sudakov. Graph products, Fourier analysis and spectral techniques. *Geom. Funct. Anal.*, 14(5):913–940, 2004.

[11] Noga Alon and Eyal Lubetzky. Uniformly cross intersecting families. *Combinatorica*, 29(4):389–431, 2009.

[12] Brian Alspach, Heather Gavlas, Mateja Šajna, and Helen Verrall. Cycle decompositions. IV. Complete directed graphs and fixed length directed cycles. *J. Combin. Theory Ser. A*, 103(1):165–208, 2003.

[13] R. A. Bailey. *Association Schemes*, volume 84 of *Cambridge Studies in Advanced Mathematics*. Cambridge University Press, Cambridge, 2004.

[14] Eiichi Bannai and Tatsuro Ito. *Algebraic Combinatorics I*. Benjamin/Cummings Publishing, Menlo Park, CA, 1984.

[15] I. Bárány. A short proof of Kneser's conjecture. *J. Combin. Theory Ser. A*, 25(3):325–326, 1978.

[16] Zs. Baranyai. On the factorization of the complete uniform hypergraph. In *Infinite and Finite Sets*, volume 10 of *Colloq. Math. Soc. János Bolyai*, pages 91–108. North-Holland, Amsterdam, 1975.

[17] Mohammad Bardestani and Keivan Mallahi-Karai. On the Erdős–Ko–Rado property for finite groups. *J. Algebraic Combin.*, 42(1):1–18, 2015.

[18] V. M. Blinovskiĭ. An intersection theorem for finite permutations. *Problemy Peredachi Informatsii*, 47(1):40–53, 2011.

[19] V. M. Blinovsky. Remark on one problem in extremal combinatorics. *Problems Inform. Transmission*, 48(1):70–71, 2012.

[20] A. Blokhuis. On subsets of GF(q^2) with square differences. *Nederl. Akad. Wetensch. Indag. Math.*, 46(4):369–372, 1984.

[21] A. Blokhuis, A. E. Brouwer, A. Chowdhury, P. Frankl, T. Mussche, B. Patkós, and T. Szőnyi. A Hilton–Milner theorem for vector spaces. *Electron. J. Combin.*, 17(1):Research Paper 71, 12 pp. (electronic), 2010.

[22] Aart Blokhuis, Andries Brouwer, Tamás Szönyi, and Zsuzsa Weiner. On q-analogues and stability theorems. *J. Geom.*, 101(1–2):31–50, 2011.

[23] Béla Bollobás. *Combinatorics*. Cambridge University Press, Cambridge, 1986.

[24] J. A. Bondy and Pavol Hell. A note on the star chromatic number. *J. Graph Theory*, 14(4):479–482, 1990.

[25] Peter Borg. Cross-intersecting families of partial permutations. *SIAM J. Discrete Math.*, 24(2):600–608, 2010.

[26] Peter Borg. Cross-intersecting sub-families of hereditary families. *J. Combin. Theory Ser. A*, 119(4):871–881, 2012.

[27] Peter Borg and Imre Leader. Multiple cross-intersecting families of signed sets. *J. Combin. Theory Ser. A*, 117(5):583–588, 2010.

[28] R. C. Bose, W. G. Bridges, and M. S. Shrikhande. A characterization of partial geometric designs. *Discrete Math.*, 16(1):1–7, 1976.

[29] Benjamin Braun. Symmetries of the stable Kneser graphs. *Adv. in Appl. Math.*, 45(1):12–14, 2010.

[30] Arne Brøndsted. *An Introduction to Convex Polytopes*, volume 90 of *Graduate Texts in Mathematics*. Springer, New York, 1983.

[31] A. E. Brouwer, A. M. Cohen, and A. Neumaier. *Distance-Regular Graphs*. Springer, Berlin, 1989.

[32] A. E. Brouwer, C. D. Godsil, J. H. Koolen, and W. J. Martin. Width and dual width of subsets in polynomial association schemes. *J. Combin. Theory Ser. A*, 102(2):255–271, 2003.

[33] A. E. Brouwer and W. H. Haemers. Structure and uniqueness of the (81, 20, 1, 6) strongly regular graph. *Discrete Math.*, 106/107:77–82, 1992.

[34] A. E. Brouwer, James B. Shearer, N. J. A. Sloane, and Warren D. Smith. A new table of constant weight codes. *IEEE Trans. Inform. Theory*, 36(6):1334–1380, 1990.

[35] Andries E. Brouwer and Tuvi Etzion. Some new distance-4 constant weight codes. *Adv. Math. Commun.*, 5(3):417–424, 2011.

[36] Andries E. Brouwer and Willem H. Haemers. *Spectra of Graphs*. Springer, New York, 2012.

[37] A. R. Calderbank and P. Frankl. Improved upper bounds concerning the Erdős–Ko–Rado theorem. *Combin. Probab. Comput.*, 1(2):115–122, 1992.

[38] Peter J. Cameron. *Projective and polar spaces*, volume 13 of *QMW Maths Notes*. Queen Mary and Westfield College, School of Mathematical Sciences, London, 1991.

[39] Peter J. Cameron and Priscila A. Kazanidis. Cores of symmetric graphs. *J. Aust. Math. Soc.*, 85(2):145–154, 2008.

[40] Peter J. Cameron and C. Y. Ku. Intersecting families of permutations. *European J. Combin.*, 24(7):881–890, 2003.

[41] Andrew D. Cannon, John Bamberg, and Cheryl E. Praeger. A classification of the strongly regular generalised Johnson graphs. *Ann. Comb.*, 16(3):489–506, 2012.

[42] Michael Scott Cavers. *The Normalized Laplacian Matrix and General Randic Index of Graphs*. ProQuest LLC, Ann Arbor, MI, 2010. Doctoral dissertation, University of Regina, Canada.

[43] Tullio Ceccherini-Silberstein, Fabio Scarabotti, and Filippo Tolli. *Harmonic Analysis on Finite Groups*, volume 108 of *Cambridge Studies in Advanced Mathematics*. Cambridge University Press, Cambridge, 2008.

[44] Tullio Ceccherini-Silberstein, Fabio Scarabotti, and Filippo Tolli. *Representation Theory of the Symmetric Groups*, volume 121 of *Cambridge Studies in Advanced Mathematics*. Cambridge University Press, Cambridge, 2010.

[45] Stephen D. Cohen. Clique numbers of Paley graphs. *Quaestiones Math.*, 11(2):225–231, 1988.

[46] Charles J. Colbourn and Jeffrey H. Dinitz, editors. *Handbook of Combinatorial Designs*. Chapman & Hall/CRC, Boca Raton, FL, second edition, 2007.

[47] William J. Cook, William H. Cunningham, William R. Pulleyblank, and Alexander Schrijver. *Combinatorial optimization*. Wiley-Interscience Series in Discrete Mathematics and Optimization. Wiley, New York, 1998.

[48] Éva Czabarka and László Székely. An alternative shifting proof to Hsieh's theorem. In *Proceedings of the Twenty-ninth Southeastern International Conference on Combinatorics, Graph Theory and Computing*, volume 134, pages 117–122, 1998.

[49] D. E. Daykin. Erdős–Ko–Rado from Kruskal–Katona. *J. Combinatorial Theory Ser. A*, 17:254–255, 1974.

[50] Maarten De Boeck. The largest Erdős–Ko–Rado sets of planes in finite projective and finite classical polar spaces. *Des. Codes Cryptogr.*, 72(1):77–117, 2014.

[51] P. Delsarte. An algebraic approach to the association schemes of coding theory. *Philips Res. Rep. Suppl.*, (10):vi+97, 1973.

[52] Ph. Delsarte. Bilinear forms over a finite field, with applications to coding theory. *J. Combin. Theory Ser. A*, 25(3):226–241, 1978.

[53] M. Deza and P. Frankl. Erdős–Ko–Rado theorem – 22 years later. *SIAM J. Algebraic Discrete Methods*, 4(4):419–431, 1983.

[54] David Ellis. A proof of the Cameron–Ku conjecture. *J. Lond. Math. Soc. (2)*, 85(1):165–190, 2012.

[55] David Ellis. Setwise intersecting families of permutations. *J. Combin. Theory Ser. A*, 119(4):825–849, 2012.

[56] David Ellis, Ehud Friedgut, and Haran Pilpel. Intersecting families of permutations. *J. Amer. Math. Soc.*, 24(3):649–682, 2011.

[57] P. Erdős. Problems and results on finite and infinite combinatorial analysis. In *Infinite and Finite Sets, Vol. I*, volume 10 of *Colloq. Math. Soc. János Bolyai*, pages 403–424. North-Holland, Amsterdam, 1975.

[58] P. Erdős, Chao Ko, and R. Rado. Intersection theorems for systems of finite sets. *Quart. J. Math. Oxford Ser. (2)*, 12:313–320, 1961.

[59] Péter L. Erdős and László A. Székely. Erdős–Ko–Rado theorems of higher order. In *Numbers, Information and Complexity*, pages 117–124. Kluwer Academic, Boston, 2000.

[60] Tuvi Etzion and Sara Bitan. On the chromatic number, colorings, and codes of the Johnson graph. *Discrete Appl. Math.*, 70(2):163–175, 1996.

[61] P. Frankl. On Sperner families satisfying an additional condition. *J. Combinatorial Theory Ser. A*, 20(1):1–11, 1976.

[62] P. Frankl. The Erdős–Ko–Rado theorem is true for $n = ckt$. In *Combinatorics*, volume 18 of *Colloq. Math. Soc. János Bolyai*, pages 365–375. North-Holland, Amsterdam, 1978.

[63] P. Frankl. Multiply-intersecting families. *J. Combin. Theory Ser. B*, 53(2):195–234, 1991.

[64] P. Frankl. An Erdős–Ko–Rado theorem for direct products. *European J. Combin.*, 17(8):727–730, 1996.

[65] P. Frankl and Z. Füredi. Extremal problems concerning Kneser graphs. *J. Combin. Theory Ser. B*, 40(3):270–284, 1986.

[66] P. Frankl and Z. Füredi. Nontrivial intersecting families. *J. Combin. Theory Ser. A*, 41(1):150–153, 1986.

[67] P. Frankl and R. M. Wilson. The Erdős–Ko–Rado theorem for vector spaces. *J. Combin. Theory Ser. A*, 43(2):228–236, 1986.

[68] Peter Frankl. The shifting technique in extremal set theory. In *Surveys in Combinatorics 1987*, pages 81–110. Cambridge University Press, Cambridge, 1987.

[69] Péter Frankl and Mikhail Deza. On the maximum number of permutations with given maximal or minimal distance. *J. Combinatorial Theory Ser. A*, 22(3):352–360, 1977.

[70] Peter Frankl and Zoltán Füredi. A new short proof of the EKR theorem. *J. Combin. Theory Ser. A*, 119(6):1388–1390, 2012.

[71] Peter Frankl and Norihide Tokushige. On r-cross intersecting families of sets. *Combin. Probab. Comput.*, 20(5):749–752, 2011.

[72] Ehud Friedgut. A Katona-type proof of an Erdős–Ko–Rado-type theorem. *J. Combin. Theory Ser. A*, 111(2):239–244, 2005.

[73] William Fulton and Joe Harris. *Representation Theory*. Springer, New York, 1991.

[74] Zoltán Füredi. Cross-intersecting families of finite sets. *J. Combin. Theory Ser. A*, 72(2):332–339, 1995.

[75] Zoltán Füredi, Kyung-Won Hwang, and Paul M. Weichsel. A proof and generalizations of the Erdős–Ko–Rado theorem using the method of linearly independent polynomials. In *Topics in Discrete Mathematics*, volume 26 of *Algorithms Combin.*, pages 215–224. Springer, Berlin, 2006.

[76] The GAP Group. *GAP – Groups, Algorithms, and Programming, Version 4.7.7*, 2015.

[77] C. D. Godsil. Graphs, groups and polytopes. In *Combinatorial Mathematics (Proc. Internat. Conf. Combinatorial Theory, Australian Nat. Univ., Canberra, 1977)*, volume 686 of *Lecture Notes in Math.*, pages 157–164. Springer, Berlin, 1978.

[78] C. D. Godsil. *Algebraic Combinatorics*. Chapman and Hall Mathematics Series. Chapman & Hall, New York, 1993.

[79] C. D. Godsil. Euclidean geometry of distance regular graphs. In *Surveys in Combinatorics*, volume 218 of *London Math. Soc. Lecture Note Ser.*, pages 1–23. Cambridge University Press, Cambridge, 1995.

[80] C. D. Godsil. Eigenpolytopes of distance regular graphs. *Canad. J. Math.*, 50(4):739–755, 1998.

[81] C. D. Godsil. *Association Schemes*. 242 pp. 2010.

[82] C. D. Godsil. Generalized Hamming schemes, 14 pp. 2010. available at http://arxiv.org/abs/1011.1044.

[83] C. D. Godsil and W. J. Martin. Quotients of association schemes. *J. Combin. Theory Ser. A*, 69(2):185–199, 1995.

[84] C. D. Godsil and M. W. Newman. Independent sets in association schemes. *Combinatorica*, 26(4):431–443, 2006.

[85] Chris Godsil and Karen Meagher. A new proof of the Erdős–Ko–Rado theorem for intersecting families of permutations. *European J. Combin.*, 30(2):404–414, 2009.

[86] Chris Godsil and Karen Meagher. Multiplicity-free permutation representations of the symmetric group. *Ann. Comb.*, 13(4):463–490, 2010.

[87] Chris Godsil and Gordon Royle. *Algebraic Graph Theory*, volume 207 of *Graduate Texts in Mathematics*. Springer-Verlag, New York, 2001.

[88] Chris Godsil and Gordon F. Royle. Cores of geometric graphs. *Ann. Comb.*, 15(2):267–276, 2011.

[89] R. L. Graham and N. J. A. Sloane. Lower bounds for constant weight codes. *IEEE Trans. Inform. Theory*, 26(1):37–43, 1980.

[90] Joshua E. Greene. A new short proof of Kneser's conjecture. *Amer. Math. Monthly*, 109(10):918–920, 2002.

[91] A. Hajnal and Bruce Rothschild. A generalization of the Erdős–Ko–Rado theorem on finite set systems. *J. Combinatorial Theory Ser. A*, 15:359–362, 1973.

[92] Marshall Hall. An existence theorem for Latin squares. *Bull. Amer. Math. Soc.*, 51:387–388, 1945.

[93] Peter Henrici. *Applied and Computational Complex Analysis: Vol. 1: Power Series, Integration, Conformal Mapping, Location of Zeros*. Wiley-Interscience, 1974.

[94] Robert Hermann. *Spinors, Clifford and Cayley Algebras*. Department of Mathematics, Rutgers University, New Brunswick, NJ, 1974.

[95] A. J. W. Hilton and E. C. Milner. Some intersection theorems for systems of finite sets. *Quart. J. Math. Oxford Ser. (2)*, 18:369–384, 1967.

[96] F. C. Holroyd and A. Johnson. BCC Problem List. http://www.maths.qmul.ac.uk/~pjc/bcc/allprobs.pdf, 2001.

[97] Ferdinand Ihringer and Klaus Metsch. On the maximum size of Erdős–Ko–Rado sets in $H(2d + 1, q^2)$. *Des. Codes Cryptogr.*, 72(2):311–316, 2014.

[98] Gordon James and Adalbert Kerber. *The Representation Theory of the Symmetric Group*, volume 16 of *Encyclopedia of Mathematics and Its Applications*. Addison-Wesley, Reading, MA, 1981.

[99] Pranava K. Jha and Sandi Klavžar. Independence in direct-product graphs. *Ars Combin.*, 50:53–63, 1998.

[100] Gareth A. Jones. Automorphisms and regular embeddings of merged Johnson graphs. *European J. Combin.*, 26(3–4):417–435, 2005.

[101] Vikram Kamat. On cross-intersecting families of independent sets in graphs. *Australas. J. Combin.*, 50:171–181, 2011.

[102] Vikram Kamat and Neeldhara Misra. An Erdős–Ko–Rado theorem for matchings in the complete graph. In Jaroslav Nesetril and Marco Pellegrini, editors, *The Seventh European Conference on Combinatorics, Graph Theory and Applications*, volume 16 of *CRM Series*, pages 613–613. Scuola Normale Superiore, 2013.

[103] G. Katona. A theorem of finite sets. In *Theory of Graphs (Proc. Colloq., Tihany, 1966)*, pages 187–207. Academic Press, New York, 1968.

[104] G. O. H. Katona. A simple proof of the Erdős–Chao Ko–Rado theorem. *J. Combinatorial Theory Ser. B*, 13:183–184, 1972.

[105] Gy. Katona. Intersection theorems for systems of finite sets. *Acta Math. Acad. Sci. Hungar*, 15:329–337, 1964.

[106] Gyula O. H. Katona. The cycle method and its limits. In *Numbers, Information and Complexity (Bielefeld, 1998)*, pages 129–141. Kluwer Academic, Boston, 2000.

[107] Joseph B. Kruskal. The number of simplices in a complex. In *Mathematical Optimization Techniques*, pages 251–278. University of California Press, Berkeley, 1963.

[108] Cheng Yeaw Ku and David Renshaw. Erdős–Ko–Rado theorems for permutations and set partitions. *J. Combin. Theory Ser. A*, 115(6):1008–1020, 2008.

[109] Cheng Yeaw Ku and David B. Wales. Eigenvalues of the derangement graph. *J. Combin. Theory Ser. A*, 117(3):289–312, 2010.

[110] Cheng Yeaw Ku and Kok Bin Wong. On cross-intersecting families of set partitions. *Electron. J. Combin.*, 19(4):Paper 49, 9 pp. (electronic), 2012.

[111] Cheng Yeaw Ku and Kok Bin Wong. Solving the Ku-Wales conjecture on the eigenvalues of the derangement graph. *European J. Combin.*, 34(6):941–956, 2013.

[112] Cheng Yeaw Ku and Tony W. H. Wong. Intersecting families in the alternating group and direct product of symmetric groups. *Electron. J. Combin.*, 14(1):Research Paper 25, 15 pp. (electronic), 2007.

[113] Benoit Larose and Claudia Malvenuto. Stable sets of maximal size in Kneser-type graphs. *European J. Combin.*, 25(5):657–673, 2004.

[114] Benoit Larose and Claude Tardif. Projectivity and independent sets in powers of graphs. *J. Graph Theory*, 40(3):162–171, 2002.

[115] Nathan Lindzey. Erdős–Ko–Rado for perfect matchings. Available at http://arxiv.org/abs/ 1409.2057, 2014.

[116] L. Lovász. Kneser's conjecture, chromatic number, and homotopy. *J. Combin. Theory Ser. A*, 25(3):319–324, 1978.

[117] László Lovász. *Combinatorial Problems and Exercises*. AMS Chelsea, Providence, second edition, 2007.

[118] László Lovász and Michael D. Plummer. *Matching Theory*. AMS Chelsea, Providence, 2009. Corrected reprint of the 1986 original [MR0859549].

[119] Jia Xi Lu. On large sets of disjoint Steiner triple systems. I, II,III. *J. Combin. Theory Ser. A*, 34(2):140–146, 1983.

[120] Jia Xi Lu. On large sets of disjoint Steiner triple systems. IV, V, VI. *J. Combin. Theory Ser. A*, 37(2):136–163, 164–188, 189–192, 1984.

[121] D. Lubell. A short proof of Sperner's lemma. *J. Combin. Theory*, 1:299, 1966.

[122] W. J. Martin. Completely regular codes: a viewpoint and some problems. In *Proceedings of 2004 Com2MaC Workshop on Distance-Regular Graphs and Finite Geometry, Pusan, Korea*, pages 43–56, 2004.

[123] William J. Martin and Hajime Tanaka. Commutative association schemes. *European J. Combin.*, 30(6):1497–1525, 2009.

[124] Rudolf Mathon and Alexander Rosa. A new strongly regular graph. *J. Combin. Theory Ser. A*, 38(1):84–86, 1985.

[125] Jiří Matoušek. A combinatorial proof of Kneser's conjecture. *Combinatorica*, 24(1):163–170, 2004.

[126] Makoto Matsumoto and Norihide Tokushige. The exact bound in the Erdős–Ko–Rado theorem for cross-intersecting families. *J. Combin. Theory Ser. A*, 52(1):90–97, 1989.

[127] Brendan D. McKay, Alison Meynert, and Wendy Myrvold. Small Latin squares, quasigroups, and loops. *J. Combin. Des.*, 15(2):98–119, 2007.

[128] Karen Meagher and Lucia Moura. Erdős–Ko–Rado theorems for uniform set-partition systems. *Electron J. Combin.*, 12(1):Research Paper 40, 12 pp. (electronic), 2005.

[129] Karen Meagher and Alison Purdy. An Erdős–Ko–Rado theorem for multisets. *Electron. J. Combin.*, 18(1):Paper 220, 8 pp. (electronic), 2011.

[130] Karen Meagher and Alison Purdy. The exact bound for the Erdős–Ko–Rado theorem for *t*-cycle-intersecting permutations, Available at http://arxiv.org/abs/1208.3638, 23 pp. 2012.

[131] Karen Meagher and Pablo Spiga. An Erdős–Ko–Rado theorem for the derangement graph of $PGL(2, q)$ acting on the projective line. *J. Combin. Theory Ser. A*, 118(2):532–544, 2011.

[132] Karen Meagher and Pablo Spiga. An Erdős–Ko–Rado theorem for the derangement graph of $PGL_3(q)$ acting on the projective plane. *SIAM J. Discrete Math.*, 28(2):918–941, 2014.

[133] Karen Meagher and Brett Stevens. Covering arrays on graphs. *J. Combin. Theory Ser. B*, 95(1):134–151, 2005.

[134] L. D. Mešalkin. A generalization of Sperner's theorem on the number of subsets of a finite set. *Teor. Verojatnost. i Primenen*, 8:219–220, 1963.

[135] Aeryung Moon. An analogue of the Erdős–Ko–Rado theorem for the Hamming schemes $H(n, q)$. *J. Combin. Theory Ser. A*, 32(3):386–390, 1982.

[136] Mikhail Muzychuk. On association schemes of the symmetric group S_{2n} acting on partitions of type 2^n. *Bayreuth. Math. Schr.*, (47):151–164, 1994.

[137] Arnold Neumaier. Completely regular codes. *Discrete Math.*, 106/107:353–360, 1992.

[138] M. W. Newman. *Independent Sets and Eigenvalues*. PhD thesis, University of Waterloo, Waterloo, 2004.

[139] Stanley E. Payne and Joseph A. Thas. *Finite Generalized Quadrangles.* EMS Series of Lectures in Mathematics. European Mathematical Society (EMS), Zürich, second edition, 2009.

[140] Valentina Pepe, Leo Storme, and Frédéric Vanhove. Theorems of Erdős–Ko–Rado type in polar spaces. *J. Combin. Theory Ser. A*, 118(4):1291–1312, 2011.

[141] Alison Purdy. *The Erdős–Ko–Rado Theorem for Intersecting Families of Permutations.* Master's thesis, University of Regina, Regina, 2010.

[142] L. Pyber. A new generalization of the Erdős–Ko–Rado theorem. *J. Combin. Theory Ser. A*, 43(1):85–90, 1986.

[143] Mark Ramras and Elizabeth Donovan. The automorphism group of a Johnson graph. *SIAM J. Discrete Math.*, 25(1):267–270, 2011.

[144] B. M. I. Rands. An extension of the Erdős, Ko, Rado theorem to *t*-designs. *J. Combin. Theory Ser. A*, 32(3):391–395, 1982.

[145] D. K. Ray-Chaudhuri and Richard M. Wilson. The existence of resolvable block designs. In *Survey of Combinatorial Theory*, pages 361–375. North-Holland, Amsterdam, 1973.

[146] Paul Renteln. On the spectrum of the derangement graph. *Electron. J. Combin.*, 14(1):Research Paper 82, 17 pp. (electronic), 2007.

[147] Bruce E. Sagan. *The Symmetric Group.* Wadsworth & Brooks/Cole Mathematics Series. Wadsworth & Brooks/Cole, Pacific Grove, CA, 1991.

[148] Jan Saxl. On multiplicity-free permutation representations. In *Finite Geometries and Designs*, volume 49 of *London Math. Soc. Lecture Note Ser.*, pages 337–353. Cambridge University Press, Cambridge, 1981.

[149] A. Schrijver. Vertex-critical subgraphs of Kneser graphs. *Nieuw Arch. Wisk. (3)*, 26(3):454–461, 1978.

[150] A. Schrijver. Short proofs on the matching polyhedron. *J. Combin Theory Ser. B*, 1983.

[151] Alexander Schrijver. *Combinatorial Optimization. Polyhedra and Efficiency. Vol. A*, volume 24 of *Algorithms and Combinatorics*. Berlin, 2003. Paths, flows, matchings, Chapters 1–38.

[152] Ernest E. Shult. *Points and lines.* Universitext. Springer, Heidelberg, 2011. Characterizing the classical geometries.

[153] Emanuel Sperner. Ein Satz über Untermengen einer endlichen Menge. *Math. Z.*, 27(1):544–548, 1928.

[154] Dennis Stanton. Some Erdős–Ko–Rado theorems for Chevalley groups. *SIAM J. Algebraic Discrete Methods*, 1(2):160–163, 1980.

[155] E. Steinitz. Polyeder und Raumeinteilungen. In *Encyclopädie der mathematischen Wissenschaften, Band 3 (Geometries)*, pages 1–139. 1922.

[156] S. Sternberg. *Group Theory and Physics.* Cambridge University Press, Cambridge, 1994.

[157] Brett Stevens and Eric Mendelsohn. New recursive methods for transversal covers. *J. Combin. Des.*, 7(3):185–203, 1999.

[158] Douglas R. Stinson. *Combinatorial Designs.* Springer, New York, 2004.

[159] John Talbot. Intersecting families of separated sets. *J. London Math. Soc.*, 2003.

[160] Hajime Tanaka. Classification of subsets with minimal width and dual width in Grassmann, bilinear forms and dual polar graphs. *J. Combin. Theory Ser. A*, 113(5):903–910, 2006.

[161] Hajime Tanaka. Vertex subsets with minimal width and dual width in *Q*-polynomial distance-regular graphs. *Electron. J. Combin.*, 18(1):Paper 167, 32 pp. (electronic), 2011.

[162] Claude Tardif. Graph products and the chromatic difference sequence of vertex-transitive graphs. *Discrete Math.*, 185(1–3):193–200, 1998.

[163] Charles B. Thomas. *Representations of Finite and Lie Groups.* Imperial College Press, London, 2004.

[164] Norihide Tokushige. On cross *t*-intersecting families of sets. *J. Combin. Theory Ser. A*, 117(8):1167–1177, 2010.

[165] Norihide Tokushige. A product version of the Erdős–Ko–Rado theorem. *J. Combin. Theory Ser. A*, 118(5):1575–1587, 2011.

[166] J. H. van Lint and R. M. Wilson. *A Course in Combinatorics*. Cambridge University Press, Cambridge, second edition, 2001.

[167] M. Vaughan-Lee and I. M. Wanless. Latin squares and the Hall-Paige conjecture. *Bull. London Math. Soc.*, 35(2):191–195, 2003.

[168] Jun Wang and Sophia J. Zhang. An Erdős–Ko–Rado-type theorem in Coxeter groups. *European J. Combin.*, 29(5):1112–1115, 2008.

[169] Li Wang. Erdős–Ko–Rado theorem for irreducible imprimitive reflection groups. *Front. Math. China*, 7(1):125–144, 2012.

[170] Helmut Wielandt. *Finite Permutation Groups*. Translated from the German by R. Bercov. Academic Press, New York, 1964.

[171] Mark Wildon. Multiplicity-free representations of symmetric groups. *J. Pure Appl. Algebra*, 213(7):1464–1477, 2009.

[172] Richard M. Wilson. The exact bound in the Erdős–Ko–Rado theorem. *Combinatorica*, 4(2-3):247–257, 1984.

[173] D. R. Woodall, I. Anderson, G. R. Brightwell, J. W. P. Hirschfeld, P. Rowlinson, J. Sheehan, and D. H. Smith, editors. *16th British Combinatorial Conference*. 1999. *Discrete Math.* 197/198.

[174] Koichi Yamamoto. Logarithmic order of free distributive lattice. *J. Math. Soc. Japan*, 6:343–353, 1954.

[175] Huajun Zhang. Primitivity and independent sets in direct products of vertex-transitive graphs. *J. Graph Theory*, 67(3):218–225, 2011.

[176] Günter M. Ziegler. *Lectures on Polytopes*, volume 152 of *Graduate Texts in Mathematics*. Springer, New York, 1995.

[177] Paul-Hermann Zieschang. *Theory of Association Schemes*. Springer Monographs in Mathematics. Springer, Berlin, 2005.

Index

Printed in the United States
by Baker & Taylor Publisher Services